Cardiovascular Physiology II

Publisher's Note

The *International Review of Physiology* remains a major force in the education of established scientists and advanced students of physiology throughout the world. It continues to present accurate, timely, and thorough reviews of key topics by distinguished authors charged with the responsibility of selecting and critically analyzing new facts and concepts important to the progress of physiology from the mass of information in their respective fields.

Following the successful format established by the earlier volumes in this series, new volumes of the *International Review of Physiology* will concentrate on current developments in neurophysiology and cardiovascular, respiratory, gastrointestinal, endocrine, kidney and urinary tract, environmental, and reproductive physiology. New volumes on a given subject generally appear at two-year intervals, or according to the demand created by new developments in the field. The scope of the series is flexible, however, so that future volumes may cover areas not included earlier.

University Park Press is honored to continue publication of the *International Review of Physiology* under its sole sponsorship beginning with Volume 9. The following is a list of volumes published and currently in preparation for the series:

Volume 1: **CARDIOVASCULAR PHYSIOLOGY** (A. C. Guyton and C. E. Jones)
Volume 2: **RESPIRATORY PHYSIOLOGY** (J. G. Widdicombe)
Volume 3: **NEUROPHYSIOLOGY** (C. C. Hunt)
Volume 4: **GASTROINTESTINAL PHYSIOLOGY** (E. D. Jacobson and L. L. Shanbour)
Volume 5: **ENDOCRINE PHYSIOLOGY** (S. M. McCann)
Volume 6: **KIDNEY AND URINARY TRACT PHYSIOLOGY** (K. Thurau)
Volume 7: **ENVIRONMENTAL PHYSIOLOGY** (D. Robertshaw)
Volume 8: **REPRODUCTIVE PHYSIOLOGY** (R. O. Greep)
Volume 9: **CARDIOVASCULAR PHYSIOLOGY II** (A. C. Guyton and A. W. Cowley, Jr.)
Volume 10: **NEUROPHYSIOLOGY II** (R. Porter)
Volume 11: **KIDNEY AND URINARY TRACT PHYSIOLOGY II** (K. Thurau)
Volume 12: **GASTROINTESTINAL PHYSIOLOGY II** (R. K. Crane)

(Series numbers for the following volumes will be assigned in order of publication)

RESPIRATORY PHYSIOLOGY II (J. G. Widdicombe)
REPRODUCTIVE PHYSIOLOGY II (R. O. Greep)
ENDOCRINE PHYSIOLOGY II (S. M. McCann)
ENVIRONMENTAL PHYSIOLOGY II (D. Robertshaw)

Consultant Editor: Arthur C. Guyton, M.D., Department of Physiology and Biophysics, University of Mississippi Medical Center

INTERNATIONAL REVIEW OF PHYSIOLOGY

Volume 9

Cardiovascular Physiology II

Edited by:

Arthur C. Guyton, M.D.
and
Allen W. Cowley, Jr., Ph.D.

Department of Physiology and Biophysics
University of Mississippi Medical Center

UNIVERSITY PARK PRESS
Baltimore · London · Tokyo

UNIVERSITY PARK PRESS
International Publishers in Science and Medicine
Chamber of Commerce Building
Baltimore, Maryland 21202

Typeset by The Composing Room of Michigan, Inc.

Manufactured in the United States of American by Universal Lithographers, Inc., and The Maple Press Co.

Library of Congress Cataloging in Publication Data

Guyton, Arthur C
   Cardiovascular physiology II.

   (International review of physiology; v. 9)
   Includes index.
   1. Cardiovascular system. 2. Blood—Circulation.
I. Cowley, Allen W., joint author. II. Title.
III. Series. [DNLM: 1. Cardiovascular system—
Physiology—Period. W1 IN834F v. 9 etc. / WG102 C2672]
QP1.P62 vol. 9 [QP102] 599'.01'08s [599'.01'1]
ISBN 0-8391-1058-8   76-12990

Consultant Editor's Note

It is now two years since the first series of the *International Review of Physiology* appeared. This new review was launched in response to unfulfilled needs in the field of physiological science, most importantly the need for an in-depth review written especially for teachers and students of physiology throughout the world. It was not without trepidation that this publishing venture was begun, but its early success seems to assure its future. Therefore, we need to repeat here the philosophy, the goals, and the concept of the *International Review of Physiology*.

The *International Review of Physiology* has the same goals as all other reviews for accuracy, timeliness, and completeness, but it also has policies that we hope and believe engender still other important qualities often missing in reviews, the qualities of critical evaluation, integration, and instructiveness. To achieve these goals, the format allows publication of approximately 2,500 pages per series, divided into eight subspecialty volumes, each organized by experts in their respective fields. This extensiveness of coverage allows consideration of each subject in great depth. And, to make the review as timely as possible, a new series of all eight volumes is published approximately every two years, giving a cycle time that will keep the articles current.

Yet, perhaps the greatest hope that this new review will achieve its goals lies in its editorial policies. A simple but firm request is made to each author that he utilize his expertise and his judgment to sift from the mass of biennial publications those new facts and concepts that are important to the progress of physiology; that he make a conscious effort not to write a review consisting of an annotated list of references; and that the important material that he does choose be presented in thoughtful and logical exposition, complete enough to convey full understanding, as well as woven into context with previously established physiological principles. Hopefully, these processes will continue to bring to the reader each two years a treatise that he will use not merely as a reference in his own personal field but also as an exercise in refreshing and modernizing his whole store of physiological knowledge.

A. C. Guyton

Contents

Preface

In the first edition of this volume of *Cardiovascular Physiology* published two years ago, the chapters were designed both to review the recent literature and to give an overall statement of present knowledge in the cardiovascular field. Now we come to a new edition — new goals, new purposes, and new needs. The chapters are different: they mainly fill in areas covered in less depth in the first edition, yet still overlapping in many scientific subjects where important progress has been made.

The reader will find in this volume, as also in the first edition, major emphasis on the integrative aspects of circulatory physiology. This is partly caused by design but mainly by the fact that this is where the forefronts of cardiovascular research are actively advancing. In the final chapter, we have made a special attempt to weave a basic design for overall function of the circulation, to provide a skeleton onto which the many bits and pieces of circulatory physiology can be appended.

To make this volume of maximum benefit to both teachers and students, the authors were specifically requested to make their chapters instructive, not only to present new information but also to weld this into common meaning with principles already generally accepted by circulatory physiologists. And, further, the authors themselves were chosen carefully for their abilities both as research workers and teachers.

The massive amount of work that goes into preparation of a review requires exceptional commitment and dedication by the authors, and repayment for their contribution is provided mainly by their knowledge that they are sharing with the reader a wealth of experience and understanding. However, for what it is worth, the Editors wish also to express their admiration for each author's devoted work.

It is also important to thank the Editorial Staff and Management of University Park Press for their efforts that have given this new *International Review of Physiology* its successful start and also for their foresight in establishing a publishing format that allows the Review to remain abreast of the times and yet to delve deeply into the conceptual basis of physiology.

A. C. Guyton
A. W. Cowley, Jr.

International Review of Physiology
Cardiovascular Physiology II, Volume 9
Edited by Arthur C. Guyton and Allen W. Cowley
Copyright 1976   University Park Press   Baltimore

1
Microrheology of Erythrocytes, Blood Viscosity, and the Distribution of Blood Flow in the Microcirculation

H. SCHMID-SCHÖNBEIN
Technical University, Aachen Germany

Supported by grants from Deutsche Forschungsgemeinschaft Schm 84/1–7.

INTRODUCTION

The influence of blood viscosity in the normal and abnormal physiology of the circulatory system has long been neglected. In light of the overwhelming influence of vasomotor influences ($V \doteq r^4$) and the small influence of viscosity ($V \doteq 1/\eta$) this is understandable in studies of the overall behavior of the macrocirculation.

However, negligence of the "viscosity factor" in microcirculatory physiology and pathophysiology is not permissible. For obvious geometrical reasons, the motion of blood in the arterioles, the capillaries, and the venules cannot be understood without reconciliation of the flow behavior of the blood cells and the plasma on one hand, and the interaction between blood vessel wall and vessel content on the other hand. This obvious fact has greatly stimulated the study of the flow properties of blood, which are long known to deviate from those of simple, Newtonian fluids.

PRESENT STATE OF BLOOD RHEOLOGY

In the late 1960s a consensus was reached by the majority of workers in the field about the factors governing blood fluidity in vitro. A number of extensive reviews have been written and should be consulted (Bicher (1); Braasch (2); Charm and Kurland (3); Chien (4); Cokelet (5); Dintenfass (6); Gabelnick and Litt (7); Larcan and Stoltz (8); Merrill (9); Schmid-Schönbein and Wells (10); Schmid-Schönbein (11); Skalak (223); Wells (12)).

As discussed elsewhere (10,11) it is now generally accepted that the so-called "anomalous viscosity" of blood in vitro is caused by the presence of *red blood cells* and must be related to two distinct red cell properties, namely, their *deformation* in rapid flow and their *aggregation* in slow flow. It was further agreed that the presence of white blood cells and isolated platelets under normal conditions has a rather minute effect on blood "viscosity" in large blood vessels. This notwithstanding, it goes without saying that the sticking of the platelets and leukocytes to the endothelium, to endothelial defects, or to other platelets (formation of a hemostatic plug) must have profound effects on the actual blood flow in the paracapillary bed. These effects have largely evaded exact hydrodynamic quantification under in vitro conditions and preliminary results about the physical forces involved in leukocytes sticking (Atherton and Born (13); G.W. Schmid-Schönbein, Fung, and Zweifach (14)) have only recently become available.

The factors found to govern blood fluidity in vitro have always been familiar with students of the living microcirculation. The unique deformability of the mammalian red blood cell has been described and appreciated in a more qualitative manner by the earliest students of microvascular beds.

Many details of the cell deformation only became evident when adequate techniques, such as high speed cinematography, made it possible to resolve rapidly flowing blood with high magnification. The phenomenon of intravascular aggregation with concomitant blood flow retardation, on the other hand, can be observed with much less technological and optical expenditure. Some of the most consistent findings of experimental pathologists studying the microcirculation after injury are the retardation of venular blood flow, the formation of intravascular red cell aggregates, the hemoconcentration, and the eventual stagnation of blood flow.

This series of events termed "prestasis" and "stasis" has been described unanimously by all investigators from the pioneering work by Cohnheim (15) up to the most recent reports. Experimental pathologists, especially in the German literature (e.g., Rickert and Regendanz (16); Nordmann (17); Illig (18); Weber (19)), have strongly emphasized the significance of "stasis" in the inflammatory reaction. Likewise, intravascular aggregation and flow retardation were observed in hemorrhagic, endotoxin, burn and tourniquet shock, and also following local or general flow retardation (for a review see Knisely (20)).

At present it is impossible to decide to what extent and by what mechanism or mechanisms an alteration in the blood mechanical properties might give rise to altered flow in the microcirculation. Admittedly, the coincidence of the phenomena observed in vitro and those in vivo suggests a close relationship. However, a warning against simplistic and overoptimistic expectations is in order. While alterations of the microrheological properties of the human blood—often correlated to "stasis" in the human microcirculation—in the paracapillary bed of the nail fold, conjunctiva bulbi (Knisely (20)) are qualitatively well documented, an unequivocal evaluation of possible rheological factors in the natural history (aetiology and pathogenesis) of the vascular response to injury calls for at least three steps: 1) quantification of the *flow properties* of blood elements followed by an analysis of their flow behavior under well defined artificial flow conditions in vitro[1]; 2) a description of the complex natural *flow*

---

[1] The theoretical problems encountered when measuring "properties" of complex materials in rheology have been treated by Scott-Blair (222). It is a priori clear that testing

conditions in the microcirculation of various organs as defined by dimensions and architectural arrangements of the different classes of vessels as well as by a quantification of the forces of flow; 3) predictions about the in vivo *flow behavior* of the complex fluid "blood" in the complex vascular beds under varying flow conditions. Only this sequence of investigations will eventually allow a precise appreciation of the causes and consequences of intravascular flow or no-flow phenomena.

RHEOLOGICAL PROBLEMS IN THE MICROVASCULATURE

The development of highly structured and big mammalian organisms, capable of adapting within wide ranges to the change in metabolic needs and to a highly variable environment, is intimately linked to the design of an effective circulatory and respiratory system. Each single parenchymal cell of this macrocosm is, just like its earliest ancestors, the unicellular inhabitant of the antidiluent waters, nourished by diffusion from its pericellular space. Life is linked to mechanisms keeping constant not only the *milieu intérieur* of each cell, but just as much by the *milieu extérieur*–the peri- or extracellular space. Oxygen transport is a most sensitive and critical task. Efficiency of the O_2 supply is based on the optimization of both diffusive transport (miniaturization and ramification of the microvascular tree) as well as of the convective transport (motion of highly concentrated red cell suspensions through narrow tubes with a minimum of frictional energy loss) and a minimum of space requirement for the vascular compartment of the tissue. Fick's law described the simple fact that transport rates by diffusion are directly proportional to concentration gradients (dC/dx), to the exchange area (A), and inversely proportional to the square of the distance between the reservoir (the blood capillaries) and the recipient of a solute (mostly the parenchymal cells).

As elaborated by Burton (22) the high energy requirements of perpetually active mammalian cells at $37°C$, in connection with the given solubility and/or diffusivity of the oxygen and terrestrian p_{O_2}, requires that $10-20 \mu m$ should be the maximum distance for diffusive transport of gases and solutes, a requirement indeed realized (Handbook of Biological Data (23)). In order to comply with these requirements, the tissue must be perforated with a set of capillaries. Due to the fourth power law of Poiseuille (24), nature had the choice to either minimize the hydraulic resistance (by supplying large-bore capillaries) or to remove most of the parenchymal cells at the cost of vascular space. As shown in Figure 1 (Burton (22)), a maximum of $20 \mu m$ intercapillary distance means that fairly large volume requirements are needed to accommodate these capillaries. The selection of 20, 10, or $5 \mu m$ capillaries leads to pronounced differences. With a $20 \mu m$ capillary diameter some 70% of the tissue space would be taken up by

of bio-materials as a rule requires comparatively large forces and deformations, so that in most tests the properties are affected by the testing procedure. Therefore, the obtained information describes "processes" rather than "properties." When the term "blood viscosity" is used hereafter in this paper, it means "the quantity computed from viscometric measurements." As the processes accompanying blood flow differ from instrument to instrument, so does the quantity obtained.

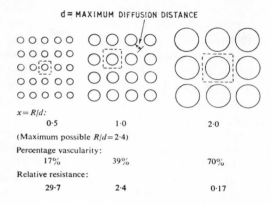

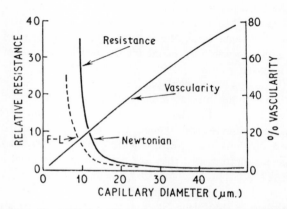

Figure 1. Perforation of a solid tissue block by a set of capillaries with a maximum distance of 20 μm. Effect of capillary diameter on vascularity (% of tissue space occupied) and on flow resistance for Newtonian fluids and for blood exhibiting the Fahraeus-Lindqvist effect (FL) (taken from Burton (22)).

blood vessels. Using 10 μm, capillaries would still require some 40% of the tissue space but would at the same time increase hydraulic resistance by a factor of 16. The choice of 5 μm capillaries only requires that some 17% of the tissue space is occupied by blood vessels, but this is paid by the 256-fold increase in vascular resistance, not counting the greater total lengths of branches, nor bends, tapers, and confluxes of blood vessels. The unusual deformability of the red blood cells and their minute effect on the flow of plasma effectively reduced the hydraulic resistance in capillaries to about 30% of the value that would be required when perfusing such capillaries with a Newtonian fluid having the "viscosity" of blood in large tubes.

The alternative, namely, the direct dissolution of the oxygen carrying hemo-protein in the plasma, has obviously not been proven successful in the phy-logeny. A direct dissolution of 32 g% hemoglobin in plasma would have had the inevitable consequence of increased colloid-osmotic pressure (producing untoward effects on transcapillary exchange) and plasma viscosity. Schmidt-Nielsen and Taylor (25), not taking into account the Fahraeus effect (26), have compared whole blood and hemolysate viscosity and have measured a hemolysate viscosity of 3.3 cP, or about 250% of plasma viscosity. In the vast majority of blood vessels, however, the Fahraeus effect *cannot* be neglected; therefore, contrary to the opinion of these authors, these findings must be taken as indication that the design of a transport cell for hemoglobin produced significant hemodynamic benefits (see below).

In order to understand the efficiency of the mammalian O_2 transport system, it must be regarded both as the consequence of a phylogenetic development and as the consequence of a day by day selection process. Statistically, only red cells capable of adapting to the given microvessels—and only microvessels capable of accommodating circulating erythrocytes—have a chance of long-term survival.

It is well established that in the process of erythropoiesis the red cells are released from the bone marrow at the time of the extrusion of the erythroblast nucleus (27). As a result of this, the cell loses proportionally more volume than surface area. At the same time, the mature erythrocyte acquires its unusual deformability, which is primarily based on the relative surplus of surface area (160 μm^2) for its given volume (90 μm^3) (see Fung (28), especially page 29), as well as on the fluidity of its contents, i.e., the lack of structured cell organelles. Leblond et al. (29) have taken measurements of cell deformability of the different stages of maturing red cells and have shown that it increases from the erythroblast to reticulocyte and thence to mature erythrocyte. The spleen appears to play a supplementary role in selecting the fittest erythrocytes. As the reticulocytes are less deformable, they are preferentially trapped in the splenic pulp (e.g., Refs. 27, 30–34, 207, 224, 229), where presumably they either await maturation or are being phagocytosed. Removal of the aged red cells also appears to take place in the spleen (Bergentz and Danon (206)), although many details of this latter process remain unclear. Most workers suppose that red cells are removed when they lose their deformability (33–37). Thus, the circulating, mature red cells are the result of a perceptual mechanical selection process. These simple facts are of great significance not only hemodynamically, but also in a practical and heuristic respect. By definition of their very presence in the circulating blood, red cells are distinguished to be normally deformable and suitable to pass even the most restricted channels patent in the microcirculation. In other words, whenever red cells are taken from an artery or a vein for rheological analysis, the population obtained is highly selected with respect to its mechanical properties, and its adequate capabilities (see below) are only a natural consequence of this selection.

Sobin (38) has made comparative measurements of the minimum capillary diameters and the dimension of the erythrocytes in various species. In all of those species that have highly deformable red blood cells (see below), the capillary diameter is in the order of 3–5 μm. Little is known about the factors leading to this appropriation. However, it is often assumed that endothelial sprouting is induced by a lack of oxygen and/or biochemical sequelae thereof; it would consequently appear quite logical to assume that the growth stimulus is removed whenever the lumen of the developing endothelial tube is just large enough to allow the passage of the erythrocytes as oxygen carriers. The more deformable the red cells, the smaller the critical capillary diameter, the more effective the O_2 exchange system.

DESCRIPTION OF FLOW CONDITIONS IN VIVO

Hydrodynamic Analysis of the Vascular Bed

Quantitative information regarding the angioarchitectonic arrangement of the different blood vessels in the organs of the body is mostly missing. The anatomical knowledge about the wall structure of the different vessels in the micro- and macrocirculation is of little help in this respect. For a detailed hydrodynamic analysis, data of the following kind would be required: (a) diameter (minimum-maximum), taper, curvature and branching of single vessels (arteries, arterioles, true capillaries, venules, veins, arterioarterial, arteriovenous, and venovenous shunts); (b) distribution of individual vessel cross-sectional area, total cross-sectional area of any class of vessels, effects of vasoconstriction and vasodilatation on these parameters; (c) intravascular pressure and its pulsation, pressure drop (ΔP) and pressure gradient ($\Delta P/l$) as well as transmural pressure under normal and various pathological conditions.

From the data presently available, the following pertinent conclusions can be drawn.

1. The majority of blood vessels (see Table 1) have diameters below 50 μm and these are in the same order of magnitude as the diameter of the blood cells. The true, nutrient capillaries in most mammals are smaller than the erythrocytes of the same species. According to preliminary data presented by Sobin (38), the capillaries of species with relatively nondeformable but small erythrocytes (goat, sheep, and chicken) are slightly larger than the resting diameter of the respective erythrocytes.

2. It is obvious that a generalized or idealized concept of the "microcirculation" would be entirely misleading. The angioarchitectonic arrangement varies from organ to organ and is adapted to the physiological task of the respective organs, i.e., based on a specific arrangement of the microvascular bed. Not only the normal circulation and function, but to the same extent the pathological circulation must therefore be subject to the organ-specific vascular arrangement and its control by metabolic, nervous, or hormonal control. Furthermore, differences in angioarchitectonics and control mechanisms might explain some of the well documented species differences and different organ-specific responses to localized or generalized circulatory disturbances.

3. Our present knowledge about the intravascular flow dynamics mostly stems from the observation of two-dimensional vascular beds as they are found in only a few limited, often highly specialized tissues. Extrapolation to other organs with three-dimensional vessel arrangement must be executed with great care (39).

4. The classical concepts of macrocirculatory physiology and pathophysiology

Table 1. Angioarchitectonics, geometry, and hydraulics of an arbitrary vascular bed

Type of vessel	Diameter (mm)	Number	Total cross-sectional area (mm²)	Length (mm)	Fraction of total volume (%)	Hindrance $l/n \cdot d^4$ (mm⁻³)	Pressure gradient $\Delta P/l$ (mm Hg mm⁻¹)[a]	Intra-vascular pressure (mm Hg)	Wall shear stress $(\Delta P \cdot d/4l)$
Aorta	10	1	0.8×10^2	4×10^2	2.0	4×10^{-2}	0.0075	100	24.9
Large arteries	3	40	3×10^2	2×10^2	4.0	6.2×10^{-2}	0.0215	97	21.4
Main artery branches	1	600	5×10^2	10^2	3.4	1.7×10^{-1}	0.129	92.7	42.8
Terminal branches	0.6	1.800	5×10^2	10	1.7	4.3×10^{-2}	0.330	79.8	65.7
Small arteries	0.019	4×10^7	1.1×10^4	3.5	2.7	6.6×10^{-1}	5.97	76.5	37.6
Arterioles	0.007	4×10^8	1.5×10^4	0.9	1.0	9.9×10^{-1}	32.1	55.6	74.6
Capillaries	0.0037	1.8×10^9	1.8×10^4	0.2	0.3	6.8×10^{-1}	89.6	25.1	110.1
Postcapillary venules	0.0073	5.8×10^9	2.5×10^5	0.2	3.6	1.2×10^{-2}	1.90	4.5	4.6
Venules	0.021	1.2×10^9	3.7×10^5	0.1	25.6	4.4×10^{-3}	0.3	4.1	2.1
Small veins	0.037	8×10^7	8×10^4	3.4	18.6	2.3×10^{-2}	0.5	3.8	6.1
Main venous branches	2.4	600	2.7×10^3	10^2	18.6	5.0×10^{-3}	0.004	2.1	3.2
Large veins	6.0	40	1.1×10^3	2×10^2	15.2	4.9×10^{-3}	0.002	1.7	4.0
Vena cava	12.5	1	1.2×10^2	4×10^2	3.4	1.7×10^{-3}	0.003	1.3	13.5

Based on data of Mall (1888) and Wiedeman (1963).

[a] Assuming = 1.5 cP.

are not a priori applicable to the analysis of microcirculatory hydrodynamics. The analogy between flow resistance across an organ and electrical resistance as described by Ohm's law is quite meaningless in this respect. The measurement of the so-called "peripheral resistance" can only present a lumped value in the behavior of all blood vessels. Such macrocirculatory measurement cannot substitute the necessary analysis of individual vessel flow dynamics. At present such analysis is admittedly often impossible. It must be stressed, however, that the tacit assumption that all blood vessels respond in a quantitatively similar fashion to stimuli or altered flow properties of the blood is untenable, as is any assumption of a kind of "average behavior" of the vessels in the microcirculation (see Fung (40) and Fung and Zweifach (209) for a more detailed discussion of the stochastic nature of the microcirculation). In order to gain an order of magnitude estimate of the hydraulics of the macro- and microcirculation, there have been a number of attempts (Goldsmith and Mason (41); Chien (42); Schmid-Schönbein and Wells (43)) to calculate shear rates and/or shear stresses in vessels of the microcirculation.

All of these are based on the limited quantitative information on angioarchitectonics available in the literature. In 1888 Mall (44) injected and fixed the canine small intestine and its mesentery. He carefully measured vessel lumina and counted the vessels of the different types. The state of the injection technique available to Mall makes it questionable whether or not capillary and postcapillary vessels were actually filled; as Mall himself admitted, he had excluded from his analysis frequent postcapillary accumulation of venules (*Venenbällchen*). Furthermore, after serial sectioning of the injected specimen, Mall was unable to estimate the vascular lengths. Some 40 years later, Schleier (45) "completed" Mall's measurements by estimating vascular lengths from the schematic drawings in Mall's original paper. Mall's data are quoted in practically all textbooks of physiology throughout the world (often as treated by Green (46)). Important physiological concepts, e.g., that of the venous capacitance system, are based on these archaic findings. Wiedeman (47) analyzed the vascular bed in the wings of unanesthetized bats, and her findings differ considerably from those of Mall. Schmid-Schönbein and Wells (43) and Chien (42) have independently used this vascular bed for a theoretical computation of the distribution of shear rates in the different vessel classes; assuming an arbitrary flow rate, the average shear rates rose steadily from the arteries ($\bar{U} = 69.2$ s^{-1}) to the capillaries (271 s^{-1}), followed by a sharp fall in the postcapillary venules (2.4–11.2 s^{-1}). Computation of minimum shear rate values in venules corresponds to the frequent observation that in vivo red cell aggregation occurs preferably in venules of both the normal and pathological microcirculation.

Keeping in mind the pitfalls of such an approach, Schmid-Schönbein and Devendran (48) have gone one step further and have combined the data of Mall and Wiedeman, obtaining a full vascular network which lends itself to a hydraulic analysis—Table 1 shows details. From the data on the number of vessels, lengths and radii, the total cross-section, total and fractional volume capacity as well as the hydraulic hindrance $(1/n \cdot d^4)$ was computed for all classes of vessels: this value shows a steady increase from the arteries to the capillaries, followed by a conspicuous fall in the postcapillary region. By assuming perfusion of this vascular bed with a Newtonian fluid with the viscosity of plasma (1.5 cP), it becomes possible to calculate the pressure drop along the different vessels. The greatest ΔP is found along the small arteries and arterioles, resulting in an intravascular pressure of 25 mm Hg in the capillaries, with only small ΔP in the venules and veins. The pressure gradient $(\Delta P/l)$, however, and thence the shear stress at the wall $(\Delta P \cdot d/4l)$ has its maximum in the capillaries (110 dyn/cm²), again followed by a steep drop in the venules to values around 4 dyn/cm². The shear stresses in the microcirculation can fall below normal due to a number of factors: fall in arterial or rise in venous pressure, steeper precapillary pressure gradients due to vasoconstriction or increased blood viscosity anywhere along the vascular tree. Shear rates in the venular part drop critically whenever general vasoconstriction affects the arterioles more than it does the venules; when capillaries but not *venules* are shut off from the circulation, or when arteriolar vasoconstriction is associated with venular dilatation (Schmid-Schönbein and Wells (43)).

Most recently, direct hydraulic mapping of the cat mesentery and omentum was performed by Zweifach (49) and Lipowski and Zweifach (50); these studies have produced data that agree favorably with the computed values, listed in Table 1. This is surprising when considering that the model computed by Devendran (48) is entirely artificial, that very crude data were used, and that many circulatory phenomena (pulsating flow and pressure, branching, taper, vasomotor effects) were neglected in the theoretical treatise. It should be stressed, however, that the distribution of total cross-sectional areas, total vascular volumes and thus of shear rates in the different classes is valid only for those vascular beds that contain abundant postcapillary venules. It is therefore valid for the omentum and mesentery, presumably for the conjunctival and subcutaneous tissues (e.g., the vascular bed in implanted cutaneous tissue chambers). The angioarchitectonics of muscle capillaries is almost certainly entirely different from the present model (see Gross (51) and Gross and Intaglietta (212)). A number of basic conclusions necessary from the understanding of blood rheology in vivo can nevertheless be drawn from the following data:

1. The wall shear stresses in all parts of the macro- and microcirculation appear to be in the order of 10–50 dyn/cm², a shear stress sufficiently high to force the blood into rapid flow where apparent blood viscosity is low (see p. 8).

2. As the wall shear stresses $(\Delta P/d/4l)$ are a function of the driving pressure $(\Delta P = P_{art} - P_{ven})$, it is immediately obvious that any reduction of the upstream pressure (due to fall in cardiac output, arterial stenosis, or arteriolar constriction) as well as an increase in downstream pressure (due to cardiac failure, venous occlusion, or thrombosis) will reduce the wall shear stresses in any vascular segment. Whether or not *vasomotion* will reduce or increase the wall shear in any vessel will greatly depend upon the effect of *vasoconstriction* on

pressure gradient and radius of the respective individual vessel, as well as on the total cross-sectional area of all vessels of one kind. Endarteriolar *dilatation* will increase wall shear stress provided that upstream dilatation has increased both the intravascular pressure and the total available pressure gradient along the endarteriole. Endarteriolar vasoconstriction can result in either rise or fall of wall shear stresses, depending on the change in ΔP or Δd.

3. While under resting conditions the wall shear stresses in the pre- and post-capillary segments appear to be at least of the same order of magnitude; the response to a flow disturbance in the high pressure system and in the low pressure system is vastly different. In the high pressure (the precapillary system), the normal wall shear stresses are only a fraction of the maximum possible shear stresses. In case an intravascular obstruction should occur in the high pressure system, the full arteriovenous pressure gradient $(P_{art} - P_{ven})$ would immediately build up along the affected vascular segment and it would increase both the normal (pressure) and the shear (flow) forces. As a result, the obstruction would be flushed away, unless it is structurally capable of withstanding such high shear forces.

In the low pressure system, quite to the contrary, any obstruction would have a far lesser effect. When situated in the major proximal veins, ΔP and thus wall shear stresses would be reduced; when situated in the smaller, distal veins and venules, wall shear stresses could only be elevated to a small extent. When, however, the obstructed postcapillary segment is neighbored by collaterals that can shunt the flow around the obstructed segment, the wall shear stresses along the occluded segment can be varied by the flow dynamics in the shunting vessels. Whenever many of them are open, and when therefore flow velocity in any single one is low, wall shear stress in the affected vessel may fall further on. As is discussed in more detail on page 44, a drop of postcapillary wall shear stresses in branching microvascular beds with many collaterals can explain the pronounced effects of red cell aggregates in vivo, despite the fact that in vitro red cell aggregates can only withstand shear stresses between 2 and 20 dyn/cm^2.

MEANING AND MEASUREMENT OF THE "VISCOSITY OF BLOOD"

The well known variability of blood viscosity as a function of flow rate, vascular diameter, and hematocrit value makes it a priori impossible to ascribe one "equivalent viscosity" to the blood flow in the complicated vascular bed. Any attempts to correlate certain states of high or low flow to a value of "apparent blood viscosity" as measured in a rotational or capillary viscometer is without base (see footnote on p. 16). It is much more meaningful to examine the microrheological factors causing the "anomalous" viscosity in vitro: such analysis will have to start from the obvious fact that blood is a rather concentrated suspension of particles (mostly erythrocytes), which themselves *change their physical characteristics in response to mere changes in shear stresses*. In addition, biochemical factors and changes in the composition of the plasma and the blood cells affect the physical behavior of the cells and their response to flow forces.

While admittedly such changes may be detected by means of viscometry, much more sensitive tools have recently become available to study directly the microrheological properties of red cells and red cell aggregates (see below).

BLOOD VISCOSITY IN VITRO

Reported values of apparent blood viscosity in the literature vary between values similar to that of plasma (Barras (52); Gerbstädt et al. (53); Braasch and Jenett (54)) in glass capillaries to maximum values of 57 P (Dintenfass (6)). As quoted in most textbooks of physiology, apparent blood viscosity in rapidly perfused large-bore blood vessels depends on the hematocrit level, i.e., the volume fraction of red cells. But even this seemingly well established fact is by no means universally applicable as it only applies for a limited range of forces and vessel diameters. The complexity of the hematocrit-viscosity relationship becomes evident when considering the extremes of prestatic flow in large, and rapid flow in narrow, tubes. In the former, viscosity values of up to 5.9 poises can be measured (62% hematocrit in a burn patient with hemoconcentration at 1.2 s^{-1} (Schmid-Schönbein, unpublished), whereas in the latter, similar hematocrit levels only lead to a "viscosity" 1.3 times that of plasma (52, 54, 55).

There have been numerous attempts to express the shear rate-shear stress relationship of human blood in terms of "blood viscosity equations." Most of these attempts are based on rather unrealistic theories or assumptions about the factors underlying the shear rate dependency of blood "viscosity." Often they are mathematical descriptions within a limited range of the shear stress-shear rate relationship for viscometric flow (power law fluids). A few examples follow: the so-called Casson equation (56) is used most commonly (see Charm and Kurland (3); Aroesty and Gross (57); Gross and Aroesty (58); Merrill (9); Oka (60); Scott-Blair (61)) and it is based on the assumption that the shear thinning behavior (apparent viscosity falling with rising shear rates) is merely the consequence of red cell aggregates, which are formed in slow flow and are dispersed in high flow. As discussed extensively elsewhere (Schmid-Schönbein and Wells (10)) and most recently in connection with a study of blood viscosity in different species (Schmid-Schönbein et al. (62)), the equation was originally proposed for nonaggregating blood. Scott-Blair (61) applied the equation to Kümin's (63) data on bovine blood, in which red cell aggregation does not occur at all, and for rapid blood flow in narrow tubes, where shear thinning is a consequence of red cell deformation (62).

Dintenfass (64) has based his "blood viscosity equation" on the far more realistic model of the red cell as a fluid drop. Consequently, blood is taken as an emulsion, i.e., as a dispersion of fluid drops in a continuous fluid phase; the viscosity of such emulsions was treated theoretically by Jeffrey (65) and Taylor (66, 67) and later by Mason and Bartok (68), Goldsmith and Mason (41), Gauthier et al. (69), Kline (216), and others. These equations need refinement to satisfy the unique micromechanical properties of the mature mammalian erythrocytes (see below), but are, at least for high shear flow, much more useful than the Casson equation, which must be considered obsolete after the establishment of red cell fluidity as the main factor governing blood flow at high shear rates.

The most useful and generally applicable theory of blood "viscosity" was presented by Chien (70), who assumed that the shear-dependent viscosity of blood (apparent viscosity falling with increasing shear rate) in rotational viscometers corresponds to a shear dependence of the effective cell volume. By measuring the relative apparent viscosity of blood (apparent viscosity of blood divided by plasma viscosity) it becomes possible to quantify the effect that a given volume fraction of red cells exerts on the flow of the plasma alone (see below). This reasoning also takes into account that in the microvasculature a flow of two distinct phases (cells and plasma) rather than bulk blood flow is taking place. Particles increase the viscosity of the medium in which they are dispersed because they disturb the mutual sliding of fluid lamellae in the medium. Einstein (71) established the fundamental equation for the flow of suspensions which is based on the following argument. Through the presence of particles, the fluid lamellae in between them have to pass more rapidly past each other to produce a given flow rate or shear rate. The shear stress necessary to produce this same rate of shear, and hence the computed apparent viscosity, is increased over the viscosity of the medium alone by a factor that depends on the viscosity of the medium (η_o), the volume fraction (H), and a shape factor (K).

$$\eta_s = \eta_o + \eta_o \cdot K \cdot H \qquad (1)$$

Einstein's formula $(\eta_{rel} = \eta_s/\eta_o = 1 + K \cdot H)$ (71) is valid only for dilute suspensions of rigid particles, whereas in concentrated suspensions it has to be modified to account for the interaction between the particles (227). Irrespective of these interactions, the relative viscosity rises when the particles aggregate into chains or networks. This rise is due first to an increased shape factor (K) and second to the augmented effective volume fraction (H) of particles since part of the volume of the continuous phase is immobilized between aggregates. Particle networks bridging all fluid elements and connecting the boundaries may abolish all laminar sliding: apparent viscosity then rises to infinity.

The formation of rouleaux by red cells, and the secondary aggregation of rouleaux into three-dimensional networks, have quite an analogous effect: apparent blood viscosity between about 1.0 and 0.1 s^{-1} increases very steeply with falling shear rate. Below 0.1 s^{-1} the apparent viscosity of "whole blood" is no longer measurable. As a consequence of phase separation effects (separation of plasma and red cell aggregates, complicated by settling) there is no constant shear stress reading for a given shear rate, but rather a complicated change of shear stress with time, which in turn strongly depends on the previous history of shearing (Figure 2). The time course is different from blood sample to blood sample (9, 10, 72, 73); in the same blood sample it depends upon the geometry of the viscometer, and upon the fact whether the measurement is taken following a preceding period of high

14   Schmid-Schönbein

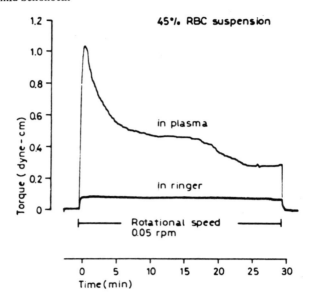

Figure 2. Transient in "apparent viscosity of blood" when measurements are attempted at a shear rate below 0.1 s^{-1}. Due to phase separation of aggregates and plasma, there is a constant decay of shear stress (and thus "apparent blood viscosity" despite of constant shear rate (taken under Usami and Chien (201)).

shear (while the previously dispersed cells rejoin into aggregates) or low shear (while the previously formed small aggregates increase in size, condense, or begin to settle (73, 74)). These difficulties notwithstanding, a number of authors (72, 75) have attempted to compute "apparent viscosity of blood" from shear stress measurements taken at shear rates below 0.1 s^{-1}. As can be seen from Figure 3, the computed quantity either remains constant, decreases, or increases. It is clear, therefore, that, contrary to previous contentions (10, 59, 76), whole blood in large-bore viscometers does *not* exhibit a yield shear stress. The red cell aggregates, of course, do withstand quite high shear stresses (4, 77, 78). At shear rates below 0.1 s^{-1}, shearing exclusively takes place in a plasma sleeve (72) that separates the red cell aggregates from the viscometer walls. The elastic components (79, 211, 218) in each red cell aggregate thus remain undetected by simple viscometry; this of course does not mean they are not there. Increasing the shear rates above 1.0 s^{-1} leads to progressive aggregate dispersion. However, unlike particles that can elastically maintain their shape, the dispersed red cells become deformed and aligned, and they participate in flow much like fluid drops (see p. 32). Consequently, their flow behavior can be approximated by equations describing the viscosity of emulsions, in which not only the volume fraction of the dispersed droplets but also their viscosity (η_i) determines the disturbance of the flow of continuous phase (η_o). Taylor (66, 67) modified Einstein's (71) equation as follows:

$$\eta_{emuls} = \eta_o + \eta_o \cdot K \cdot H \cdot T \tag{2}$$

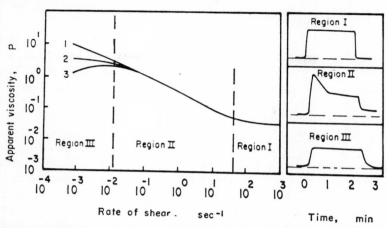

Figure 3. "Apparent viscosity" of whole human blood as a function of the rate of shear as measured in a Weissenberg rheogeniometer. At shear rates below 0.01 s⁻¹, the data are not consistent. *Insert*: Shear stress for a given shear rate as a function of shear time. High shear, steady state rapidly reached (region I); low shear, transients in time due to phase separation (region II); creeping shear, steady state rapidly reached, presumably by shearing in plasma (from Copley, Huang, and King (72)).

where

$$T = (\eta_i/\eta_o + 0.4) / (\eta_i/\eta_o + 1.0).$$

This formula again does not consider droplet interaction nor shear rate dependency. Dintenfass (63) and later in more details Schmid-Schönbein and Wells (10, 80) and Goldsmith (69, 81) established that red cells behave like fluid droplets with the notable exception that red cells are deformed much more easily than ordinary fluid drops. Consequently, the relative viscosity of red cell dispersions can be even lower than predicted by Taylor's equation.

Barras (52) and Gerbstädt (53) have shown that in capillaries between 50 and 10 μm diameter the Fahraeus-Lindqvist phenomenon becomes so pronounced that blood viscosity in rapid flow approaches that of plasma viscosity alone, or relative viscosity approaches 1.0. The various factors involved are all rooted in the unusual deformability or fluidity of erythrocytes (see Figure 17, p. 34). Due to red cell fluidity, strong axial migration occurs in arteriolar size vessels (Goldsmith and Beitel (82)), and bullet-shape deformation takes place in even smaller vessels. Consequently, the shape factor changes, the effective hematocrit is reduced when cells travel much faster than plasma (83, 84), the marginal plasma layer lubricates the axial stream, the hemodynamic effect of which becomes largely independent of the number of cells transported (52, 54, 55), and their aggregation has only a small effect on viscosity (Skalak et al. (85)).

Relative viscosity of blood, written in its most basic form

$$\eta_{rel} = \frac{\eta_{susp}}{\eta_{plasma}} = 1 + K \cdot T \cdot H \tag{3}$$

can vary between infinity, extremely high values, and 1.0 because H is no constant and K and T vary as a function of shear stress and tube radius.

This variability has in fact been documented frequently by viscometric results obtained on whole blood in rotational viscometers. Figure 4 shows, as an example, the effect of various shear stresses on the one hand, and of various hematocrits on the other, on relative apparent viscosity of human blood. In all samples of normal blood,[2] relative apparent viscosity is minimum at high shear stresses ($>$ 5 dyn/cm^2) due to deformation of the dispersed cells. With decreasing shear stresses, relative apparent viscosity increases. This increase is highly dependent upon hematocrit and aggregate properties. Under pathological conditions, these viscosity profiles are frequently shifted to higher values for a given hematocrit (see p. 17). However, even when correcting for a given hematocrit value and plasma viscosity, one cannot determine the reason for the increased blood viscosity: it may be intensified red cell aggregation or severed cell deformability. It was therefore necessary to develop independent methods to quantify the kinetics of red cell aggregation and to quantify red cell deformability (see p. 20).

SMALL-BORE TUBES—THE FAHRAEUS EFFECT

Chien's (70) theory accounts for the experimental results obtained in rotational viscometers under various experimental conditions, whereby after appropriate changes in the composition of the plasma or the cells the relative apparent viscosity of red cell suspensions was reduced by procedures inducing red cell deformation (Schmid-Schönbein et al. (88), Chien (4), Dintenfass (64), Meiselman and Cokelet (89)). In such instruments, the actual red cell volume fraction is invariant; only the hydrodynamically effective volume fraction (factor H and K in equation 3) varies as a function of the shear rate.

When measuring blood viscosity in capillaries below 300 μm in diameter, not only the hydrodynamically effective but also the actual volume fraction of red cells are inconstant. This is the consequence of the phase separation between red cells and plasma, which occurs in these vessels due to the axial migration of the red blood cells. While axial migration, i.e., the movement of particles from the region near the vessel wall to the vessel axis, is a general phenomenon in *dilute* suspensions (221), rate and extent of axial migration in the *concentrated* suspension of red cells are unique. As shown by Goldsmith and Mason (41), and more recently by Devendran et al. (90, 91), the pronounced axial migration of red cells is a function of cell deformability and cell aggregation. Axial migration is most pronounced in rapidly flowing suspensions of strongly aggregating, highly flexible red cells, and least pronounced in rigidified cells. When using

---

[2] When red cell deformability is severed, the apparent blood viscosity may *increase* with *increasing shear rate* ("dilatancy") and shear time ("rheopexy"). This phenomenon was studied extensively for analytical reasons (see p. 18) but may also occur in vivo since it is seen in blood with high osmotic pressure as it occurs in the vasa recta of the mammalian kidney in antidiuresis (Schmid-Schönbein et al. (86)).

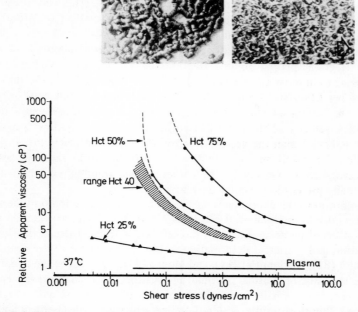

Figure 4. Relative apparent viscosity ($\eta_{blood}/\eta_{plasma}$) of human blood as a function of shear stress and hematocrit value in viscometric flow (couette and cone plate-viscometer). The pronounced non-Newtonian viscosity is the consequence of two rheological properties: red cell aggregation below 1 dyn/cm^2, red cell deformation, orientation, and membrane rotation above 10 dyn/cm. Transition occurs between 1 and 10 dyn/cm^2 and is strongly affected by such factors as temperature, hematocrit level, and plasma protein composition.

totally fixed red blood cells, Devendran (91) not only failed to procure an axial migration of red cells, but a blockade of the capillary entrance occurred caused by cells sitting at the walls of the entrance segment.

The phenomenon of axial migration, which was first observed in vitro by Poiseuille (24) and in vivo by Eberth and Schimmelbusch (92), has strong hemodynamic consequences, which were fully appreciated and measured by Fahraeus and Lindqvist (26) in their classic papers. The axial migration has two effects:

1. A lubricating layer of low viscosity plasma is created, which in itself is likely to reduce the viscous energy dissipation.
2. As the cells move in the rapidly flowing axial part of the velocity profile, the red cell flux per unit time is far higher than the plasma flux (210): as a consequence, the red cells reside a much shorter time period in the blood vessels—or, in other words, relatively fewer cells than plasma are present in the rapidly perfused blood vessels.

Due to the more rapid erythrocyte flow rate, the actual volume fraction of the red cells in tubes of diameters below 300 μm progressively decreases, until it reaches a minimum, presumably at the capillary level. This was already fully appreciated by Fahraeus, and later investigators of this effect (e.g., Barbee and Cokelet (93)) have only supplemented his original ideas.

In order to transfer a given amount of red cells and plasma (e.g., red cell volume fraction, plasma volume fraction 0.55) from one large-bore reservoir (the aorta) to another (the vena cava), only the flow fraction (450 ml/min red cell flow per 1,000 liters/min of blood flow) in all blood vessels must be equal to the volume fraction of red cells in the reservoirs. In all of those vessels, in which the average velocity of the red cells is higher than the plasma flow rate, the flow fraction is higher than the volume fraction by a factor inversely proportional to the ratio of red cell and plasma flow rate. If this ratio thus increases with progressively decreasing radius, the actual volume fraction, i.e., the volume of cells actually present in the tube, decreases.

In addition to the dependency of the actual volume fraction on tube size and flow rate, the hydrodynamical effect of this reduced volume fraction may be further affected by arrangement of the cells and cell aggregates in the flow field. Orientation and perpetual deformation of blood cells, as they were described by Goldsmith (81), Hochmuth (214), and Monro (94), may reduce the effect of cells on the shearing of plasma. Normal aggregation, or even more so enhanced aggregation, may actually assist the formation of very dense axial cores. Fully developed plug flow results, with a flattened velocity profile (Berman and Fuhro (95), Gaehtgens (96), G.W. Schmid-Schönbein and Zweifach (220)), where shear only takes place in the marginal plasma zone. Therefore, the presence of red cell aggregates in the axial plug of rapidly flowing blood is by no means a model of increased viscosity; it may on the contrary reduce "viscosity" (Gaehtgens and Uekermann (97); Skalak et al. (85); Devendran (90)). However, the relationship among red cell rheology, axial migration, formation of marginal plasma layers, and the reduction of the actual and the hydrodynamically active volume fraction only applies to rapidly flowing blood.

When the blood flow is retarded, and thence the shear stresses are reduced, most of these hydrodynamically favorable effects of cell aggregation no longer take place in creeping flow: the rate and extent of axial migration are strongly reduced, as suspected by Bayliss (98), later shown by cinematography (Merrill et al. (59)) and by quantitative measurements of the width of the marginal layer as a function of shear stress (Devendran and Schmid-Schönbein (91)). In addition, at these same low rates of shear, the suspension of red cells becomes aggregated, and when reaching across planes of shear these aggregates further increase the apparent viscosity of blood.

When tested in small-bore tube viscometer, this effect of aggregation is much more difficult to quantify than in rotational viscometers; only a few reports on blood viscosity in small bore (< 300 μm diameter) are available in the literature, due to the technical difficulties involved. The data available from Barbee and Cokelet (93) in 29 μm tubes, and from Devendran et al. (90) in 45-μm tubes agree in that at shear stresses below 5 dyn/cm^2 the apparent viscosity of normally aggregating blood in capillary viscometers is considerably lower than that measured in rotational viscometers on comparable blood samples. Moreover,

Devendran et al. (90) found that the viscosity of nonaggregating samples of red cells in 45-μm glass capillaries was actually higher than that of aggregating samples (Figure 5). This finding is directly opposite to the behavior found in rotational viscometers, where nonaggregating samples notoriously exhibit lower values of apparent viscosity than aggregating samples at shear stresses less than 2 dyn/cm² (Figure 5). Devendran et al. (90), simultaneously measuring shear stresses and observing the microrheological behavior, was able to find the solution for these two puzzling paradoxes, as he observed not only a better axial migration of the aggregating sample, but in addition rapid sedimentation of the cells and cell aggregates to the bottom of the glass tubes. As fairly long intervals pass before a steady state measurement of flow in narrow glass tubes under extremely low driving pressures is possible, sedimentation leads to phase separation of the suspension before an actual measurement of the apparent viscosity can be taken. This effect is most pronounced in horizontal or slanted glass tubes, but also occurs in vertical tubes (unpublished observation) perfused by creeping human blood.

For these reasons, the influence of red cell aggregates on apparent blood viscosity in prestatic flow in narrow tubes remains unsettled. The measurements of Burton (99) and Haynes (55), so frequently cited in textbooks of physiology, have evaded all these complexities. The values measured, however, do not apply to "blood" and should therefore be abandoned. Haynes was apparently aware of the sedimentation difficulties so that he did not use whole blood, but rather red cells suspended in protein-free ACD solutions. In such medium, there is neither aggregation nor normal shape adaptation of red cells (see section "Blood Viscosity in Vivo"), since human and canine red cells in ACD solution are invariably

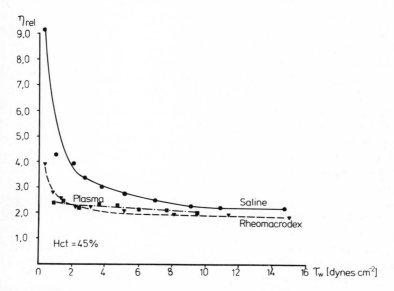

Figure 5. Relative apparent viscosities of red blood cell suspensions (η_{susp}/η_{pl}) as functions of wall shear stress (tube diameter).

transformed into crenated discs and spheres (Schmid-Schönbein and von Gosen, unpublished observation).

Another word of caution must be phrased against the so-called "inversion of the Fahraeus effect" (as introduced by Dintenfass (100)), who performed his measurement in wide slit-like channels with heights between 4 and 12 μm and not in cylindrical tubes. While it is interesting that he also finds a reduction in apparent viscosity with decreasing height of the channel, the upswing of the plotted curves does not seem to reflect actual measurements but rather indicates that the flow through these channels was totally or partially blocked. In the description of his experiments, the author concedes the possibility that emboli (platelet aggregates, etc.) may be responsible for these results. As viscosity by definition is "internal friction of a fluid," all factors leading to a reduction of channel dimension or to blockade due to external friction between solid particles and vessel walls should not be explained in terms of "viscosity." The interesting possibility that such an "inversion" of the Fahraeus effect actually exists therefore awaits experimental clarification under more closely controlled experimental conditions.

ANALYSIS OF BLOOD MICRORHEOLOGY:
OBSERVATION, PHOTOMETRY, AND VISCOMETRY

Following microrheological studies by Goldsmith and Mason (41), the study of the red cell flow behavior in the present author's laboratory was largely centered on the microscopic observation of the cells while subjected to quantifiable shear stresses in a "rheoscope" chamber (101, 102) (Figure 6). Subsequently, photometric and viscometric methods were developed in order to monitor objectively and simultaneously the subjective observations. Figure 6 depicts schematically such a rheoscope. Figure 7 shows photomicrographs of blood in stasis, under conditions of slow flow and rapid flow, as well as normal and enhanced red cell aggregation.

At shear stresses above 3 dyn/cm^2, normal red cells are monodispersed, but not biconcave; instead they are perpetually elongated and oriented with their major axis parallel to flow. Hydrodynamic analysis of this deformation has earlier led to the conclusion that the red cell membrane is in perpetual tank-tread-like rotation around the cell fluid content during such deformation (see below), rendering the red cell properties akin to those of a fluid droplet. Evidence demonstrating that such membrane rotation actually takes place is meanwhile available (see p. 32).

Upon mere reduction of shear stresses, the red cells are seen to aggregate into primary rouleaux and secondary rouleaux structures, a phenomenon also causing an increase in apparent viscosity. There are strong quantitative as well as qualitative (Figure 8) differences between the normal aggregation as found in the blood of human subjects and that of most animal species, and the enhanced red cell aggregation as found in disease, which can be mimicked by the addition of aggregating colloids in vitro (Schmid-Schönbein et al. (103–105); Volger (106–108)). These differences cannot readily be detected by rotational viscometers

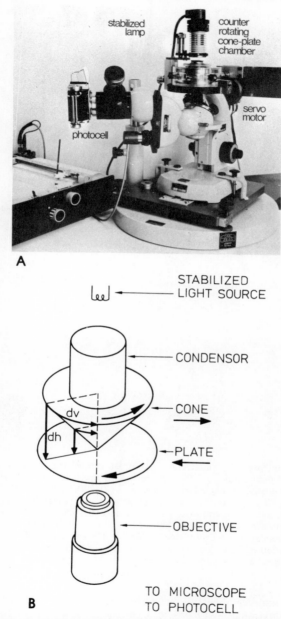

Figure 6. Schematic representation of the essential parts in the counterrotating rheoscope chamber. The light from a voltage-stabilized incandescent lamp (or from a strobe lamp) is collected in the rotating condenser. The image of the blood layer under study is projected into ocular or cameras. With the help of a beam splitter, a part of the light is directed to a selenium barrier layer photocell, the photo voltage of which is recorded on a d.c. compensation recorder. The photocell circuitry is identical with the one described in Ref. 104.

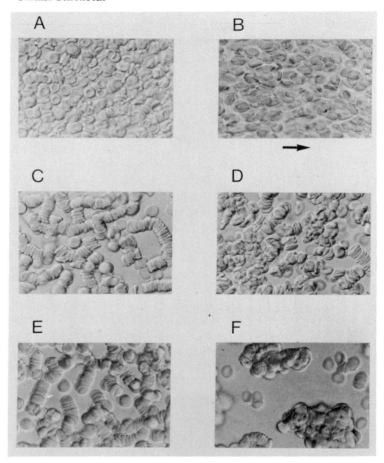

Figure 7. Photomicrographs (interference contrast optics, magnification 100 × 6.3) of human RBC and dextran-induced RCA in the "rheoscope" chamber. (A) 1000 mg% Dx 70: (stasis) single rouleaux, no secondary aggregation. (B) 1000 mg% Dx 70, 400 s⁻¹: Irregular deformation of RBC in flow; *arrow*: direction of flow. (C) 1000 mg% Dx 110. Stasis, continuous network due to end-to-side attachment. (D) 1000 mg% Dx 110; flow at 7 s⁻¹; short RCA, plasma gaps within and between aggregates. (E) 1500 mg% Dx 250; stasis, primary and secondary aggregation due to side-to-side and end-to-side attachment; hemispheric deformation of RBC at the end of a rouleaux. (F) 1500 mg% Dx 500; flow at 7 s⁻¹; large "agglomerates" with no gaps within, but large gaps between RCA.

(Figure 9). A number of techniques have been developed to measure separately the factors operational in the non-Newtonian viscosity of the blood. The critical description of these methods is beyond the scope of the present treatise and the reader is referred to the original communications describing the apparatus for red cell (109) and platelet (110) aggregometry, for measuring the deformability of red cells in suspension (111), and of individual red cells (112) by monitoring their ability to pass 5-μm pores.

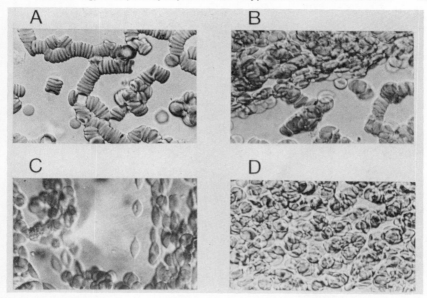

Figure 8. Rheoscopical pictures of pathological red cell aggregates in viscometric flow. (Interference contrast optics, 100 × objective). *A*, network of normal human red cell aggregates (RCA): primary aggregation into typical rouleaux, secondary aggregation due to end-to-side attachment. *B*, pathological RCA at 10 s^{-1} (myeloma) at the free ends of rouleaux, erythrocytes are drawn into hemispherical caps, irregular clumping of red cells. *C*, meshes of pathological red cell aggregates at 20 s^{-1}; extended red cell chains bridging plasma gaps (cholangitis with disseminated intravascular coagulation). *D*, pathological red cell aggregates at high shear flow (230 s^{-1}) (myocardial infarction); incomplete hydrodynamic dispersion: "flocs" and short rouleaux persist, individual cells are irregularly deformed.

PHOTOMETRIC AGGREGOMETRY

Rheophotometric techniques (109, 113–115, 226), i.e., the measurement of light transmission of blood as a function of varying flow rates or shear stresses, has been used extensively for hemorheological measurements. We utilized it to quantitate the shear stresses (dyn/cm^2) that are required to keep red cell aggregates dispersed, and to measure the velocity of the aggregate formation, i.e., the half-time and rate constant of aggregate formation following hydrodynamic dispersion. These techniques make use of the well known fact that both cell aggregation and cell deformation or orientation increase light transmission (for a detailed discussion see Ref. 109).

The measurement of the mechanical integrity of the aggregates in flow, i.e., their ability to withstand the hydrodynamic forces of flow, is possible by determining the shear rate corresponding to the maximum optical density (or minimum light transmission) of blood. As both cell aggregation (at low shear rates) and cell orientation (at high shear rates) distinctly increase the light transmission, the shear rate of minimum light transmission corresponds to a state where cells are neither dispersed nor oriented (Figure 10). The transition

24   Schmid-Schönbein

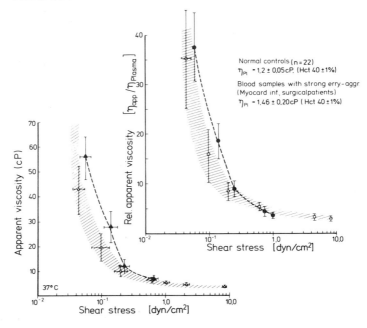

Figure 9. Apparent viscosity and relative apparent viscosity of blood as a function of shear stress: comparison between normal and pathological blood samples. Hematocrit value standardized to 40 ± 1%.

(τ_{Tmin}) between these two rheological states requires high forces whenever the adhesive forces between cell aggregates increase (Figure 11). The determined shear stress of hydrodynamic disaggregation is equivalent to a critical shear stress at which the flow of blood changes from that of a (low viscosity) emulsion into that of a (high viscosity) suspension.

As both aggregation and dispersion are entirely reversible processes, the rate of aggregate formation can be determined repeatedly. For this purpose, a rapidly flowing cell suspension, in which all cell aggregates are dispersed, is abruptly brought to full stop. The aggregates re-form spontaneously and the rate of this formation is followed photometrically (104). From such records (Figure 12) the half-time and thus the rate constant of the aggregation process can be measured (Schmid-Schönbein et al. (104)). In normal blood the half-time of aggregate formation (3–5 s) is much longer than in pathological blood, where the half-time is sufficiently short (0.5–1.5 s) to allow aggregate re-formation in venules, despite their dispersion in the true capillaries (Table 2).

It has long been known that under pathological conditions red cell aggregates have a tendency to form dense and rather irregular cell clumps, which have been called "agglomerates" (Ruhenstroth-Bauer (116)) or "agglutinates" (20, 117). The rheoscope has shown that this type of enhanced red cell aggregation is also entirely reversible, but requires higher shear stresses for dispersion. It is also the consequence of a primary aggregation into typical rouleaux. However,

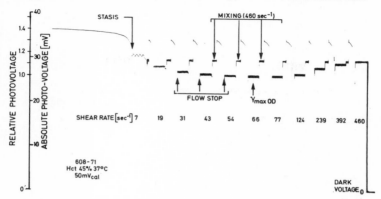

Figure 10. Light transmission of human blood as a function of shear rate. Original tracing. Steady state photovoltage is recorded starting from high shear rates (460 s⁻¹, right-hand side). Between each episode of shearing, the blood sample is mixed (shearing at 460 s⁻¹), minimum light transmission occurs at 66 s⁻¹. Light transmission increases steeply after flow stop.

under conditions of slow flow, these pathological cell aggregates attach in a side-to-side fashion, forming continuously growing cell aggregates in spite of a constant flow velocity. As a result, after a few seconds of low shear, the red cell

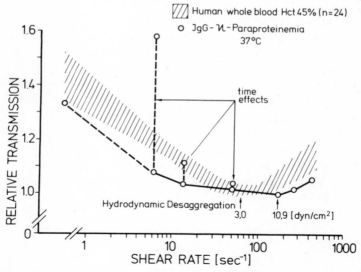

Figure 11. Schematic representation of the light transmission of human blood as a function of shear rate. Relative transmission. Photovoltage at any given shear rate divided by minimum light transmission (corresponding to random cell orientation and aggregate dispersal). Shaded area = normal controls: A shear rate of 55 ± 18 s⁻¹ is required for hydrodynamic disaggregation, which is equivalent to a shear stress of 3.0 dyn/cm². Myeloma blood: 10.9 dyn/cm² are required for hydrodynamic dispersion ($T_{T_{min}}$). Note strong time effects, compared to Fig. 13.

26   Schmid-Schönbein

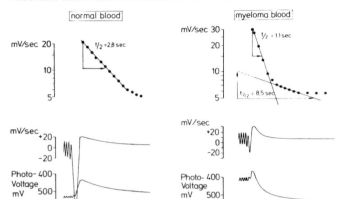

Photometric determination of the Kinetics of RCA-Formation in Stasis (Hct 45%, 37°C)

Figure 12. Changes of the light transmission of human blood during aggregate formation. *Left-hand side:* Simultaneously plot of a "syllectogram" (V) of a normal blood sample, first derivative ($\dot{I}=dV/dt$) of the syllectogram and log $\dot{I} = dV/dt$ decays exponentially, $t_{1/2}$ = 2.8 s. *Right-hand side:* Pathological blood sample (myeloma): dV/dt shows double exponential decay: $t_{1/2}$ = 1.1 sec, $t_{2/2}$ = 8.5 s.

aggregates in flow are much more pronounced than they would have been in full stasis; in other words, slow flow induces rather than disperses red cell aggregates (as in normal human blood). The growth of red cell aggregates (Figure 13) is caused by a continuous uptake of individual cells and small cell aggregates and, furthermore, by a kind of elastic recoil of cell aggregates accompanied by the formation of large, cell-free plasma gaps *between* aggregates, whereas the plasma immobilized *within* the aggregates is reduced. This latter factor, in turn, has pronounced effects on the optical properties of cell suspensions and has been

Table 2. Flow behavior of red cell aggregates in blood from normal subjects, diabetic and myeloma patients, and gravidas at term

	Shear stress of hydrodynamic disaggregation (dyn/cm^2)	Half-time of aggregate reformation $t_{1/2}$ (s)
Normal controls	2.56 ± 0.06	3.6 ± 0.4
Diabetics	3.13 ± 0.15	1.68 ± 0.3
Myeloma	6.51 ± 4.88	1.13 ± 0.81
Pregnancy at term	6.32 ± 3.8	0.94 ± 0.3

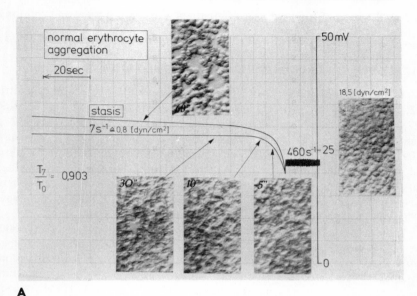

A

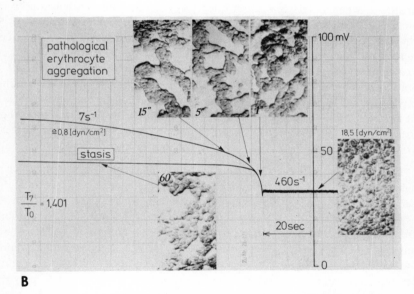

B

Figure 13. *A*, kinetics of *physiological* red cell aggregation in slow flow and in stasis. Photomontage of two experiments: 1) Photometric tracing of the light transmission when flow is abruptly switched from rapid (460 s^{-1}) flow to stop. 2) Photometric tracing obtained when rapid flow is switched abruptly to slow flow (7 s^{-1}). 3) Microphotographs taken in the rheoscope chamber are synchronized to certain parts of the photometric curves. This mode of data presentation clearly shows that RCA is most pronounced *in stasis* (FSAR > 1.0). *B*, kinetics of pathological RCA: Photomontage as explained above. The aggregation is *more* pronounced in slow flow than in stasis (FSAR > 1.0) (myeloma blood).

used as the basis of a simple test differentiating between normal and abnormal aggregation (102, 118).

FACTORS OPERATIONAL IN RED CELL AGGREGATION

Ultrastructural (Chien and Jan (119); Rowlands and Skibo (120)), viscometric (42, 59, 70, 74, 121–126), and aggregometric techniques (62, 77, 78, 106–109, 127), have elucidated the effect of plasma proteins (59, 70, 106, 118, 127), their concentration (62) (and thus the effect of hemodiluting anticoagulants (109), plasma concentration (103) (ultrafiltration (103)) of red cell deformability, and membrane pliability (10, 87, 128) (formation of a maximum adhesive area (70)) on red cell aggregation. Hypothermia strongly increases shear resistance of red cell aggregates (124); hyperthermia up to 47°C is of no effect, and above 49°C it abolishes red cell aggregation (123).

There is a considerable interindividual variability in aggregate properties, but the variation in disease goes far beyond these, as was to be expected. Grouped results from three groups of patients are listed in Table 2. This variation is the consequence of elevated concentrations of fibrinogen, but also of nonclottable serum proteins (126), of which α_2-macroglobulins (103, 127) and IgM (127) have been identified, whereas IgG is without effect; α_2-macroglobulin conspicuously produced the clumping type of red cell aggregation as described above (see p. 24), whereas fibrinogen and IgM exclusively cause network aggregation, i.e., continuous rouleaux structures. Model experiments with dextran-induced red cell aggregation (which closely resembles natural red cell aggregation in all its morphological and rheological aspects) strongly suggest that one of the variables responsible for either end-to-side or side-to-side attachment is molecular size of the bridging colloids (Volger et al. (108)). The complicated effect of the red cell surface charge has been actively studied and has been the subject of a recent symposion (see Bibl. Anat., Vol. **11**, 1972). It is clear to date that:

1. In normal human blood, red cell aggregation occurs in the presence of high molecular proteins despite the fact that the red cell surface charge is maintained (121, 122, 129) by these high molecular weight colloids.
2. When increasing the electrophoretic mobility (by reduction of the ionic strength of the medium, which decreases the extent of the cloud of counter ions) red cell aggregation is less pronounced (121, 122, 129). Only at extremely low ionic strengths, red cell aggregation is enhanced despite grossly accelerated electrophoretic mobility. This is caused by spontaneous formation of polymers in the colloids promoting aggregation (Volger et al. (106)).
3. When reducing the electrophoretic mobility by the action of enzymes (neuraminidase, trypsin, chymotrypsin, pronase), the tendency to aggregation is (as a rule) enhanced (121, 129) (for a detailed discussion see Schachtner et al. (130)).

There are surprising, well established (131) species differences not only between bovine (nonaggregating) and equine (excessively aggregating) blood, but also between man and the conventional laboratory animals used in intravital

Table 3. Red cell aggregation kinetics in various
mammals, Hct 33%, room temperature

	γT_{min} (s^{-1})	$t_{1/2}$ (s)
Weak aggregation		
Cattle	0	—
Rabbit	0.6	60
Guinea-pig	0.9 ± 0.4	60
Hamster	1.1 ± 1.0	180
Rat	29.8 ± 21.8	5.3 ± 3.5
Cat	46.8 ± 35	29 ± 9
Man	56 ± 32	2.4 ± 0.77
Strong aggregation		
Dog	111 ± 82	4.25 ± 1.7
Pig	180	1.5
Horse	470	2.7

microscopic studies. Table 3 summarizes studies by von Gosen (132). The tendency to aggregation in the species conventionally used as laboratory animals for intravital microscopy studies (rat, hamster, guineapig, rabbit) is far less pronounced than that found in man, whereas the aggregation in cat and dog blood is similar to man. This is important in the evaluation of data on intravascular aggregation obtained in experiments using these species.

It should be noted that as a rule we found conspicuously low hematocrit levels in "aggregating" species and/or individual animals. This seems to reflect a general (or compensatory) principle, as for these strongly aggregating species (cat and horse) hematocrit levels below 35% are reported standard (see Table 4, compiled from Ref. 23). Strongly aggregating human blood samples in chronic disorders are also present, as a rule, at a reduced hematocrit level (unpublished observation).

SHAPE AND DEFORMATION
OF RED CELLS: CONFORMATIONAL INSTABILITY

The occurrence and critical significance of perpetual red cell deformation to the perfusion of nutrient capillaries are generally accepted. However, it is frequently overlooked that red cells in larger blood vessels are also deformed in flow. The critical role of perpetual red cell deformation, which greatly enhances whole blood fluidity in large blood vessels, was attributed on theoretical grounds to a rotation of the erythrocyte membrane, leading to a transmission of shear stresses into the cell interior, where flow of the liquid cell content is enforced (Schmid-Schönbein and Wells (10, 133). The advent of the counter-rotating rheoscope chamber (102) allowed the microcinematographic recording of this type

Table 4. Comparison of microrheological properties of red blood cells and red cell volume fractions in different species (3)

Species	Type of erythrocyte	%
Camel	Ovalocytes	29[a]
Goat	Rigid microcytes	33[a]
Pigeon	Nucleated erythrocytes	35[a]
Horse	Pronounced RCA	33
Cat	Pronounced RCA	40
Pig	Pronounced RCA	40
Dog	Shear compliant RBC	45
Rat	Shear compliant RBC	46
Hamster	Shear compliant RBC	49
Man	Shear compliant RBC	42 (♀), 47 (♂)
Pregnant ♀ at term	Pronounced RCA	37

[a]Plasma trapping?

of flow adaptation of the red cell. Figures 14 and 15 show a sequence of frames from a film taken in a rheoscope, an apparatus in which the red cells remain stationary to the observer in the region between counter-rotating cone and plate. Changes in the rotational speeds of these subject the cells to quantifiable shear stresses that are easily manipulated. Figure 14 depicts prints of consecutive frames of a 16-mm motion picture taken at 25 frames per second. The cells are suspended in a dextran solution of 60 cP, and are subjected to a shear rate of 20 s^{-1} (resulting in a shear stress of 12 dyn/cm^2 acting on the interphase between the cell surface and the continuous phase). The large open arrow in the second picture of the upper row (see Figure 15) indicates the direction of shear; as can be seen, all red cells, which are biconcave at rest, are deformed into ellipsoids and are aligned so that their major axis is parallel to flow. The precipitated fibrin particles are also seen, one such particle being attached to the trailing end of the red cell in the first frame shown in the upper panel (small arrow). During the next 640 ms (16 frames, panels 1 and 2, Figure 14) that particle is gradually moving to the leading end of the erythrocyte (small arrow in the last frame of panel 2) and then back to the trailing end of the erythrocyte (last frame of panel 4). Following this, the particle begins to move again towards the leading end (last frame in panel 5). This back-and-forth motion occurs over and over again, while the cell remains always in its aligned state and never tumbles over. The back-and-forth motion of a particle attached to the erythrocyte surface together with invariant alignment of the cell can exclusively be explained by the mechanism of membrane rotation. No other mechanism could possibly explain the co-existence of particle rotation and maintenance of cell orientation. The membrane itself, of course, evaded detection by light microscopy.

The last frame in panel 3 of Figure 14 shows two erythrocytes, each with a particle attached (small fat arrow). In that frame one cell has its satellite protein

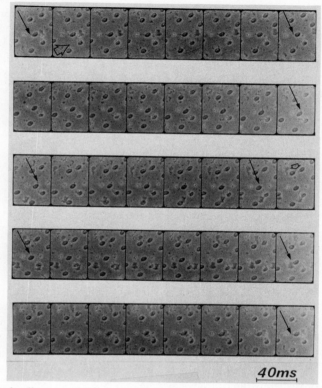

Figure 14. Red cells suspended in an isotomic dextran solution (=60 cP) and subjected to viscometric flow (20 s^{-1}).

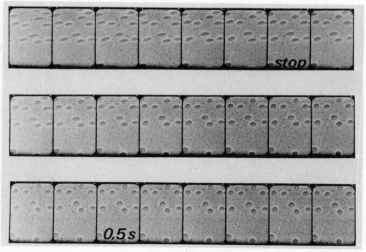

Figure 15. Red cells suspended in dextran solution. Transient from viscometric flow (25 s^{-1}) to full stop.

precipitate in front, the other in back position. The particle rotation can also be seen in this pair of cells that is followed in the preceding and subsequent scene, although they are occasionally not in focus. Transients between the ellipsoid (aligned) and round (biconcave) cell shape are seen in Figure 15. Such transients occur very rapidly according to the change in shear rate. Figure 15 shows that such transients from viscometric flow with a shear rate of 25 s^{-1} to full stop under the present conditions are followed by resumption of biconcavity within about 600 ms. An accidental scratch (diagonal white line) in the surface of the transparent cone indicates the moment of flow stop (frame 7, top panel). The slow recoil within 600 ms is indicative of small elastic forces, presumably generated by the bending rigidity of the erythrocyte membrane as discussed by Canham (134) and Bull and Brailsford (135).

The orientation and deformation of red blood cells in flow also occur when red cells are suspended in plasma; under the conditions of crowding in suspensions with normal hematocrit, the deformation is considerably more irregular. Continuous membrane rotation and transmission of shear stress into the interior of the cell must be invoked to explain the macrorheological behavior (4, 6, 64, 70, 101, 108, 136) of red cell suspensions, especially the inverse relationship between continuous phase (plasma) viscosity and relative apparent blood viscosity (64, 78, 89, 108) (see p. 14). Bull (137) has also clearly shown (Figure 16) that a similar type of membrane motion can occur when individual cells are fixed to a surface and a membrane marker is moved by tangential flow forces. After these forces cease, the marker remains in its new position; thus there is no indication of elastic recoil into the original position, despite the fact that after cessation of the stress the cell again assumes a biconcave shape.

The well known surplus of the surface area for the given volume, in connection with the set of physical properties listed above, helps to explain the unusual shape of the red cell, as well as its rheological behavior. The orientation, deformation, and tank treading of the erythrocytes seen in the rheoscope under artificial but well defined conditions of viscometric flow are only one example of many other forms of flow adaptation that the erythrocyte perpetually undergoes in vivo. This is the consequence of the fact that: (a) the cell in vivo actually is always subjected to shearing, extensional and also compressive forces. Not blood as such, but the flowing blood, with crowding by vessel walls and by other blood cells, is the natural habitat of the erythrocytes; and (b) the cell yields to these forces due to its *inherent conformational instability*. The cell interior behaves like a fluid; but due to the relative surplus of surface area for the given volume the erythrocyte as a whole combines physical characteristics of a fluid drop with those of a free fluid (105). Ordinary fluid drops, owing to the surface forces acting between the suspended and the suspending phase, become spherical at rest; due to the surface tension they have quite substantial conformational stability. In spherical drops deformation can only be achieved provided that the external forces are high enough to produce elastic surface strain against the effect of interfacial stress. The erythrocyte is deformed without membrane strain; the only forces that have to be overcome are the small but finite bending

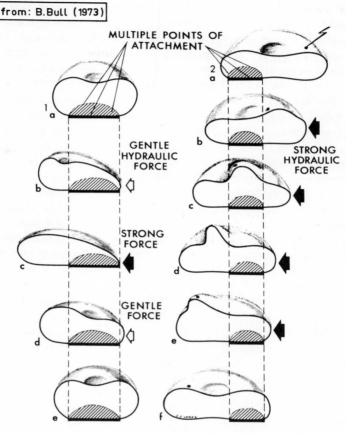

from: B.Bull (1973)

MULTIPLE POINTS OF
ATTACHMENT

GENTLE
HYDRAULIC
FORCE

STRONG
HYDRAULIC
FORCE

STRONG
FORCE

GENTLE
FORCE

Figure 16. The formation of a single dimple and an asymetrical double dimple. *1*) The membrane on a cup cell when distorted by mild hydraulic force will roll as far as permitted by the extent of the fixation point. This moves the dimple to the left. More vigorous hydraulic force (solid arrow) stretches the membrane, causing the dimple to disappear. When hydraulic force (still in the same direction) is decreased, the dimple reappears in its original position relative to the fixation point but on a new segment of membrane. When further decrease in hydraulic force occurs, the dimple stays spatially fixed but the membrane rolls under it. *2*) When firmly attached the membrane of the discoid cell will roll, relocating dimples and annulus 90° around the cell. In addition, if the hydraulic force is strong, it will cause a wavelike segment of the annulus to traverse the top of the cell. Cells at the midpoint of this process thus have two dimples on the same side and three hemi-equators (from Bull (137)).

rigidity of the membrane (28, 105, 134, 135, 138), the membrane surface viscosity (64), and the low viscosity of the cell content (10). Compliance to both tangential and normal forces is thus the reason for the *conformational instability of the erythrocytes*, which is responsible for its resting shape and for its ability to adapt to forces of flow. The biconcave shape of the erythrocyte is assumed only under exceptional circumstances—namely, in the absence of forces of blood flow. Much to the confusion of the medical public, one such exception is the erythrocyte lying on a coverslip under a microscope, where it is once in a way not subjected to any external physical forces and thus bending rigidity of the membrane enforces the peculiar biconcave resting shape. Previous speculations about the functional significance of erythrocyte biconcavity in connection with O_2 diffusion should be abandoned. The conformational instability, nevertheless, greatly facilitates oxygen uptake and release. The perpetual motion of red cells in the peripheral and pulmonary capillaries induces intracellular mixing and thus convective transport of O_2, hemoglobin and oxyhemoglobin, as shown by Zander et al. (139, 140).

The conformational instability, the deformability in flow, or the shear compliance of the red blood cells is the basis for "autoadaptation" of the red blood cells (Schmid-Schönbein (105, 141)) which minimizes viscous energy dissipations in all vessels, as it minimizes the inevitable disturbance of plasma flow created by particles. Examples of the flow adaptation are as follows (Figure 17): (a) orientation, deformation, and tank tread-like rotation of the cells when flowing in large blood vessels; (b) a rapid axial migration and consequent

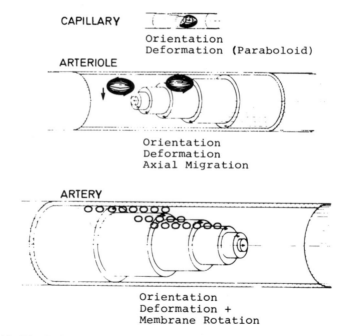

Figure 17. Rheological consequences of the red cell deformability in different classes of vessels. *Artery:* Orientation, deformation, and perpetual membrane rotation induce "fluid drop"-like behavior of individual erythrocytes, turn whole blood into an emulsion of low apparent viscosity. *Arteriole:* Initial membrane rotation, accompanied by rapid axial migration: formation of a lubricating plasma layer, reduction of effective hematocrit: Fahraeus effect. *Capillary:* Familiar parachute-like deformation: Reduction of effective cell diameter, creation of lubricating layer; very pronounced Fahraeus effect.

Fahraeus effect in small arteries and arterioles; (c) parachute-like adaptation and reduction of effective cell diameter when flowing in nutrient capillaries; (d) motion of the cells per diapedesis when either leaving the bone marrow sinusoids or when passing through slits in the endothelium (142).

The deformations of erythrocytes observed in the microcirculation have been misinterpreted in the past as modification of a disc shape (3, 143). Cinematographic records of flow in vivo (Figure 18) as well as photomicrographs of deformed cells in rapidly frozen lung segments (Figure 19) clearly demonstrate that practically all shapes and not only deformed discs occur. Hutchins, Goldstone, and Wells (144) have published photomicrographs showing the often observed fact that the parachute red cell shape is maintained for long periods of time in stagnant capillaries (Figure 20). Lighthill's (145) and Fitzgerald's (146,

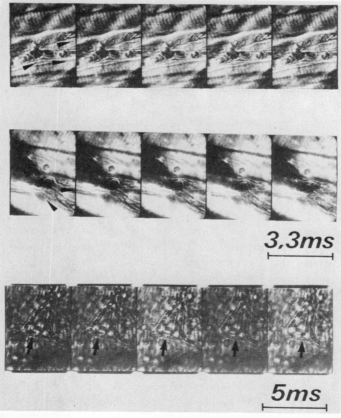

Figure 18. Flow adaptation of normal mammalian RBC in narrow and highly bent capillary intersections as seen from high speed motion pictures. *Upper panel:* Rat cremaster muscle (courtesy of Dr. R.E. Wells, Harvard Medical School, and used by permission). *Lower panel:* Dog mesentery: note passage of elongated RBC around "tissue post" while maintaining alignment to flow. This behavior is highly suggestive of "tank treading" (courtesy of Dr. Ted Bond, University of Texas Medical School; used by permission).

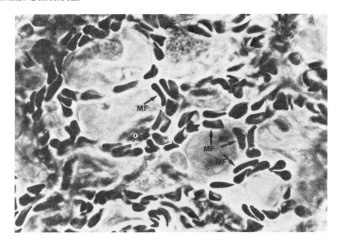

Figure 19. Deformation of mammalian red blood cells in the alveolar microcirculation. Quick-freeze technique retains cell shapes. Irregular shapes resembling fluid drops, but not folded discs (from Miyamoto and Moll (202)).

147) assumptions that external flow forces are prerequisite for the maintenance of parachute shape are not borne out by facts.

Besides facilitating blood flow itself by adapting in the cylindrical blood vessels of all calipers, the red cells also guarantee even and proportionate distribution of oxygen carriers under the adverse "traffic" conditions prevailing in narrow capillary intersections. The significance of this fact in the maintenance of adequate and homogeneous supply to tissues cannot be overestimated. In a theoretical treatise, Fung (28) has earlier argued that in response to the set of forces acting in the bifurcation, the red cells are preferentially moved into the more rapidly perfused capillary. In his argument, the compliance of the cells to the forces of flow is neglected. High speed motion pictures of cells passing intersections have, however, demonstrated that the normal erythrocytes are being strongly deformed and consequently pass such intersections with surprising ease, while abnormal erythrocytes (and leukocytes) often become lodged here (Figure 21).

DISTURBED RED CELL DEFORMABILITY: CONFORMATIONAL STABILITY

Many forms of hemolytic anemias have been found associated with abnormal shape of resting red cells, but also with reduced deformability. This fact has aroused considerable interest in blood rheology among hematologists, and has prompted a number of recent reviews (34, 37, 148–150) and symposia (148, 149). The detailed discussion of red cell membrane mechanics and erythrocyte micromechanics is beyond the scope of the present review. As both survival and flow adaptation are two consequences of the *rheological competence* of the erythrocyte, it is immediately obvious that serious defects should not be com-

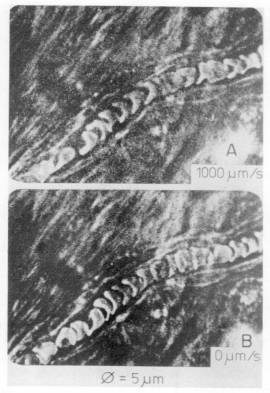

Figure 20. Parachute-like deformation of erythrocytes in rapid flow and in the first seconds after flow stop. The weak energy stored in bending the *membrane* is insufficient to induce rapid resumption of biconcave shape (from Hutchins et al. (144)).

patible with the survival of either the erythrocyte or the affected individual. The described physical properties of the erythrocytes can thus be regarded as the results of a perpetual physical selection of those cells fitted best to the in vivo flow conditions. Whenever so-called "shape changes" of the erythrocyte occur, and can be observed in the erythrocyte resting on a cover slip, they are to be regarded as indications of generalized or localized "conformational stability." Such shape changes are thus indicative of a higher extent of energies stored in the membrane (Bull (138)) or in the cell content. The two most easily understandable forms of conformational stability are the *sickle cell* (elastic tensions in the polymerized S-hemoglobin) and the *true spherocyte* (the build-up of surface tension). In either case, the observed rigidity in vitro and in vivo is understandable and additional assumptions about other abnormalities are not necessary. Less drastic changes, as seen in "echinocytes," "stomatocytes," "acanthocytes," or "codocytes" (see Bessis and Weed (149)), are indicative of a localized conformational stability, of a mechanical alteration within the membrane or on either side of it. Such conformational stability can, but it need not, disturb the

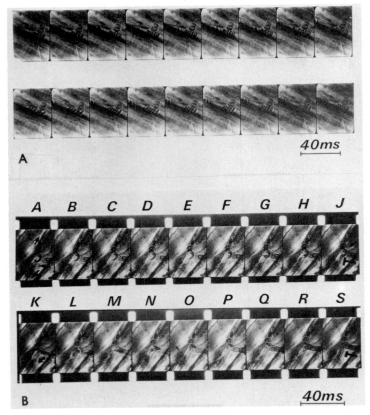

Figure 21. *A,* behavior of rigidified red blood cells (glutaraldehyde fixed cells injected into rat abdominal aorta) when reaching the microcirculation (rat cremaster muscle). Note clogging of several cells when they reach capillary intersections. *B,* behavior of sickle cells in the microcirculation (irreversibly sickled human RBC injected into the rat aorta and observed in the microcirculation of the rat cremaster muscle). The sickled cells become logged in the capillary bifurcation *(A-J),* are freed occasionally *(K-P),* and eventually lead to permanent flow stop due to size-up of many cells *(Q-S)* (from a film kindly supplied by Drs. J. Goldstone and R.E. Wells, Harvard Medical School, Boston, Mass.; shown by permission).

shear compliance of the affected red blood cell. Severe forms of "erythrocyte sclerocytosis" (150) in the peripheral blood are relatively seldom; failure to find them in the circulating blood does not mean they do not occur. On the contrary, the design of the peripheral vascular tree (small diameters, narrow branches, etc.) as well as the great susceptibility of the erythrocytes to physical, thermal, toxic, or biochemical changes (33, 123, 151, 152) creates a considerable risk of vascular obstruction by rigidified erythrocytes. It is conceivable that these enter a capillary in a normal rheological state, are secondarily affected by local changes, rapidly lose their deformability, and become permanently lodged in the capillary. A typical, generally accepted event of this type is the "sickle cell crisis," thought to be brought about by red cells sickling in situ. It should be noted in this context that the microrheology of circulating blood in sickle cell patients is *not* severely altered (133, 153).[3]

BLOOD VISCOSITY IN VIVO

Attempts to gain information about blood viscosity in vivo date back to the times of Poiseuille (24), whose interest in this problem prompted the experiments that led to the discovery of the classic Poiseuille-Hagen law. Many authors have attempted to assess blood viscosity in entire organs by exchange perfusion; blood and Newtonian reference fluids were compared and viscosity values were computed from the ratio of the flow rates and/or resistances. The validity of such measurements stands and falls with the confidence that both individual vessel diameter and total number of perfused vessels remain *absolutely invariant* in such experiments. Although such confidence is *not warranted,* it must be accepted for the time being that *blood viscosity in the vascular bed* is *low.* Whittacker and Winton (154), later Levy and Share (155), more recently Benis et al. (156) as well as Djojosiguto et al. (157) and Bäckström et al. (158) in Folkow's laboratory, Gaehtgens and Uekermann (97), and Hint (203) have unequivocally reported low "viscosity" values. The interpretation of Whittacker and Winton's data, who used cell-free plasma as a reference fluid, has to be modified (due to a kinetic energy correction, Bassenge et al. (204) and Benis et al. (156)). Even when measured under conditions of reduced flow (Bäckström et al. (158)), the apparent viscosity as measured in isolated, dilated vascular beds was found *lower* than the apparent viscosity as measured in rotational viscometers, even when there was evidence of vascular blockade (see below).

There is much less information available on blood viscosity in individual blood vessels. Kurland et al. (159) measured blood viscosity in a rat tail artery (218–380 μm ϕ) and found values similar to those in a viscometer. Meiselman et al. (160), in a quasi ex vivo outflow technique, found a Fahraeus effect. Recent data by Lipowski and Zweifach (161) are most relevant in this respect. By hydraulic mapping of the microcirculation, these authors measured an apparent viscosity of 1 cP in a 10-μm capillary at 30 mm/s, but 20 cP in the same capillary when the velocity was reduced to 0.2 mm/s. These data clearly corroborate the idea that a very pronounced Fahraeus effect occurs in capillaries, but only in the presence of high shear forces.

Our knowledge of the Fahraeus effect in small blood vessels, as well as the hydrodynamic analysis of the vascular bed, makes all findings in vascular beds very plausible. Caused by their large number, and their narrow diameters, the hydrodynamic hindrance of the circulation resides in those paracapillary vessels, in which fortuitously the blood viscosity is equal to or only slightly higher than that of plasma. This statement is of course not new—it only repeats the explanation given by Whittacker and Winton for their experimental results.

---

[3] Since the completion of this manuscript, Usami *et al.* (231) have reported a marked decrease of HbSS RBC deformability (as measured by their ability to pass through 5-μm pores) when ambient p_{O_2} was lowered below 100 mm Hg.

BLOOD VISCOSITY AND HEMODYNAMICS

The value of blood viscosity as measured in conventional viscometers only applies to the very few large blood vessels, in which flow dissipates a small fraction of the total energy necessary for the perfusion of a total vascular bed. Whether or not blood viscosity in these large vessels is higher or lower than normal by a few percent is therefore not very relevant. Furthermore, the hematocrit level only affects the value of apparent blood viscosity in these few large, but not in the majority of the small, blood vessels (22, 54, 85, 162). If, conversely, in conventional viscometers elevated viscosity is paralleled by hematocrit changes, this may have no significance at all to the entire vascular bed, as long as flow forces are normal. (It appears hopeless at the moment that the viscosity component and geometrical components responsible for the "peripheral vascular resistance" might ever be clearly separated.)

When scrupulously corrected for plasma viscosity, hematocrit, and temperature, the data obtained in viscometers do, nevertheless, reflect the *rheological potentials* of suspensions of red blood cells under varying flow forces. These potentials are responsible on the one hand for the favorable flow behavior of blood when it is rapidly streaming, but on the other hand they can turn into the cause of prolonged stagnation of blood flow in microscopic blood vessels under conditions of circulatory insufficiency.

Whenever relative apparent blood viscosity at high shear stresses (above 5 dyn/cm^2) is normal, this indicates that in large vessels blood actually flows as an "emulsion" of perpetually deforming hemoglobin "droplets." When the same red cells approach the microvasculature, axial migration, reduction of effective hematocrit, and their assumption of the typical bullet shape minimizes the viscosity of blood (by maximizing the Fahraeus effect) so that in the rapidly perfused true capillaries it is only slightly higher than or equal to plasma viscosity.

Considered in its relevance to dynamics of the *normal* circulation, the viscometric detection of high apparent blood viscosity due to red cell aggregation in prestatic flow (> 0.5 dyn/cm^2) therefore represents not an actual but rather only a *virtual* property of blood. This virtual property may certainly become quite an effective property under conditions of *pathological* hemodynamics. Whenever the flow is chronically retarded—irrespective of the factor causing the flow retardation—the hemodynamic effects of apparent blood viscosity may become the *critical* and *limiting* factor governing flow in the affected vessels. Depending on actual pressure gradient, vasomotor reserve, local hematocrit level, etc., the effect of elevated viscosity due to aggregation may then be far more pronounced than predicted by rotational viscometry.

Resulting from locally obtaining low shear stresses in single vessels or vessel provinces, these may be clogged entirely by cellular material. Such a blockade was actually recorded by Djojosiguto et al. (157) who performed simultaneous measurements of the "viscosity" in vivo and the transcapillary exchange.

The material responsible for blockage in narrow capillaries may be large leukocytes, or small platelet aggregates in arterioles, or it may be red cell aggregates in any blood vessel with reduced blood flow. Leukocytes and platelet aggregates may act as *functional microemboli,* because their mechanical properties are sufficient to withstand the forces acting in the normal microcirculation. On the other hand, the mechanical properties of the red cell aggregates (see p. 24) only allow them to resist the forces of the disturbed, the hypoperfused microvasculature. For all of their reversibility and for all of their frailty that allows normal flow forces to clear the affected vessel, the red cell aggregates cannot arrest blood flow. They can lead to prolonged vascular obstruction in a major fraction of microvessels under conditions of sustained hypoperfusion, especially if accompanied by local or generalized changes in hematocrit or aggregating tendency (see p. 43). Whether a flow retardation is caused by rheological or by hemodynamic reasons, the consequences mainly derive from the highly variable flow behavior of both red cells and their aggregates. Even if there is no aggregation axial migration no longer occurs (9, 82, 91), the beneficial effects of the lubricating plasma layer and the reduced effective hematocrit are lost; the red cells might even establish contact with the capillary wall (Lighthill (145, 163); Fitzgerald (146)). If the individual red cells are unable to display their adequate flow properties through mere lack of adequate shear forces, the concentrated suspension of red cells exerts a highly effective resistance against the mutual motion of fluid lamellae (and thence flow). Blood then assumes the properties of a conventional suspension.

When in addition the red cells are aggregated, the blood assumes the properties of a "reticulated" suspension. At the unusually high volume fraction near 50%, all mesh-work-like, structured suspensions are flow resistant and/or exhibit elastic properties (218). In blood, a two-phase suspension, cell aggregates as such exert hemodynamic effects only provided: (*a*) that they reach across planes of shear; (*b*) that they withstand the forces acting between these planes of shear; (*c*) that they are kept from evading these forces (as in the case of axial migration or settling); and (*d*) that they impede the motion of fluid lamellae past each other.

In rotational viscometers (Couette, viscometric flow, see p. 12), the motion away from the shear forces (shear-induced phase separation) is governed by parameters different from those in perfused natural or glass capillaries, where axial migration and sedimentation lead to shearing mainly in the lubricating plasma layer. As its viscosity is low, phase separation obscures the fact that the cell aggregates (comprising a nonsheared portion) have lost their fluidity.

A state of irreversible stagnation may thus be easily reached by changes in rheological behavior. Associated with a reversible loss of blood fluidity, it may trigger a sequence of effects with positive feedback. Flow retardation, formation of red cell aggregates, and increase in viscosity may lead to further decrease in velocity, unless compensated by an increase in cardiac output, arterial pressure, or diameter of the affected vessel or vessels immediately upstream or downstream.

RELATIVE ROLE OF VASOMOTOR AND VISCOSITY FACTORS

Reduction of flow forces alone does not necessarily lead to critically reduced flow rate, nor to elevated viscosity. Hypoperfusion is immediately followed by the sequence of events known as *autoregulation,* i.e., a myogenic or a metabolically controlled vasodilation tending to re-establish blood flow (negative feedback). Whether or not the circulation deteriorates by the described chain of reactions therefore depends on the relative gain of the positive feedback loop by increased viscosity and that of the compensating negative feedback loop by metabolic autoregulation. This interdependence is schematically depicted in Figure 22. Very obviously, under all physiological conditions flow velocity is high and viscosity is low and only slightly shear dependent (low gain of the viscosity loop). At the same time, the vasomotor reserve is high (because the arteriolar vessels are under tone and can greatly improve flow velocity by very limited vasodilation, a factor not only restoring flow but also improving the fluidity of the blood). Therefore, the physiological *regulation* of the peripheral blood supply is doubtlessly dominated by vasomotor factors and blood viscosity factors play a negligible role: high gain of vasomotor loop.

However, in grossly pathological conditions of circulatory deficiency, the situation may be totally different. The reduced flow velocity has several different consequences: hypoperfusion leads to chronic hypoxia, vasodilator metabolites are produced by the parenchymal cells, and they paralyze the vascular smooth muscle. This may proceed up to the point where the vasomotor reserve is exhausted. The reduced linear flow velocity in any single blood vessel, on the other hand, not only increases blood viscosity but now shifts the viscosity profile to its steep portion (high gain of the positive feedback viscosity loop). In other words, under conditions of chronic or prolonged circulatory deficiency, the gain of the vasomotor negative feedback loop goes to zero, while at the same time the gain of the viscosity positive feedback goes to its maximum. Consequently, the deficient circulation is *limited* by the actual level of apparent viscosity in the respective blood vessels. Hematocrit level and aggregation tendency affect the critical shear rates below which aggregation occurs, but even more so the gain of the viscosity loop (position and steepness of the viscosity profile). Thus, it is easy to understand why hemodilution (164, 165) (reduction of hematocrit, reduction of adhesive plasma proteins) or defibrination (166–169) has such pronounced beneficial effects under conditions of circulatory deficiency.

One additional principal difference between vasomotor control and viscous limitation of perfusion follows from the same argument. While vasomotor shutdown of blood vessels (notoriously small arterioles and their peripheral end, the functional precapillary sphincter (Folkow (170); Johnson (171)) are highly effective in producing a rapid but short-lasting flow stop, metabolic autoregulation will paralyze the constriction and soon restore flow. Any kind of maintained flow retardation and/or stasis must, therefore, be caused by an intravascular obstacle. The physiologically and particularly the pathologically en-

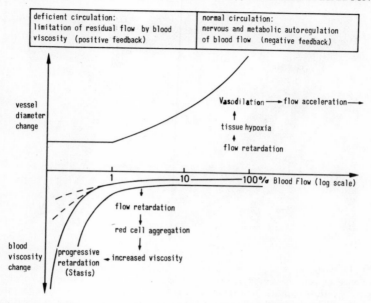

deficient circulation: limitation of residual flow by blood viscosity (positive feedback)	normal circulation: nervous and metabolic autoregulation of blood flow (negative feedback)

vessel
diameter
change

Vasodilation ⟶ flow acceleration ⟶

↑

tissue hypoxia

↑

flow retardation

1 ————————— 10 ——————— 100% Blood Flow (log scale)

↓

flow retardation

↓

red cell aggregation

↓

blood
viscosity
change

progressive
retardation → increased viscosity
(Stasis)

Figure 22. Schematic representation of the effects of "blood viscosity" changes and of diameter changes on blood flow (for explanation see text).

hanced cell aggregates are likely candidates to be held responsible for such a state of maintained intravascular stagnation. Thus, rheological changes *are not likely to initiate but to sustain* arrest of flow.

RHEOLOGY OF INTRAVASCULAR RED CELL AGGREGATION AND DISSOCIATED MICROVASCULAR PERFUSION: STASIS AND SHUNTING

It has long been established (20, 117, 172–178) that both in vivo and in vitro red cell aggregation depends on the equilibrium between the shearing forces of the blood flow and the adhesive forces acting between red cells. Reduction of flow forces (Schmid-Schönbein and Gaehtgens (176); Gaehtgens and Schmid-Schönbein (96)) as they occur in vivo following arterial clamping, in the course of hemodynamic reflexes involving precapillary vasoconstriction (orthostasis, cold pressure tests) is notoriously accompanied by phenomena of intravascular aggregation in the venules of the bulbar conjunctiva (see Figure 23). Also, an increase in the adhesive properties of red cell aggregates produces the phenomenon of intravascular aggregation in most vessels, even at normal perfusion pressure. Slight pressure reduction enhances the effect, an increasing number of vessels become occluded by erythrocyte masses, which either move extremely slowly or are fully stagnant—only to resume motion upon re-establishment of normal perfusion pressure.

The microscopic analysis of the microcirculation (see Figure 18) reveals that

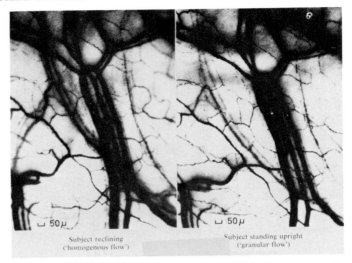

Subject reclining
('homogenous flow')

Subject standing upright
('granular flow')

Figure 23. Intravital microscopy of erythrocyte aggregates in the human bulbar conjunctiva. Influence of general hemodynamics: *Left-hand side:* Patient sitting; upon subjective observation the flow appears homogeneous. Flashlight illumination reveals aggregates and plasma gaps. *Right-hand side:* Patient standing upright; general flow retardation, the aggregation is more pronounced and can be observed even by subjective microscopic observation (from Bibl. Anat. 10, 1969, Basel: Karger).

the red cells very quickly change their shape and also that red cell aggregates are formed and dispersed within a fraction of a second. In other words we can assume that the blood in vivo is capable of changing its physical properties as quickly as in vitro (see p. 29). However, normal red cell aggregates in human blood are extremely shear labile. This frailty notwithstanding, they are frequently observed in the microcirculation, particularly in the venules, but also in the arterioles. Prolonged stasis or even flow reversal is not an uncommon finding in the microcirculation. The venules as a predilectory site for intravascular red cell aggregation with concomitant retardation have several hemodynamic peculiarities that might predispose them for low shear régimes.

The mere fact that in most organs the total cross-sectional area of all venules is considerably higher than that of all other classes of vessels is presumably most important. While under normal conditions this fact is advantageous, under pathological conditions it bears the risk that venular shear stresses are subcritical, while the shear stresses in all other blood vessels are above the critical level necessary to keep red cell aggregates dispersed. The walls of most of the venules appear to contain very little, if any, vascular smooth muscle and they exhibit a high permeability. On the other hand, the arterioles immediately upstream are strongly invested with vascular smooth muscle and appear to have very low permeability ("gradient of vascular permeability"). A general vasoconstriction, in which venules are spared due to their lack of effector muscle cells, might also lead to a preferential fall of venular shear stresses (Schmid-Schönbein and Wells (43)) and an increase in viscosity there (Figure 24). As a consequence, a shift in

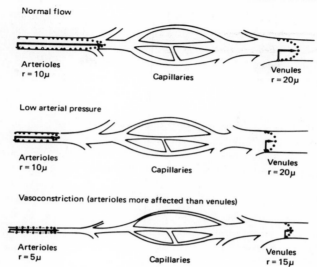

Figure 24. Schematic model describing the effect of arteriolar constriction on the velocity gradients in the postcapillary venules. *Normal flow:* shear rates much lower in venules than in arterioles. *Low arterial pressure* (in the absence of arteriolar constriction): shear rates reduced in all parts of the microcirculation. *Vasoconstriction* (arterioles more affected than venules): overproportional reduction of shear rates in the venules. The ultimate of this be the combination of low arterial pressure with strong vasoconstriction of the arterioles.

the equilibrium between precapillary and postcapillary resistance occurs, a factor discussed by Chien (42). In those venules that do possess vascular smooth muscle, the preferential loss of venular tone, as it occurs early in hemorrhagic shock when arteriolar tone is still high (Goldstone et al. (175); Hutchins et al. (144)), has a similar effect. If loss of tone is associated with enhanced permeability, local hemoconcentration occurs with a 3-fold effect: (*a*) increase in hematocrit, (*b*) increase in plasma protein concentration, and (*c*) increase in the shear resistance of red cell aggregates (103). Blood that slowly enters such a venular segment with quite normal rheology might undergo secondary changes that turn it into a highly concentrated, highly adhesive mass of red cells which strongly resists flow. Furthermore, since the static network of cell aggregates has elastic properties (and thus stops red cell flow) it does leave free spaces that allow plasma to flow in orthograde direction, thereby producing local hemoconcentration.

The comparison of the data on shear resistance of red cell aggregates in vitro and shear forces acting in vivo (Table 1, Lipowski and Zweifach (161)) makes it highly unlikely that red cell aggregates produce arteriolar plugging by acting as microemboli, as earlier suggested by Knisely (20) and Gelin and Zederfeldt (174). Since whole blood yields to forces as low as 0.1 dyn/cm^2, and since moreover even the most pronounced red cell aggregates are incapable of withstanding shear forces higher than 20 dyn/cm^2, one has to accept the fact that it is a priori unlikely that cell aggregates should ever be able to resist the kind of shear forces in such vessels (as they were computed or measured), especially when these are positioned *in series* to the "main arteriovenous pathway." Even a venule positioned *in series* will be unlikely to become blocked by even the most adhesive red cell aggregates observed so far,[4] because following such blockade an increasing portion of the arteriovenous pressure gradient will build up streams of the occluded venule and would tend to flush away the cell aggregates—unless there is a bypass in parallel to the affected vessel.

In the microcirculation of most tissues, however, vessels can actually be bypassed, as the vast majority of them are positioned in *parallel* to the main arteriovenous pathway. Here, a definition of the term "main arteriovenous pathway" is in order. We define as such "in-series vessel" the one channel connecting artery and vein that has the least hydraulic resistance. Very obviously, such a vessel could be, but must not be, anatomically defined; moreover, it need not always be specifically the same vessel. Vessels may take turns in this function, adapting to spontaneous, hormonal, metabolic, or nervous changes in vascular tone. Very obviously, long, narrow, nutritive capillaries are unlikely candidates for this function, whereas any type of short arteriovenous pathway, especially when dilated, is predisposed to operate as a main arteriovenous channel.

The old controversy about the absence or presence of anatomically pre-formed "shunt" vessels, as well as the endless discussion about the organization of microvascular beds with "thoroughfare" and/or "preferential channels" (Chambers and Zweifach (179)) has been based too much on anatomical arguments (Hammersen (213); Illig (18); Staubesand (180); Tischendorf (228)). The simple consequences of Hagen-Poiseuille's law make it unnecessary to search for anatomical channels to explain physiological "shunting." Any arteriovenous channel, irrespective of its diameter, which is endowed with smooth muscles, and is normally kept under tone, may become the "main arteriovenous path-way." It always may serve this function, or it may assume this function whenever the vascular smooth muscle is relaxed. The hemodynamic potentials of vasodilation at the capillary level are depicted schematically in Figures 25 and 26 and are based on the simple fact that in the microcirculation large relative changes in vessel diameter occur, and these affect flow as described by the law of Hagen-Poiseuille. The hemodynamic and metabolic situation of circulatory deficiency, as defined on page 42, not only creates the bypass for a vessel occluded by an aggregate, but may likewise give rise to the rheological conditions (flow with progressively decreasing shear rate) that initiate the progressive formation of red cell aggregates within capillary networks. Venules admittedly are the most likely sites for low shear stress flow in parallel vessels, but other microvessels may be subject to a similar shift in forces. The highly flow-dependent viscosity of the blood, the general fall in shear stresses during hypoperfusion in combina-

---

[4] The red cell aggregates found in certain extreme cases of disseminated intravascular coagulation may be an exception to this rule. In such cases, red cell aggregates have been found to resist shear stresses higher than 27 dyn/cm^2 (unpublished observation).

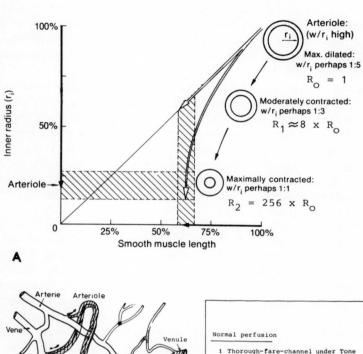

A

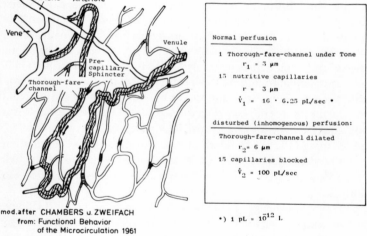

B

Figure 25. *A,* effect of smooth muscle contraction on vascular diameter and resistance in microscopic blood vessels. Resting state = 75% maximum muscle length. Dilatation on 25%: inner diameter increases to 40%, resistance decreases by factor 8. Maximum constriction: inner diameter decreases to 25%, resistance increases by factor 32. *B,* schematic representation of the hemodynamic potentials of a thoroughfare-channel, when normally kept under tone by vascular smooth muscle: a microvascular module containing 15 true capillaries and 1 thoroughfare-channel is assumed to be perfused by 1 pL/sec. Following obstruction of nutrient capillaries and simultaneous dilatation of thoroughfare-channels, the same flow rate is accomplished by a dilated thoroughfare-channel, which is turned into a "preferential channel" (modified after Chambers and Zweifach (179)).

tion with dilation, and the preferential reduction of shear stress in parallel vessels must be held responsible for the frequent observation of heterogeneous, nonuniform perfusion of the microcirculation seen under various diseased states, i.e., flow stagnation in one vessel, normal or even increased blood flow in its immediate neighbors.[5]

In cases of acute hypoperfusion, a nonuniform flow might be explained by nonuniform action of vasomotor factors, e.g., critical closure of precapillary sphincters. In cases of prolonged or even chronic hypoperfusion, the strongly variant flow properties of blood are much more likely to cause this effect. As red cell aggregates can clearly resist finite forces without yielding (see p. 25), they can be said to possess a yield shear stress and this would easily explain the frequent dissociation of the microcirculation into perfused and nonperfused vessels without resorting to critical closure or coagulation. The fluid dynamic principles that apply are depicted schematically in Figure 26.

Flow in one branch (A) leads to a pressure gradient ($P_1 - P_2$) which also acts along the branch (B or B'). However small this pressure gradient, it must lead to flow. If, however, the perfusing fluid possesses a yield shear stress, flow in B may come to stop as long as the ratio $\Delta P \cdot d/4l$ is below the yield value. Since, in turn, the pressure gradient along vessel A is inversely proportional to the mean flow velocity and the fourth power of the vessel diameter, it is easy to understand the existence of "preferential channels," i.e., such vessels which can dilate after losing muscular tone, especially when slowly perfused, have such a small viscous energy dissipation that wall shear stresses in B and B' also fall below the "yield shear stress" of the blood elements, giving rise to permanent stagnation.

For this potential to impede flow, we (141) coined the German term *kollaterale Viskositätserhöhung*, very appropriate but not translatable into English; hence we propose the term *collateral blood viscidation*[6] or "collateral loss of blood fluidity." Many epiphenomena of low flow states, such as shunting (174, 181), red cell trapping (182–184), disseminated activation of the coagulation system (185), and hidden acidosis (186, 187), can thus be explained on the basis of the bipotential flow properties of the red blood cells, and their effect on blood flow behavior.

The present knowledge about the factors governing blood rheology in vitro (plasma viscosity, hematocrit value, and transition between aggregated and deformed states as a function of the prevailing shear stresses) might supply the basis for a future evaluation of the alterations in the intravascular flow phenomenon paraphrased by the term "sludging." The controversy about "blood sludging" has arisen from the fact that this phenomenon on the one hand is regularly seen to be associated with other objective signs of severe microvascular de-

---

[5] Since the completion of this manuscript, Gaehtgens *et al.* ((1976) *Pflügers Arch.* 361: 183–190) have shown conclusively that after microembolization functional shunting (with increased venous p_{O_2}) occurs through vessels smaller than 10 μm in diameter.

[6] According to Webster's dictionary, the term "viscidity" describes the *state* of being sticky and viscous. Both features are assumed by slowly moving blood; the *process* which makes the blood sticky and viscous might therefore be called "viscidation."

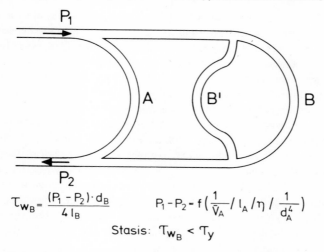

$$\tau_{w_B} = \frac{(P_1 - P_2) \cdot d_B}{4 \, l_B} \qquad P_1 - P_2 \approx f\left(\frac{1}{\overline{V}_A} \middle/ l_A \middle/ \eta \middle/ \frac{1}{d_A^4}\right)$$

$$\text{Stasis:} \quad \tau_{w_B} < \tau_y$$

Figure 26. Schematic representation of the concept of "collateral blood viscidation" in branching and collaterals of the microcirculation. Flow over branch A causes pressure drop $(P_1 - P_2)$, which is the cause of flow over B and B' in true fluids. After generalized reduction of flow forces (e.g., in low flow states) $P_1 - P_2$ decreases (as a function of velocity in A, length of A as compared to B, viscosity of blood in A and especially after dilation in A). Wall shear stress in B or B' (τ_{wR}) is directly proportional to $P_1 - P_2$. Spontaneously arrested blood will not resume flow if τ_{wB} is below the "yield stress" of red cell aggregates: collateral blood viscidation.

ficiency in shock, burn, endotoxinemia, and freezing (for review see Knisely (20); more recently Stalker (188); Davis (189); Gelin and Löfström (190); Merlen (219); Piovella (191); Shoemaker and Iida (182); Stallworth (225)), in severe forms of myocardial insufficiency (Bloch (192)), diabetes (Ditzel (173)), immunological incidents, etc. On the other hand, the same phenomenon has been observed over and over again in subjects who are neither critically ill nor show evidence of localized circulatory disturbance (193, 194). The paradox is easily explained by the pronounced effect of hematocrit: the hemodynamic effects of red cell aggregates above all depend upon such factors as hematocrit value, absolute velocity, class of vessel, etc. Even the most pronounced forms of aggregation are of little hemodynamic significance if associated with anemia. Under these conditions large red cell aggregates can easily form (78); however, even larger plasma gaps come into existence. As a consequence the aggregates can more easily be discriminated under a microscope, while on the other hand they are kept from exerting pronounced effects on blood flow.

Quite to the contrary, normal aggregation, when associated with a strong local reduction of shear stresses and elevated hematocrit level, has very pronounced effects on flow and may give rise to the phenomenon of *kollaterale Viskositätserhöhung* as described above. Obviously, the combination of strong aggregation and high hematocrit is critical and fortunately they rarely occur together.

Anyone who has ever observed the microcirculation in low flow states has seen the dissociation between normal perfusion and total stasis in blood vessels immediately adjacent to each other (e.g., Kulka (217)). The presently available microscopic techniques, however, make it extremely difficult to record these changes objectively as very high magnification is necessary to decide with certainty the flow status in an individual vessel. With such high magnifications, only a very small area with few vessels can be observed at one time. Special techniques have been developed (Schmid-Schönbein and Driessen (195)) to record the nonuniformity of perfusion as a consequence of intravascular cell aggregation. On a macroscopic scale, a nonuniform, patchy perfusion is frequently observed in organs (see a textbook on pathology) or extremities (196) that are underperfused. Moreover, in tissues which do not allow direct visualization, a number of authors have reported physiological changes in shock or other low flow states that can only be explained by assuming that an increasing number of nutrient capillaries are actually excluded permanently from the circulation by intravascular cell aggregates.[7]

The studies of Appelgren (166, 181) provide a very detailed analysis of the changes in nutrient capillary perfusion in skeletal muscle as they occur in hemorrhage, regional hypotension, as well as following experimental alteration of blood rheology. By measuring the clearance of two topically administered radioactive tracers (^{133}Xe, Na^{131}I), they were able to differentiate clearly between overall blood flow (which is proportional to the ^{133}Xe clearance) and nutrient capillary perfusion. The latter is measured by the clearance of NaI. Therefore, NaI clearance strongly depends upon a number of perfused capillaries, as this substance diffuses slowly and is only removed when patent capillaries are in close proximity to the deposit of the injected tracer. On the other hand, Xe, a lipid-soluble gas, diffuses so easily that nonnutrient channels (shunts) will also remove it quite effectively from the deposit. The ratio of the fractional disappearance rates measures the "effectiveness" of nutrient perfusion. Reduced ^{131}I clearance, indicating a reduction of the number of perfused capillaries, is found in many forms of shock and is not necessarily associated with a reduced ^{133}Xe clearance. Nutrient capillaries were found to be excluded from the circulation whenever either the flow *conditions* produced aggregation (hemorrhage or regional hypotension) or when they resulted in altered flow *properties* (hemoconcentration or enhanced red cell aggregation).

Hemodilution with low molecular weight dextran solution, following which the rheological properties of the blood are less shear rate dependent (Gelin (197); Schmid-Schönbein et al. (198)) ameliorates the nutrient capillary perfusion under all experimental conditions studied by these authors. As a result, the ratio

---

[7]*Note Added in Proof:* Since the completion of the manuscript, the thesis of Lipowski from Zweifach's laboratory has appeared in which a pronounced reduction of the fraction of perfused vessels as a function of decreasing arteriovenous pressure gradients in the cat mesentery was established.

of the fractional clearance rates of the two substances is returned towards normal.

Basically, the same results were found by Backström et al. and they confirm by an independent method the frequent subjective observation of improved intravascular flow following the administration of low molecular weight dextran. When the prestatic increase in viscosity is avoided by disaggregating measures, even when the general flow forces are considerably reduced (Schmid-Schönbein, Gaehtgens, and Hirsch (176)), all capillaries remain evenly perfused.

The present concept is also supported by experiments in which tissue oxygenation was correlated to the rheological changes in the blood. The intensive studies by Messmer and co-workers (136, 164) on hemodilution are especially noteworthy in this context. Messmer and co-workers improved blood rheology by its isovolemic exchange against dextran 60. In spite of a very marked reduction in oxygen-carrying capacity (hematocrit 20%!) a better and more even oxygen supply to skeletal muscle, liver, kidney, pancreas, and small intestine was found. A typical experiment of Messmer is shown in Figure 27 indicating a shift to high p_{O_2} values following improved blood fluidity.

In similar experiments, Bicher et al. (199), using oxygen microelectrodes in the brain, found a highly significant decrease in the reoxygenation time (defined as the time required to reach control p_{O_2} after a short period of anoxic anoxia), following the administration of a disaggregating drug. Appelgren (166) has recently also found similar flow improvement by defibrinogenation, thereby substantiating claims by Ehrly (168, 169), who correlated measurable decrease in "viscosity" following the reduction of fibrinogen level with the subjective improvement of peripheral perfusion in patients suffering from chronic arterial occlusions.

The improvement of blood rheology by reinfusion of saline rather than whole blood in experimental hemorrhagic shock has been shown to interfere with its irreversibility and also with disseminated intravascular coagulation (101). It is impossible at present to establish the cause-and-effect relationship of flow retardation and disseminated intravascular coagulation (see footnote on p. 46). It is, however, very likely that the two phenomena are closely interconnected and may potentiate each other. Flow retardation favors coagulatory processes while, on the other hand, the tendency to aggregations of various blood elements is enhanced following general activation of the coagulation system (coagulation in the sense of fibrin formation as well as in the sense of platelet activation). It is logical to assume that hemodilution and improved blood rheology also interfere with the positive feedback of aggregation, rise in "viscosity," stasis, and coagulation. Many details, however, remain to be clarified.

The low flow state was so far considered only under aspects of intravascular changes of the blood. It is very obvious, however, that in severe states of hypoperfusion other components of the microcirculation are equally affected.

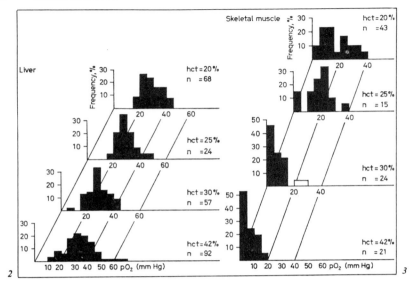

Figure 27. Frequency distribution of p_{O_2} values in liver and skeletal muscle is obtained from measurements by two multiwire electrodes in relation to the degree of hematocrit during normovolemic hemodilution (from Messmer et al. (164)).

In particular, vessel wall factors, i.e., the integrity of the endothelial layer, are important for normal behavior of the microcirculation. It can be deduced from Branemark's studies (172) that the microcirculatory derangement with cell aggregation phenomena is fully reversible by mere hemodynamic normalization. They become irreversible when endothelial damage is inflicted at the same time as intravascular aggregation. This notwithstanding, rheological factors are nevertheless pivotal to the eventual fate of the deranged microcirculation. As discussed above, rheological factors can *maintain* hypoperfusion. In doing so, they not only interfere with metabolite supply but also with catabolite removal. Therefore, they must be held responsible for the side effects of stagnant hypoxia (200). Since catabolites of anoxic parenchymal cells exert negative effects on red cell deformability (as discovered by Teitel (151), and later confirmed repeatedly) prolonged stagnant hypoxia turns initially reversible changes of blood rheology into irreversible ones.

SUMMARY

The normal rheological behavior of the red blood cells is prerequisite for the survival of the red cells and also for the functioning of microcirculation. Severe alterations of red cell deformability are incompatible with life. When compensated by anemia, even relatively severe rheological incompetence of individual red cells is tolerable.

Functional loss of red cell deformability is widely known to occur under conditions of sustained hypoperfusion, and disseminated stagnation of blood in the paracapillary bed occurs. The resulting capillary occlusion does not necessarily reveal itself in grossly reduced flow rates or increased "total peripheral resistance," since it is compensated by shunting through microscopic anastomoses. The biological significance of the phenomenon of red cell aggregation (collateral loss of blood fluidity, "collateral blood viscidation") is related to hemodynamics only on the level of individual capillaries. Since the compensatory potentials of vasomotor factors at this level are also very high, the collateral viscidation is not only facilitated but by the same token partially compensated. Therefore, unless complicated by a defect in the macrocirculation, the biological significance of blood rheology seen under the aspects of entire organs is not primarily related to hemodynamics, but to diffusive transcapillary exchange. As a consequence of collateral blood viscidation, diffusion takes place under suboptimal conditions. The available surface area for exchange is reduced, the diffusion distances are increased. In themselves, these changes are no acute threat to the survival of the entire individual. In combination with other defects, they are capable of sustaining prolonged states of flow arrest. Since the rheological properties of blood can be easily manipulated, sustained circulatory deficiencies can be avoided or treated by improving blood fluidity.

ACKNOWLEDGMENTS

It is a pleasure to acknowledge the help of my collaborators Drs. H. J. Klose, E. Volger, G. Gallasch, J. von Gosen, W. Schlick, H. Schachtner, and H. Rieger and the extensive and profound discussions with Priv. Doz. Dr. Paul Teitel (who also examined the manuscript), Prof. F. Hammersen, and Prof. K. Messmer.

Their critique was invaluable in the development of the concepts described in this paper. Miss Ch. Stollman's untiring secretarial help in the preparation and completion of this manuscript and F.J. Kaiser's photographic skill are deeply appreciated.

REFERENCES

1. Bicher, H.I. (1972). Blood Cell Aggregation in Thrombotic Processes. Charles C. Thomas, Springfield, Ill.
2. Braasch, D. (1971). Red cell deformability and capillary blood flow. Physiol. Rev. 51:679.
3. Charm, S.E., and Kurland, G.S. (1974). Blood flow and Microcirculation. J. Wiley and Sons, New York, London, Sydney, Toronto.
4. Chien, S. (1972). The present state of blood rheology. In K. Messmer and H. Schmid-Schönbein (eds.), Hemodilution, Theoretical Basis and Clinical Application, p. 1. Karger, Basel, New York.
5. Cokelet, G.R. (1972). The rheology of human blood. In Y.C. Fung, N. Perrone, and M. Auliker (eds.), Biomechanics: its Foundations and Objectives. Prentice-Hall, Englewood Cliffs, N.J.
6. Dintenfass, L. (1971). Blood Microrheology—Viscosity Factors in Blood Flow, Ischaemia and Thrombosis. Butterworths, London.

7. Gabelnick, H., and Litt, M. (1973). Rheology of Biological Systems. Charles C. Thomas, Springfield, Ill.
8. Larcan, A., and Stoltz, J.-F. (1970). Microcirculation et hémorhéologie. Masson, Paris.
9. Merrill, E. W. (1969). Rheology of blood. Physiol. Rev. 49:863.
10. Schmid-Schönbein, H., and Wells, R.E. (1971). Rheological properties of human erythrocytes and their influence upon the "anomalous" viscosity of blood. Erg. Physiol. Biol. Chem. Exper. Pharmacol. 63:147.
11. Schmid-Schönbein, H. (1974). Zelluläre Physiologie der Mikrozirkulation: Ausbildung von Risikofaktoren als Folge optimaler Anpassungfähigkeit. In R. Ahnefeld and K. Messmer (eds.), Die Mikrozirkulation I, p. 1. Springer-Verlag, Heidelberg.
12. Wells, R.E. (1973). The rheology of blood. In B. W. Zweifach and L. Grant (eds.), The inflammatory process, 2nd Ed. Vol. II. Academic Press, New York, London.
13. Atherton, A., and Born, G.V.R. (1972). Quantitative investigations of the adhesiveness of circulating polymorphonuclear leucocytes to blood vessel walls. J. Physiol. 222:447.
14. Schmid-Schönbein, G.W., Fung, Y.C., and Zweifach, B.W. Vascular endothelium—leucocyte interaction: Sticking shear force in venules. Abstr. Microvasc. Res. In press.
15. Cohnheim, J. (1867). Über Entzündung und Eiterung. Virchows Arch. 40:1.
16. Rickert, G., and Regendanz, P. (1921), Beiträge zur Kenntnis der örtlichen Kreislaufstörungen. Virchows Arch. Path. Anat. 231:1.
17. Nordmann, M. (1933). Kreislaufstörungen und Patholog. Histologie. Theodor Steinkopff, Dresden and Leipzig.
18. Illig, L. (1961). Die terminale Strombahn. Capillarbett und Mikrozirkulation. Band 10 der Reihe: Pathologie und Klinik Springer, Berlin.
19. Weber, H.W. (1955). Zur Begriffsbestimmung der Stase. Klin. Wschr. 33:387.
20. Knisely, M.H. (1965). Intravascular erythrocyte aggregation (blood sludge). In W.F. Hamilton and P. Dow (eds.), Handbook of Physiol. Sect. 2, Vol. III, Washington, D.C.
21. Scott-Blair, G.W. (1958). The importance of the sigma phenomenon in the study of the flow of blood. Rheol. Acta 1:123.
22. Burton, A.C. (1969). The mechanics of red cell in relation to its carrier function. In G.E.W. Wolstenholme and J. Knight (eds.), Ciba Symposium: Circulatory and Respiratory Mass Transport. Churchill, London.
23. Handbook of Biological Data (1971). Respiration and circulation, F.A.S.E.B., Washington, D.C.
24. Poiseuille, J.L.M. (1841). Recherches sur les causes du mouvement du sang dans les vaisseux capillaries. Mém. Acad. Roy. Sciences Inst. France, Sciences, Math. et Phys. 7:105.
25. Schmidt-Nielsen, K., and Taylor, G.R. (1968). Red blood cells: why or why not. Science 162:274.
26. Fahraeus, R., and Lindqvist, T. (1931). The viscosity of blood in narrow capillary tubes. Amer. J. Physiol. 96:562.
27. Bessis, M. (1973). Living Blood Cells and Their Ultrastructure. Springer-Verlag, New York, Heidelberg, Berlin.
28. Fung, Y.C. (1966). Theoretical considerations of the elasticity of red cells and small blood vessels. Fed. Proc., 25:1761.
29. Leblond, P.F., Lacelle, P.L., and Weed, R.I. (1971). Cellular deformability: a possible determinant of the normal release of maturing erythrocytes from the bone marrow. Blood 37:40.
30. Groom, A.C., Song, S.H., Lim, P., and Campling, B. (1971). Physical characteristics of red cells collected from the spleen. Can. J. Physiol. Pharmacol. 49:1092.
31. Groom, A.C., Song, S.H., and Campling, B. (1973). Clearance of red blood cells from the vascular bed of skeletal muscle with particular reference to reticulocytes. Microvasc. Res. 6:51.
32. Jandl, J.H., Simmons, R.L., and Castle, W.B. (1961). Red cell filtration and the pathogenesis of certain hemolytic anemias. Blood 18:133.
33. Teitel, P., and Nicolau, C.T. (1964). Physico-chemical and metabolic factors influencing the erythrocyte theology. In C. Nicolau (ed.), Molecular Biology and Pathology, Bucharest.

34. Weed, R.I. (1970). The importance of erythrocyte deformability. Amer. J. Med. 49:147.
35. Danon, D. (1968). Reversible deformability and mechanical fragility as a function of red cell age. Hemorheology, Proc. 1st Int. Conf. Reykjavik, 1966. Pergamon Press, Oxford, New York.
36. Danon, D. (1968). Biophysical aspects of red cell ageing. Bioliotheca Haematol. Proc. 11th Cong. Int. Soc. Blood Transf. Sydney, 29:178.
37. LaCelle, P.L., and Weed, R.I. (1972). The contribution of normal and pathological erythrocytes to blood rheology. Progr. Hematol. 7:1.
38. Sobin, S.S. (1972). The comparison of erythrocyte and capillary diameters in various mammalian species. Paper read to the American Society of Microcirculation, Atlantic City.
39. Lamport, H. (1964). Vascularisation compared in thin sheets and blocks of tissue. 2nd Europ. Conf. Microcirc. 1962, Pavia. Bibl. Ant. Vol. 4 Karger, Basel, New York.
40. Fung, Y.C. (1973). Stochastic flow in capillary blood vessels. Microvasc. Res. 5:34.
41. Goldsmith, H.L., and Mason, S.G. (1967). The microrheology of dispersions. In F.R. Eirich (ed.), Rheology, Theory and Applications. Academic Press, New York.
42. Chien, S. (1970). Blood rheology and its relation to flow resistance and transcapillary exchange, with special reference to shock. Adv. Microcirculation 2:89.
43. Schmid-Schönbein, H., and Wells, R.E. (1959). Quantification of the dynamics of red cell aggregation. Bibl. Anat. 10:45.
44. Mall, F. (1888). Die Blut- und Lymphwege im Dünndarm des Hundes. Ber. Sachs. Ges. Akad. Wiss. 14:151.
45. Schleier, J. (1918). Der Energieverbrauch in der Blutbahn. Pflügers Arch. 173:172.
46. Green, H.D. (1944). Circulation: Physical principles. In O. Glasser (ed.), Medical Physics Year Book Publ., Chicago.
47. Wiedeman, M. (1963). Dimensions of blood vessels from distributing artery to collecting vein. Circ. Res. 12:375.
48. Schmid-Schönbein, H., and Devendran, Th. (1972). Blood rheology in the micro-circulation. Pflügers Arch. (Suppl.) 336:84.
49. Zweifach, B.W. (1974). Quantitative studies of microcirculation structure and function, I. Circ. Res. 34:843.
50. Lipowsky, H.H., and Zweifach, B.W. (1975). In vivo study of "apparent viscosity" and vessel wall shear stress in cat mesentery. Abstr. 1st World Congr. Microcirc., Toronto, 1975.
51. Gross, J.F. The significance of pulsatile microhemodynamics. In B.W. Zweifach (ed.), Microcirculation. In press.
52. Barras, J.P. (1968). L'écoulement du sang dans les capillaires. Helv. Med. Acta, 48, 118.
53. Gerbstädt, H., Vogtmann, C.H., Rüth, P., and Schöntube, E. (1966). Die Schein-viskosität von Blut in Glaskapillaren kleinster Durchmesser. Naturwissenschaften 53:526.
54. Braasch, D., and Jenett, W. (1969). Erythrocyte flexibility, hemoconcentration and blood flow resistance in glass capillaries with diameters between 6 and 50 microns. Bibl. Anat. 10:109.
55. Haynes, R.H. (1962). The viscosity of erythrocyte suspensions. Biophysic. J. 2:95.
56. Casson, N. (1959). A flow equation for pigment-oil suspensions of the printing ink type. In C.C. Mill (ed.), Rheology of Disperse Systems. Pergamon Press, New York, London, Paris, Los Angeles.
57. Aroesty, J., and Gross, J.F. (1972). The mathematics of pulsatile flow in small vessels. I. Casson theory. Microvasc. Res. 4:1.
58. Gross, J.F., and Aroesty, J. (1972). Mathematical models of capillary flow: a critical review. Biorheology 9:225.
59. Merrill, E.W., Cokelet, G.C., Britten, A., and Wells, R.E. (1963). Non-Newtonian rheology of human blood. Effect of fibrinogen deduced by substraction. Circ. Res. 13:48.
60. Oka, S. (1973). Pressure development in a non-Newtonian flow through a tapered tube. Biorheology 10:207.

56   Schmid-Schönbein

61. Scott-Blair, G.W. (1959). An equation for the flow of blood, plasma and serum through glass capillaries. Nature 183:613.
62. Schmid-Schönbein, H., von Gosen, J., and Klose, H.J. (1973). Comparative microrheology of blood: Effect of disaggregation and cell fluidity on shear thinning of human and bovine blood. Biorheology 10:545.
63. Kümin, K. (1949). Bestimmung des Zähigkeitskoeffizienten μ für Rinderblut bei Newton 'schen strömungen in verschieden weiten Röhren und Kapillaren bein physiologischer Temperatur. Inaug. Diss. Schweiz, Freiburg.
64. Dintenfass, L. (1968). Internal viscosity of the red cell and a blood viscosity equation. Nature 219:956.
65. Jeffery, G.B. (1922). The motion of ellipsoidal particles immersed in a viscous fluid. Proc. Roy. Soc. (London) 102A:162.
66. Taylor, G.I. (1932). The viscosity of a fluid containing small drops of another fluid. Proc. Roy. Soc. (London) 138A:41.
67. Taylor, G.I. (1934). The formation of emulsions in definable fields of flow. Proc. Roy. Soc. (London) Ser. 146A:501.
68. Mason, S.G., and Bartok, W. (1959). In C.C. Mills (ed.), Rheology of Disperse Systems, Chapt. II. Pergamon Press, Oxford.
69. Gauthier, F.J., Goldsmith, H.L., and Mason, S.G. (1972). Flow of suspensions through tubes-X. Liquid drops as models of erythrocytes. Biorheology 9:205.
70. Chien, S. (1970). Shear dependence of effective cell volume as a determinant of blood viscosity. Science 168:977.
71. Einstein, A. (1906). Eine neue Bestimmung der Moleküldimensionen. Ann. Physik 19:289.
72. Copley, A.L., Huang, C.R., and King, R.G. (1973). Rheogoniometric studies of whole human blood at shear rates from 1000 to 0.0009 sec^{-1} Part I-Experimental Findings. Biorheology 10:17.
73. Usami, S., King, R.G., Chien, S., Skalak, R., Huang, C.R., and Copley, A.L. (1975). Microcinephotographic studies on red cell aggregation in steady and oscillatory shear. In Proc. 2nd Int. Congr. Biorheology, Jerusalem, 1974. Biorheology, 12:87.
74. Schmid-Schönbein, H., Gaehtgens, P., and Hirsch, H. (1968). On the shear rate dependence of red cell aggregation in vitro. J. Clin. Invest. 47:1447.
75. Chmiel, H., Effert, S., and Methey, D. (1973). Rheologische Veränderungen des Blutes beim akuten Herzinfarkt und dessen Risikofaktoren. Dtsch. Med. Wschr. 98:1641.
76. Charm, S.E., and Kurland, G.S. (1962). A comparison of the tube flow behavior and shear stress-shear rate characteristics of canine blood. Trans. Soc. Rheol. 6:25.
77. Schmid-Schönbein, H., Volger, E., and Klose, H.J. (1972). Microrheology and light transmission of blood. II. The photometric quantification of red cell aggregate formation and dispersion in flow. Pflügers Arch. 333:140.
78. Schmid-Schönbein, H., Gallasch, G., von Gosen, J., Volger, E., and Klose, H. J. (1976). Red cell aggregation in blood flow. I. New methods of quantification. Klin. Wschr. 54:149.
79. Thurston, G.B. (1972). Viscoelasticity of human blood. Biophys. J. 12:1205.
80. Schmid-Schönbein, H., and Wells, R.E. (1969). Fluid drop-like transition of erythrocytes under shear. Science 165:288.
81. Goldsmith, H.L. (1971). Deformation of human red cells in tube flow. Biorheology 7:235.
82. Goldsmith, H.L., and Beitel, L. (1970). Axial migration of red cells in tube flow. Fed. Proc. 29:319.
83. Fung, C.C. (1969). Blood flow in the capillary bed. Biomech. 2:353.
84. Henry, J.P., and Meehan, J.P. (1971). The circulation. An integrative physiologie study. Year Book Medical Publisher, Inc.
85. Skalak, R., Chien, P.H., and Chien, S. (1972). Effect of hematocrit and rouleaux on apparent viscosity in capillaries. Biorheology 9:67.
86. Schmid-Schönbein, H., Wells, R.E., and Goldstone, J. (1973). Effect of ultrafiltration and plasma osmolarity upon the flow properties of blood: a possible mechanism for control of blood flow in the renal medullary vasa recta. Pflügers Arch. 338:93.

87. Chien, S., Usami, S., Dellenback, R.J., and Bryant, C.A. (1971). Comparative hemo-rheology-hematological implications of species differences in blood viscosity. Biorheology 8:35.
88. Schmid-Schönbein, H., Wells, R.E., and Goldstone, J. (1971). Fluid drop-like behavior of erythrocytes—disturbance in pathology and its quantification. Biorheology 7:227.
89. Meiselman, H.J., and Cokelet, G.R. (1973). Blood rheology. Adv. Microcirc. 5:32.
90. Devendran, T., Kline, K.A., and Schmid-Schönbein, H. (1973). Capillary viscometry of erythrocyte suspensions in various media. Proc ASME, 73-WA/Bio 35.
91. Devendran, T., and Schmid-Schönbein, H. (1975). Axial concentration in narrow tube flow for various RBC suspensions as function of wall shear stress. Pflügers Arch. 355:R 19.
92. Eberth, C.J., and Schimmelbusch, C. (1888). Die Thrombose nach Versuchen und Leichenbefunden. Verlag von Ferdinand Enke, Stuttgart.
93. Barbee, J.H., and Cokelet, G.R. (1971). Prediction of blood flow in tubes with diameter as small as 29 μ. Microvasc. Res. 3:17.
94. Monro, P.A.G. (1969). Progressive deformation of blood cells with increasing velocity of flowing blood. Bibl. Anat. 10:99.
95. Berman, H.J., and Fuhro, R.L. (1968). Effect of rate of shear of the velocity profile and orientation of red cell in arterioles. Bibl. Anat. 10:32.
96. Gaehtgens, P., Meiselman, H.J., and Wayland, H. (1970). Erythrocyte flow velocities in mesenteric microvessels of the cat. Microvasc. Res. 2:151.
97. Gaehtgens, P., and Uckermann, U. (1973). The apparent viscosity of blood in different vascular compartments of the autoperfused canine foreleg, and its variation with hematocrit. Bibl. Anat. No. 11, 76.
98. Bayliss, L.E. (1962). The rheology of blood. In W.F. Hamilton (ed.), Handbook of Physiology, Section 2, Circulation I. Washington, D.C.
99. Haynes, R.H., and Burton, A.C. (1959). Role of the nonNewtonian behavior of blood in hemodynamics. Amer. J. Physiol. 197:943.
100. Dintenfass, L. (1967). Inversion of the Fahraeus-Lindqvist phenomenon in blood flow through capillaries of diminishing radius. Nature, 215:1099.
101. Schmid-Schönbein, H., Wells, R.E., and Schildkraut, R. (1969). Microscopy and viscometry of blood flowing under uniform shear rate. J. Appl. Physiol. 26:674.
102. Schmid-Schönbein, H., von Gosen, J., Heinich, L., Klose, H.J., and Volger, E. (1973). A counter-rotating 'rheoscope chamber' for the study of the microrheology of blood cell aggregation by microscopic observation and microphotometry. Microvasc. Res. 6:366.
103. Schmid-Schönbein, H., Gallasch, G., Volger, E., and Klose, H.J. (1973). Microrheology and protein chemistry of pathological red cell aggregation (blood sludge) studied in vitro. Biorheology 10:213.
104. Schmid-Schönbein, H., Kline, K.A., Heinich, L., Volger, E., and Fischer, T. (1975). Microrheology and light transmission of blood, III. The velocity of red cell aggregate formation. Pflügers Arch. 354:299.
105. Schmid-Schönbein, H. Erythrocyte rheology and the optimization of mass transport in the microcirculation. Blood Cells (Vol. 2), Springer-Verlag, Heidelberg. In press.
106. Volger, E., Schmid-Schönbein, H., and Mehrishi, J.N. (1973). Artificial red cell aggregation caused by reduced salinity: production of a polyalbumin. Bibl. Anat. 11:29.
107. Volger, E., Schmid-Schönbein, H., and Klose, H.J. (1973). Rheological studies on the kinetics of artificial red cell aggregation induced by dextrans. Bibl. Anat. 11:83.
108. Volger, E., Schmid-Schönbein, H., von Gosen, J., Klose, H.J., and Kline, K.A. (1975). Microrheology and light transmission of blood, IV. The kinetics of artificial red cell aggregation induced by dextran. Pflügers Arch. 354:319.
109. Klose, H.J., Volger, E., Brechtelsbauer, H., Heinich, L., and Schmid-Schönbein, H. (1972). Microrheology and light transmission of blood. I. The photometric quantification of red cell aggregation and red cell orientation. Pflügers Arch. 333:126.
110. Klose, H.J., Rieger, H., and Schmid-Schönbein, H. A rheological method for the quantification of platelet aggregation (PA) in vitro and its kinetics under defined flow conditions. Thromboses. In press.

58   Schmid-Schönbein

111. Schmid-Schönbein, H., Weiss, J., and Ludwig, H. (1973). A simple method for measuring red cell deformability in models of the microcirculation. Blut 26:369.
112. Schlick, W., and Schmid-Schönbein, H. Measurement of single red cell deformability (Preliminary report), Blood Cells. In press.
113. Berman, H.J., and Fuhro, R.L. (1973). Quantitative red cell aggregometry of human and hamster blood. Proc. 7th Conf. on Microcirculation, Aberdeen, 1972. Karger, Basel.
114. Dognon, A. (1969). Granulométrie optique de la suspension sanguine. Nouvelle technique d'étude. C.R. 268:974.
115. Healy, J.C. (1973). Etude expérimentale des associations réversibles entre les globules rouges. Thèse de doctorat, Univ. Paris.
116. Ruhenstroth-Bauer, G. (1966). Mechanismus und Bedeutung der beschleunigten Erythrocytensenkung. Klin. Wschr. 44:533.
117. Bloch, E.H., Powell, A., Meyman, H.T., Warner, L., and Kafig, E. (1956). A comparison of the surface of human erythrocytes from health and disease by in vivo light microscopy and in vitro electron microscopy. Angiology 10:6.
118. Schmid-Schönbein, H., Gallasch, G., von Gosen, J., Volger, E., and Klose, H.J. (1976). Red cell aggregation in blood flow. I. New Methods of Quantification. Klin. Wschr. 54:159.
119. Chien, S., and Jan, K. M. (1973). Ultrastructural basis of the mechanism of rouleaux formation. Microvasc. Res. 5:155.
120. Rowlands, S., and Skibo, L. (1972). The morphology of red cell aggregates. Thrombosis Res. 1:47.
121. Brooks, D.E., Goodwin, J.W., and Seaman, G.V.F. (1970). Interactions among erythrocytes under shear. J. Appl. Physiol. 28:172.
122. Brooks, D.E., Goodwin, J.W., and Seaman, G.V.F. (1974). Rheology of erythrocyte suspensions: electrostatic factors in the dextranmediated aggregation of erythrocytes. Biorheology 11:69.
123. Schmid-Schönbein, H., Volger, E., Klose, H.J., and Wells, J. (1972). Blood microrheology and the development of stasis in the microvasculature. In A.B.E. Kovach (ed.), Neurohumoral and Metabolic Aspects of Injury. Plenum Publishers, New York.
124. Schmid-Schönbein, H., Klose, H.J., Volger, E., and Weiss, J. (1973). Hypothermia and blood flow behavior. Res. Exp. Med. 161:58.
125. Wells, R.E., Gawronski, T.H., Cox, P.M., and Perera, R.D. (1964). Influence of fibrinogen on flow properties of erythrocyte suspensions. Amer. J. Physiol. 207:1035.
126. Wells, R.E., Schmid-Schönbein, H., and Goldstone, J. (1971). Flow behavior of red cells in pathologic sera: Existence of a yield shear stress in the absence of fibrinogen. In H. Hartert and A.L. Copley (eds.), Clinical and Theoretical Hemorheology. Springer-Verlag, Heidelberg, New York.
127. Gallasch, G. (1975). Zur rheologischen Wirksamkeit isolierter Serumproteine auf die Erythrocyten-Aggregation. Inaug. Diss., Munich.
128. Seaman, G.V.F., and Swank, R.L. (1967). The influence of electrokinetic charge and deformability of red blood cells on the flow properties of its suspensions. Biorheology 4:47.
129. Chien, S. (1973). Electrochemical and ultrastructural aspects of red cell aggregation. Bibl. Anat. No. 11, Karger, Basel.
130. Schachtner, W., Schmid-Schönbein, H., and Brandhuber, M. Microrheology and light transmission of blood VI. Effects of red cell surface charge. In preparation.
131. Fahraeus, R. (1929). The suspension stability of blood. Physiol. Rev. 9:241.
132. von Gosen, J., Wells, J., and Brandhuber, M. (1974). Microrheology of red blood cells in conventional experimental animals and man. Pflügers Arch. Europ. J. Physiol. Suppl. to Vol. 347.
133. Schmid-Schönbein, H., and Wells, R.E. (1971). Red cell deformation and red cell aggregation: Their influence on blood rheology in health and disease. In Hartert and Copley (eds.), Clinical and Theoretical Hemorheology, Proc. 2nd Intern. Conf. Hemorheology, Heidelberg, 1971.
134. Canham, P.B. (1970). The minimum energy of bending as a possible explanation of the biconcave shape of the human red blood cell. J.Theor.Biol. 26:61.
135. Bull, B., and Brailsford, J.D. (1973). The biconcavity of the red cell—an analysis of several hypotheses by means of models. Blood 41:833.

136. Sunder-Plassmann, L., Kloevekorn, W.P., and Messmer, K. (1971). Blutviskosität und Hämodynamik bei Anwendung kolloidaler Volumenersatzmittel. Anaestesist 20:172.
137. Bull, B. (1973). Red cell biconcavity and deformability. A macromodel based on flow chamber observations. In M. Bessis, R.I. Weed, and P.L. Leblond (eds.), Red Cell Shape, Physiology, Pathology, Ultrastructure. Springer-Verlag, New York, Heidelberg, Berlin.
138. Bull, B.S. The importance of bending elasticity in red cell deformation. Blood Cells. In press.
139. Zander, R., and Schmid-Schönbein, H. (1972). Influence of intracellular convection on the oxygen release by human erythrocytes. Pflügers Arch. 335:58.
140. Zander, R., and Schmid-Schönbein, H. (1973). Intracellular mechanisms of oxygen transport in flowing blood. Resp.Physiol. 19:279.
141. Schmid-Schönbein, H. (1975). Blood rheology and the distribution of blood flow within nutrient capillaries. In K. Messmer and H. Schmid-Schönbein (eds.), Proc. 2nd Intern. Conf. Hemodilution, Rottach-Egern, 1974. Bibl. Haemat. 41:1.
142. Schmid-Schönbein, H. (1970). Hemorheological aspects of splenic function. In K. Lennert and D. Harms (eds.), Die Milz: Struktur, Funktion, Pathologie, Klinik, Therapie. Springer-Verlag, Berlin, Heidelberg, New York.
143. Skalak, R., and Branemark, P.I. (1969). Deformation of red blood cells in capillaries. Science 1964:717.
144. Hutchins, P.M., Goldstone, J., and Wells, R. (1973). Effects of hemorrhagic shock on the microvasculature of skeletal muscle. Microvasc. Res. 5:131.
145. Lighthill, M.J. (1968). Pressure-forcing of tightly fitting pellets along fluid filled elastic tubes. J. Fluid Mech. 34:113.
146. Fitzgerald, J.M. (1969). Mechanics of red cell motion through very narrow capillaries. Proc.Roy. Soc.London B. 174:193.
147. Fitzgerald, J.M. (1972). In D.H. Bergel (ed.), The Mechanics of Capillary Blood Flow in Cardiovascular Fluid Dynamics. Academic Press, London, New York.
148. Bessis, M. (1973). Living Blood Cells and Their Ultrastructure. Springer-Verlag, New York, Heidelberg, Berlin.
149. Bessis, M., and Weed, R.I. (eds.) (1975). Red cell rheology and deformability. Blood Cells No. 2.
150. Wells, R. (1970). Syndromes of hyperviscosity. New Eng.J. Med. 283:183.
151. Teitel, P. (1965). Disk-sphere transformation and plasticity of alteration of red blood cells. Nature 206:409.
152. Teitel, P. (1967). Le test de la filtrabilité érythrocytaire (TFE). Une méthode simple d'étude de certaines propriétés microrhéologiques des globules rouges. Nouv. Rev. Franc. Hémat. 7:195.
153. Chien, S., Usami, S., and Bertles, J.F. (1970). Abnormal rheology of oxygenated blood in sickle cell anemia. J.Clin.Invest. 49:623.
154. Whittacker, S.R.F., and Winton, R.R. (1933). The apparent viscosity of blood flowing in the isolated hindlimb of the dog, and its variation with corpuscular concentration. J.Physiol. (London) 78:339.
155. Levy, M., and Share, R.L. (1953). The influence of erythrocyte concentration upon the pressure flow relationship of the dog's hind limb. Circ.Res. 1:247.
156. Benis, A.M., Chien, S., Usami, S., and Jan, K.M. (1973). Inertial pressure losses in perfused hindlimb: a reinterpretation of the results of Whittacker and Winton. J. Appl. Physiol. 34:383.
157. Djojosugito, A.M., Folkow, B., Oberg, B., and White, S. (1970). A comparison of blood viscosity measured in vitro and in a vascular bed. Acta Physiol. Scand. 78:70.
158. Backström, P., Folkow, B., Löfving, B., Kovách, X., and Öberg, B. Evidence of plugging of the microcirculation following acute hemorrhage. Proc. VI. Conf. on Microcirculation, Allborg Bibl. Anat. II, In press.
159. Kurland, G.S., Charm, S.E., Brown, S., and Tousignant, P. (1968). Comparison of blood flow in a living vessel and in glass tubes. In A. L. Copley (ed.), Hemorheology.
160. Meiselman, H.J., Frasher, W.G., Jr., and Wayland, H. (1972). In vivo rheology of dog blood after infusions of low molecular weight dextran or saline. Microvasc. Res. 4:399.
161. Lipowsky, H.H., and Zweifach, B.W. (1974). Net work analysis of microcirculation of cat mesentery. Microvasc. Res. 7:73.

162. Devendran, T., Kline, K.A., and Schmid-Schönbein, H. (1974). Flow properties of suspensions of normal, crenated and hardened RBC in narrow, curved plastic tubes. Proc. ASME 74-WA/Bio-5.
163. Lighthill, M.J. (1972). Physiological fluid dynamics: a survey. J.Fluid Mech. 52:475.
164. Messmer, K., Sunder-Plassmann, L., Klövekorn, W.P., and Holper, K. (1972). Circulatory significance of hemodilution: Rheological changes and limitations. Adv.Microcirc. 4:1.
165. Messmer, K., and Schmid-Schönbein, H. (eds.) (1972). Hemodilution, Theoretical Basis and Clinical Application. Karger, Basel.
166. Appelgren, L., Gustavsson, L., and Myrvold, H. (1975). Flow improvement by defibrinogenation (DF). Abstr. 1st World Congr. Microcirc., Toronto, 1975.
167. Ehringer, H., Dudczak, R., Kleinberger, G., Lechner, K., and Reiterer, W. (1971). Arvin: Schlangengift als neue Therapiemöglichkeit bei Durchblutungsstörungen. Wien. Klin. Schr. 83:411.
168. Ehrly, A.M. (1972). Rheological changes due to fibrinolytic therapy. In Hemodilution, Theoretical Basis and Clinical Application. Int. Symp. 1971. Karger, Basel.
169. Ehrly, A.M. (1973). Verbesserung der Fliesseigenschaften des Blutes: Ein neues Prinzip zur medikamentösen Therapie chronisch arterieller Durchblutungsstörungen. VASA, Suppl.No. 1.
170. Folkow, B. (1964). Autoregulation in muscle and skin. Suppl. I to Circulat.Res., XIV and XV.
171. Johnson, P.C. (1967). Autoregulation of blood flow. Gastroenterology 52:435.
172. Branemark, P.I. (1971). Intravascular Anatomy of Blood Cells in Man. Karger, Basel.
173. Ditzel, J. (1959). Relationship of blood protein composition to intravascular erythrocyte aggregation (sludged blood). Acta Med.Scand. Suppl. B 164:43.
174. Gelin, L.E., and Zederfeldt, B. (1961). Experimental evidence of the significance of disturbances in the flow properties of blood. Acta Chir.Scand. 122:33.
175. Goldstone, J., Hutchins, P.M., Schmid-Schönbein, H., Urschel, C., Sonnenblick, E., and Wells, R. (1971). Correlation of microvascular and rheological factors in hemorrhagic shock. In J. Ditzel and D. Lewis (eds.), Proc. 6th Europ. Soc. Microcirc. Karger, Basel.
176. Schmid-Schönbein, H., Gaehtgens, P., and Hirsch, H. (1967). Nicht Newton'sche Viskosität des Blutes und Erythrozytenaggregation. Proc. III. Symp. Intern. Anaesthes. Poznan (Polen).
177. Thorsen, G., and Hint, H. (1950). Aggregation, sedimentation and intravascular sludging of erythrocytes. Acta Chir. Scand. Suppl. 154:1.
178. Thuranskii, K. (1957). Der Blutkreislauf der Netzhaut. Ungarische Akademie der Wissenschaften, Budapest.
179. Chambers, R., and Zweifach, B.W. (1944). Topography and function of the mesenteric circulation. Amer. J. Anat. 75:173.
180. Staubesand, J. (1974). Arterio-venöse Anastomosen. In G. Heberer, G. Rau, and W. Schoop (eds.), Angiologie. Grundlagen, Klinik und Praxis. 2. Aufl. Thieme, Stuttgart.
181. Appelgren, L. (1972). Perfusion and diffusion in shock. Acta Physiol. Scand. Suppl. 378.
182. Shoemaker, W.C., and Iida, F. (1962). Studies on the equilibration of labeled red cells and T1824 in hemorrhagic shock. Surg. Obst. Gyn. 114:539.
183. Suzuki, M., and Shoemaker, W.C. (1964). Effect of low viscosity dextran on red cell circulation in hemorrhagic shock. Surgery 55:304.
184. Wollheim, E. (1931). Blutmenge und Dekompensation des Kreislaufs. Z. Klin Med. 116:269.
185. Hardaway, R.M. (1970). The fallacy of hemorrhagic shock models in dogs. Abstr. Vol. Europ. Soc. Microcirc. VI Conf., Allborg.
186. Bergentz, S.E., Carsten, A., Gelin, L.E., and Kreps, J. (1969). "Hidden Acidosis" in experimental shock. Ann. Surg. 169:227.
187. Kessler, M., Höper, J., Schäfer, D., and Starlinger, H. (1974). Sauerstofftransport im Gewebe. In S.W. Ahnefeld, C. Burri, W. Dick, M. Halmagyi (eds.), Mikrozirkulation, Springer-Verlag, Berlin, Heidelberg, New York.
188. Stalker, A.L. (1964). Intravascular erythrocyte aggregation. Bibl. Anat. 4:108.
189. Davis, E. (1968). The conjunctival vascular micropool as a sign of proneness to ischemic heart disease. Microvasc. Res. 1:102.

190. Gelin, L.E., and Löfström, B. (1954). A preliminary study on peripheral circulation during deep hypothermia. Acta Chir. Scand. 108:402.

191. Piovella, C. (1972). In vivo observations of the microcirculation of the bulbar conjunctiva in migraneous. Res. Clin. Stud. Headache 3:277.

192. Bloch, E.H. (1972). Sludged blood, human disease and chemotherapy. Oxygen transport to tissue. In D.F. Bruley and H.I. Bicher (eds.), Pharmacology, Mathematical Studies and Neonatology. Plenum Press, New York.

193. Replogle, R.L. (1969). The nature of blood sludging, and its relationship to the pathophysiological mechanisms of trauma and shock. J. Trauma 9:675.

194. Robertson, H.S., Wolf, S., and Wolff, H.G. (1950). Blood "sludge" phenomenon in human subjects. Notes on its significance and on the effects of vasomotor drugs. Amer. J. Med. Science 219:534.

195. Schmid-Schönbein, H., and Driessen, G. A double exposure technique to assess the distribution of red cell flow velocities in all vessels of microvascular moduli. In preparation.

196. Ratschow, M. (1974). Angiologie, Grundlagen, Klinik und Praxis. 2. Auflage G. Heberer, G. Rau, W. Schoop (eds.). G. Thieme Verlag, Stuttgart.

197. Gelin, L.E., Rudenstam, C.M., and Zederfeld, B. (1965). The rheology of red cell suspensions. Bibl. Anat. 7:368.

198. Schmid-Schönbein, H., Klose, H.J., and Volger, E. (1972). Effect of colloidal plasma substitutes on microrheology of human blood. In K. Messmer and H. Schmid-Schönbein, (eds.), Hemodilution: Theoretical Basis and Clinical Application. Karger, Basel, New York.

199. Bicher, H.I., Bruley, D., Knisely, M.H., and Reneau, D.D. (1971). Effect of microcirculation changes on brain tissue oxygenation. J. Physiol. 217:689.

200. Kessler, M. (1974). Oxygen supply to tissue in normoxia and in oxygen deficiency. Microvasc. Res. 8:283.

201. Usami, S., and Chien, S. (1973). Optical reflectometry of red cell aggregation under shear flow. Proc. 7th Conf. Microcir., Aberdeen, 1972. Bibl. Anat. 11. 91. Karger, Basel.

202. Miyamoto, A., and Moll, W. (1971). Measurements of dimensions and pathway of red cells in rapidly frozen lungs in situ. Resp. Physiol. 12:141.

203. Hint, H.C. (1964). The flow properties of erythrocyte suspensions in isolated rabbit ear: the effect of erythrocyte aggregation, hematocrit and perfusion pressure. Bibl. Anat. 4:112.

204. Bassenge, E., Höfling, B., and von Restorff, W. (1975). Inertial pressure loss in hemodilution. In K. Messmer and H. Schmid-Schönbein (eds.), Intentional Hemodilution, Bibl. Haematol., 41:140.

205. Benis, A.M., and Lacoste, J. (1968). Distribution of blood flow in vascular beds: model study of geometrical, rheological and hydrodynamical effects. Biorheology 5:147.

206. Bergentz, S.E., and Danon, D. (1966). Alterations in red blood cells of traumatized rabbits. II. Preferential sequestration of old cells. Acta Chir. Scand. 132:26.

207. Fischer, J. (1970). Die Milzintigraphie als Methode zur funktionellen Milzanalyse. (Spleen scanning as a method of functional analysis of the spleen.) In K. Lennert and D. Harms (eds.), Die Milz, the Spleen. Springer-Verlag, Berlin, Heidelberg, New York.

208. Folkow, B., and Neil, E. (1971). Circulation. Oxford University Press, London.

209. Fung, Y.C., and Zweifach, B.W. (1971). Microcirculation: Mechanics of blood flow in capillaries. Ann. Rev. Fluid Mech. 3:189.

210. Gaehtgens, P., Benner, K.U., and Schickendantz, S. (1975). Inst. of Norm. and Path. Physiology, Univ. Köln, Germ. Flow velocities of RBC and their suspending medium measured in a glass capillary (I.D. 11μ) using the photometric dual slit method. Abstr. 1st World Congr. Microcirc. Toronto, 1975.

211. Goldstone, J., Schmid-Schönbein, H., and Wells, R.W. (1970). The rheology of red blood cell aggregates. Microvasc. Res. 2:273.

212. Gross, J.F., and Intaglietta, M. (1973). Effects of morphology and structural properties on microvascular hemodynamics. Bibl. Anat. No. 11, 532. Karger, Basel.

213. Hammersen, F. Vorkommen, Struktur und Funktion echter arterio-venöser Anastomosen beim Menschen. Zugleich ein Beitrag zum Begriff der sog. Kurzschlussdurchblutung. In press.

214. Hochmuth, R.M., Marple, R.N., and Sutera, S.P. (1970). Capillary blood flow. I. Erythrocyte deformation in glass capillaries. Microvasc. Res. 2:409.

215. Hutchins, P.M., Goldstone, J., and Wells, R.E. (1971). Maximum of erythrocyte deformation in mammalian capillaries. Microvasc. Res. 3:115.

216. Kline, K.A. (1972). On a liquid drop model of blood rheology. Biorheology 9:287.

217. Kulka, J.P. (1969). Injurious effects of microcirculatory impairment in inflammatory disorders. In W.L. Winters and A.N. Brest (eds.), The Microcirculation. Thomas, Springfield, Ill.

218. Lessner, A., Zahavi, J., Silberberg, A., Frei, E.H., and Dreyfus, F. (1971). The viscoelastic properties of whole blood. In H.H. Hartert and A.L. Copley (eds.), Theoretical and Clinical Hemorheology. Springer-Verlag, Berlin, Heidelberg, New York.

219. Merlen, J.F. (1972). Physiologie de la microcirculation. Folia Angiologica: intern. J. of clinical Angiology XX, 4/5.

220. Schmid-Schönbein, G.W., and Zweifach, B.W. (1975). RBC velocity profiles in arterioles and venules of the rabbit omentum. Microvasc. Res. 10:153.

221. Schofield, R.K., and Scott-Blair, G.W. (1930). The influence of the proximity of a solid wall on the consistency of viscous and plastic materials. J. Phys. Chem. Ithaca 34:248.

222. Scott-Blair, G.W. (1969). Elementary Rheology. Academic Press, London, New York.

223. Skalak, R. (1972). Mechanics of the Microcirculation: Its Foundation and Objectives, Y.C. Fung, N. Perrone, and M. Anliker (eds.), p. 457. Prentice-Hall, Englewood Cliffs, N.J.

224. Song, S.H., and Groom, A.C. (1971). Storage of blood cells in spleen of the cat. Amer, J. Physiol. 220(3):779.

225. Stallworth, J.M., Ramirez, A., Barrington, B.A., and Bradham, R.R. (1969). Hypovolemic shock: microcirculatory changes during and after specific therapy. Annals of Surgery 169:694.

226. Taylor, M. (1955). The flow of blood in narrow tubes. II. The axial stream and its formation, as determined by changes in optical density. Austral. J. Exp. Biol. 33:1.

227. Thomas, D.G. (1965). The transport characteristics of suspension. VIII. A note on the viscosity of Newtonian suspensions of uniform spherical particles. J. Coll. Sci. 20:267.

228. Tischendorf, F., Bardolini, G., and Curri, S.B. (1971). Histology, histochemistry and function of the human digital arteriovenous anastomoses (Hoyer-Grosser's organs, Masson's glomera). Microvasc. Res. 3:323.

229. Vaupel, P., Hutten, H., Wendling, P., and Braunbeck, W. (1974). Experimentelle und theoretische Untersuchungen der intralienalen Mikrozirkulation beim Kaninchen. Red. Exp. Med. 164:223.

230. Zweifach, B.W. (1961). Functional behavior of the microcirculation. Charles C Thomas, Springfield, Ill.

231. Usami, S., Chien, S., and Bertles, J.F. (1975). Deformability of sickled cells as studied by microsieving. J. Lab. Clin. Med. 86:274.

International Review of Physiology
Cardiovascular Physiology II, Volume 9
Edited by Arthur C. Guyton and Allen W. Cowley
Copyright 1976   University Park Press   Baltimore

2
Peripheral Circulation: Fluid Transfer Across the Microvascular Membrane

F. J. HADDY, J. B. SCOTT,[1] and G. J. GREGA
Michigan State University

[1] Career Development Awardee of the National Heart and Lung Institute.

64   Haddy, Scott, and Grega

INTRODUCTION

Fluid transfer across the microvascular membrane determines the partition of extracellular water between the vascular and interstitial spaces. It is therefore an important determinant of the absolute blood volume and the store of fluid in the tissue spaces. The former is important to venous return and, hence, cardiac output. The latter may be of little functional consequence in many tissues but it can greatly influence function in others, for example, lung and brain. In this literature survey, the determinants of fluid transfer across the microvascular membranes of peripheral vascular beds are briefly reviewed and then emphasis is given to the effects of naturally occurring vasoactive agents, the nervous system and several stresses upon these determinants and consequently upon fluid transfer. For fluid transfer across the microvessels of the pulmonary vascular bed, the reader is referred to the reviews by Visscher, Haddy, and Stephens (1) and Staub (2).

PHYSICAL DETERMINANTS OF FLUID
TRANSFER ACROSS MICROVASCULAR MEMBRANE

There appears to be little reason to abandon the basic tenets of Starling's hypothesis (3–5) which describes water transfer across the microvascular membrane in purely physical terms. There is no convincing evidence for active fluid flux. The relevant physical forces are the internal and external hydrostatic pressures and the internal and external colloid osmotic pressures and these can be expressed by the equation:

$$F = k \left[(P_c - P_i) - (\pi_p - \pi_i) \right]$$

where F = rate of fluid flow; k = proportionality constant, the microvascular filtration coefficient, equal to the product of microvascular permeability to filtered fluid and microvascular surface area; P_c = microvascular hydrostatic pressure; P_i = interstitial fluid hydrostatic pressure; π_p = plasma colloid osmotic pressure; and π_i = interstitial fluid colloid osmotic pressure. Filtration occurs when the algebraic sum is positive and absorption when it is negative.

It is true that measurements of these forces are difficult, and that considerable argument exists even as to the sign of interstitial fluid hydrostatic pressure in some tissues, but there is little evidence that other physical forces are directly involved in the steady state transfer value under most conditions. As will be pointed out later, changes in crystalloid osmotic pressure can directly influence transfer in the transient.

According to the classic view, some microvessels filter and others reabsorb along the entire length but, on the average, filtration occurs at the arterial end and absorption at the venous end because the transmural hydrostatic pressure gradient exceeds the transmural colloid osmotic pressure gradient at the former site, and the reverse is the case at the latter site. Only a portion of the filtered fluid is thought to return to the vascular system via the lymphatics. Recently this concept has been challenged by Intaglietta and Zweifach (6). They suggest that, in certain tissues such as mesentery and omentum, fluid efflux occurs along the entire length of the microvessel and that all of this fluid returns to the vascular system via the lymphatics. This, they feel, results from higher hydrostatic pressure and protein permeability in the venous microvessel than generally recognized. On the other hand, Wiederhielm's data on the wing of the unanesthetized bat (7) support the classic view. It seems likely that the ratio of filtration to reabsorption will vary from vascular bed to vascular bed and within the same bed from time to time due to differences in the forces outlined by Starling. For example, a normally low arterial resistance or a fall in arterial resistance will shift the site of transition from filtration to reabsorption further down the length of the microvessel due to effects on microvascular hydrostatic pressure (see section "Microvascular Hydrostatic Pressure").

Hydrostatic Pressure Gradient

The difference between the internal and external pressure provides the force for driving fluid from the microvessel to the interstitial spaces. This gradient decreases along the length of the microvessel as the internal pressure is dissipated. The absolute magnitude of the gradient will be in doubt as long as there are two polarized groups of investigators, one believing interstitial fluid pressure is subatmospheric in some tissues (8) and the other believing that it is not (5, 9). Regardless of the absolute value, large changes in the gradient can be quickly induced by altering the microvascular hydrostatic pressure.

Microvascular Hydrostatic Pressure   Assuming a constant microvascular wall compliance, mean microvascular hydrostatic pressure is set by the volume of blood in the microvessel and consequently by the relation between micro-vascular inflow and outflow. Inflow is determined by the level of the arterial pressure and the resistance to blood flow through the arteries. Outflow is determined by the resistance to blood flow through the veins and the level of the outflow or venous pressure. The mathematical expression of the influence of these four variables on mean microvascular hydrostatic pressure, $\bar{P}_c$, is:

$$\bar{P}_c = (\bar{P}_a - P_v)\frac{R_v}{R_a + R_v} + P_v$$

where $\bar{P}_a$ = mean arterial pressure; P_v = outflow pressure or venous pressure; R_v = resistance to blood flow through veins; and R_a = resistance to blood flow through arteries. From this expression, it is apparent that an increase in aortic pressure, venous resistance, or right atrial pressure will raise microvascular pressure, whereas an increase in arterial resistance will lower this pressure. However, a variable rarely operates by itself; fortunately one or more of the others frequently move in a direction to maintain a constant microvascular pressure.

Arterial Pressure   The effect of a change in arterial pressure on micro-vascular hydrostatic pressure is relatively small in many vascular beds because the resistance to flow through arteries is high and because this resistance automatically moves in a direction to antagonize it (autoregulation (10)). Consequently, microvascular inflow changes very little. A good example is the renal vascular bed, where changes in arterial pressure over a limited range have essentially no effect on glomerular capillary pressure and, consequently, on the rate of glomerular filtration (11–13). Lymph flow from a hilar lymphatic vessel also changes very little (14). Similar effects are seen in other vascular beds. In the mesenteric vascular bed of the cat, only 15% of a reduction in arterial pressure from 78 to 56 mm Hg is transmitted to venules 10–20 μm in diameter (15). In cat skeletal muscle, net transcapillary fluid movement is essentially zero over a wide range of arterial pressures (16). The constancy of microvascular pressure to a large extent results from the active arteriolar dilation but passive venous constriction also contributes (17). When arterial pressure falls or rises to levels where autoregulation is less pronounced or absent, the effect on micro-vascular pressure is greater. Even here, however, factors other than arterial pressure determine microvascular hydrostatic pressure. For example, a severe reduction in arterial pressure will passively collapse both arteries and veins, the former tending to decrease and the latter tending to increase microvascular pressure.

Arterial Resistance Secondary changes also tend to blunt the effects of alterations in arterial resistance[2] on microvascular hydrostatic pressure. For example, the rise in capillary hydrostatic pressure subsequent to arteriolar dilation is somewhat blunted by a fall in venous resistance (17). The latter occurs because the increased venous inflow raises venous pressures, thereby passively distending the veins. On the other hand, the fall in capillary hydrostatic pressure subsequent to arteriolar constriction is to some extent blunted by a rise in venous resistance due to a fall in venous inflow and hence venous transmural pressure.

Venous Resistance In some vascular beds such as forelimb and intestine, the effects of venous resistance[3] on microvascular pressure are tempered by changes in arterial resistance of the same sign. The effect of venous constriction with a tourniquet, for example, is partly blunted by automatic arteriolar constriction as arteriolar transmural pressure rises (20) (venous-arteriolar response (19–21)). This reduces microvascular inflow, thereby tempering the effect of the decrease in microvascular outflow on microvascular hydrostatic pressure. This should limit filtration. Lymph flow has been measured in the dog forelimb and hindlimb before and after constricting the veins with a tourniquet and the increment in flow is surprisingly small (22, 24). The permeability-surface area product apparently does not change (25). It would be of interest to compare the increment in lymph flow with that seen when the venous-arteriolar response is abolished or when arterial inflow is held constant with a pump.

Venous Pressure In these same vascular beds, effects of venous pressure on microvascular pressure are tempered by changes in both the venous and arterial resistance. An increase in outflow pressure, for example, both passively distends the veins and venules and actively constricts the arterioles, metarterioles, and precapillary sphincters (venous-arteriolar response) (15, 17, 20, 23, 26). Both of these secondary changes tend to reduce microvascular hydrostatic pressure and the net effect is a smaller increase than would be seen in their absence. In the feline mesenteric vascular bed, for example, only 60% of an increase in venous pressure from 7 to 16 mm Hg is reflected in venules 10–20 μm in diameter (15). Consequently, the increase in filtration and lymph flow should be limited (especially since surface area also decreases due to closure of precapillary sphincters (23, 27)).

---

[2] It is frequently stated that arterial resistance is more affected by the arterioles than by the precapillary sphincters. This is an arbitrary separation, however; arterial resistance is sensitive to radius changes down the entire length of the arterial bed. The arterial resistance can have a large effect on microvascular hydrostatic pressure; indeed, closure of the precapillary sphincter completely stops inflow into the capillary and drops capillary pressure to the prevailing venous pressure (7). The action of the sphincter is not necessarily all or none, however, graded changes in caliber are more usual (18).

[3] Venous resistance can have a large effect on microvascular hydrostatic pressure; indeed, capillary outflow can be completely stopped by elevation of venous extraluminal pressure (with a tourniquet, for example). In this vascular section, passive changes in caliber are prominent because of the great compliance of veins over the lower pressure range. Active changes are also prominent, even in the larger veins.

The secondary changes (active arteriolar constriction and passive venous dilation) also complicate the calculation of the capillary filtration coefficient (filtration in ml min^{-1} mm Hg^{-1}/100 g of tissue (28)) from the filtration seen on elevation of venous outflow pressure. In practice, this value is calculated by assigning a fixed relation between venous pressure and capillary hydrostatic pressure. Clearly, this can lead to error because the increment in capillary hydrostatic pressure for a given increment in venous pressure may vary from bed to bed, from preparation to preparation of the same bed (15), and from condition to condition in the same preparation. The preparation with the most vigorous venous-arteriolar response or the most compliant veins will have the smallest increment in capillary hydrostatic pressure for a given increase in venous pressure. When the venous-arteriolar response is naturally absent (for example, lung) or when the venous-arteriolar response or venous compliance is reduced by some experimental maneuver (for example, giving a vasoactive agent), the increment will increase.

Interstitial Fluid Hydrostatic Pressure This pressure depends upon interstitial fluid volume and tissue compliance. Interstitial fluid pressure has traditionally been described as being positive. While this is clearly the case in many tissues (kidney (29), brain, bone marrow, abdominal vicera), recent studies (8) suggest that it is normally subatmospheric in subcutaneous tissue and becomes positive only with hydration. This proposal has not been universally accepted, in part because of measurements in the bat wing (9).

Volume of Tissue Fluid All fluids in and around tissue can contribute to its interstitial fluid pressure, including fluid in the interstitial compartment, cell compartment, lymphatic bed, vascular bed, heart chambers, tubular lumen (kidney, gut), and cavity (peritoneal, pericardial, cerebrospinal). With the volume of fluid in the cell compartment constant, the volume of fluid in the interstitial compartments of simple tissues such as skin and skeletal muscle depends upon the net influx of fluid into the interstitial compartment from the microvessels and the efflux from the interstitial compartment via the lymphatics. Increased influx or decreased efflux will raise volume and hence pressure, especially in tissue with low compliance (see later). This will tend automatically to limit the increment in volume, i.e., defend against edema formation. In the first case, the increased influx from the microvessels will tend to disappear and efflux via the lymphatics will increase. In the second case, the decreased efflux via lymphatics will tend to disappear and influx from microvessels will decrease.

With the volume of fluid in the interstitial compartment constant, changes in cell volume and vascular volume (especially venous) can also affect interstitial pressure, particularly in tissues such as brain and kidney. So can changes in tubular (kidney) and cavity (brain) volumes. Shifts of fluid between the cellular compartment and the interstitial compartment, while affecting the volume of the latter, will have no effect on its pressure when the cell wall is infinitely compliant.

Tissue Compliance The bone marrow and brain are enclosed in a rigid box. The kidney has a rigid stroma which is enclosed in a fibrous capsule. Skeletal muscle is covered by facia and the abdominal viscera are retained by the abdominal wall. In these organs, effective tissue compliance is low and, consequently, for a given interstitial fluid volume, interstitial pressure is high. In the kidney, for example, it is estimated to be approximately 10 mm Hg (29). In such organs, the position relative to the heart and the venous outflow resistance become important determinants of interstitial pressure since the increased pressure in the distended veins is transmitted to the interstitial spaces (14, 29). Indeed, in the kidney, the changes in capillary and interstitial fluid hydrostatic pressures may be almost equal, resulting in little change in the capillary hydrostatic pressure gradient (29). Increased cell volume, tubular volume, luminal volume (gut), and cavity volume can also result in pressure transmission to the interstitial space.

It has been proposed that subcutaneous tissue has a nonlinear compliance, being first very low and then very high as the tissue is hydrated (8). This proposal has recently been discussed in some detail by Eliassen et al. (30) in connection with a study of the compliance of skeletal muscle.

Colloid Osmotic Pressure Gradient

This gradient opposes the hydrostatic pressure gradient. When the colloid gradient exceeds the hydrostatic gradient fluid moves from the interstitial to the vascular space. The magnitude of the colloid osmotic pressure gradient depends on the difference between the plasma and interstitial fluid colloid osmotic pressures and, hence, at least to a major extent, on the difference in the protein concentrations between the two fluids. This difference is a function of the permeability of the microvascular membrane to plasma proteins. Since permeability differs from bed to bed and indeed even along the length of the microvessel in a given bed, so does the magnitude of the colloid osmotic pressure gradient.

Plasma Colloid Osmotic Pressure This depends upon the plasma protein concentration which in turn is dependent upon the amounts of circulating proteins and water. The amount of circulating protein is influenced by protein production, protein losses, and other factors, whereas the amount of circulating water is determined by intake, output, and Starling's forces.

Plasma colloid osmotic pressure is approximately 27 mm Hg in man (the value is somewhat less in cat, dog, rabbit, and rat (31, 32)). The contribution of albumin is much greater than that of globulin because the albumin molecule is approximately one-half the size of the globulin molecule and in most species it is present in higher concentrations. Furthermore, albumin exerts a greater osmotic force than can be accounted for solely on the basis of the number of molecules dissolved in plasma (partly the result of the Gibbs-Donnan effect) and this additional force becomes greater at the high concentrations (5, 33). Consequently, in man approximately 65% of plasma colloid osmotic pressure is attributable to albumin and only approximately 15% to globulin.

Interstitial Fluid Colloid Osmotic Pressure As in plasma, this pressure depends upon macromolecular concentration which in turn is a function of the amounts of macromolecules and water in the interstitium. Macromolecular amount is, to a large extent, determined by the influx of protein from microvessels and the efflux of protein via the lymphatics. The amount of water in the interstitium is also determined by influx from microvessels and efflux via lymphatics. Influx of water is regulated by Starling's forces (see section "Physical Determinants of Fluid Transfer Across Microvascular Membrane").

The influx of protein is importantly determined by the permeability of the microvascular membrane to the protein molecules. According to pore theory developed prior to the advent of the electron microscope (5), the passage of molecules from blood to interstitium is mediated by two systems of pores. The larger molecules, including plasma proteins, cross the microvascular wall through large pores while the smaller molecules also escape through pores having a much smaller diameter. The studies of Palade, Rhodin, and others, with the electron microscope, revealed two basic types of microvessels: those with continuous endothelium, as in muscle, and those with fenestrated endothelium, as in certain viscera (others have features of both basic types) (34). They also revealed slit-like junctions between the endothelial cells and a large population of vesicles within the endothelial cells. About 60% of the vesicles open to the surfaces and they occasionally connect to form a patent transendothelial channel.

The size postulated for the large pores corresponds to the dimensions of the vesicles and fenestrae but the vesicles are not pores (and are similar in appearance to vesicles seen in many other types of cells, for example, vascular smooth muscle) and most of the fenestrae have diaphragms of unknown nature and porosity. The size postulated for the small pores corresponds roughly to the dimension of the cleft in the slit-like junction between the endothelial cells but its size is not reliably indicated by morphological techniques alone.

The morphological technique was, therefore, combined with the use of visible "probe" molecules of known dimensions (such as horseradish peroxidase) by Karnowsky and others. This technique suggests that the vesicles can transport molecules across the endothelial cell and that they may be the structural equivalent of the large pore in capillaries with continuous endothelium. It also indicates that a fraction of the fenestrae also serve this function in capillaries with fenestrated endothelium. The vesicles and the intercellular junctions may be the structural equivalent of the small pore in microvessels with continuous endothelium whereas all fenestrae, including those with more complete diaphragms, also serve this function in the fenestrated capillaries. The basement membrane seems to serve as an extra coarse filter behind the endothelium (it also apparently contributes to the rigidity of the true capillary).

It is thought that albumin and globulin escape from the microvessels via vesicles and fenestrae (5, 6, 24, 25, 35). Whether albumin also escapes via the slit-like intercellular junctions is still undecided. The vesicles and fenestrae are especially prominent in the venous portion of the capillary (34), thereby accounting for the higher permeability of this segment to proteins (6). The venous portion of the capillary also has more surface area than the arterial portion.

It is thought that vesicular transport is independent of capillary pressure (36) but certain evidence suggests that the size of the pores can increase with pressure (22, 35, 37, 38). Perhaps this is related to the fact that the venous capillary and venule distend with pressure (39, 40) (the true capillary is relatively indistensible, changing only 0.1 μm with systole (41)). Other evidence suggests that the size of the intercellular slits can be influenced by certain chemical agents, such as histamine, through rounding and contraction of the endothelial cells (40, 42). It has also been suggested that some of these agents increase vesicular transport (36).

The structure of the interstitium is still not clearly defined but it is thought that it is a two-phase system, namely, a gel-like phase composed of ground substance and a free-fluid phase which contains the proteins (9, 43, 44). In addition to the proteins, the interstitium contains other osmotically active particles such as the mucopolysaccharides of the ground substance. It is estimated that these are present in amounts sufficient to produce an osmotic pressure of 4 mm Hg. Mucopolysaccharides are immense molecules which form a network in a gel-like matrix which is sufficiently dense to exclude plasma proteins. The dense network of the ground substance also markedly impedes the transport of smaller molecules, even molecules as small as water. It does allow the passage of water, however, and this water may contribute to lymph water. If true, this presents the possibility that the concentration of plasma proteins in the free fluid of the interstitial space is actually higher than in the lymph. Considering how high the concentration of protein is in lymph (see later) and the possibility that the interstitium contains other osmotically active particles, it is difficult to escape the conclusion that the colloid osmotic pressure surrounding the microvessels is much higher than realized previously.

Other evidence also suggests that the colloid osmotic pressure of the interstitium is not negligible. Fluid sampled from subcutaneous tissue and skeletal muscle of the rat by the nylon wick method reveals colloid osmotic pressures and total protein concentrations approximately 50% of those in plasma (45, 46). Fluid from capsules implanted in the subcutaneous tissue of dog (47) and rabbit (44) contains 1.9 and 2.5 g of protein per 100 ml, respectively. Fluid sampled from subcutaneous tissue of the rabbit by the liquid paraffin cavity technique contains roughly 1.9 g of protein per 100 ml (the samples also contain mucopolysaccharides) (44). It is, however, difficult to be absolutely certain that fluids collected by the wick and paraffin methods are not contaminated with plasma or that any of the three fluids in fact contain all of the osmotically active molecules found in the interstitium. To circumvent the problems of sample size and contamination with the wick and paraffin techniques and the inconvenience of the long waiting period with the capsule technique, the value for interstitial fluid colloid osmotic pressure is often estimated from lymph draining the part in question. This fluid, like wick, capsular, and paraffin cavity fluid, frequently has a surprisingly high protein concentration (see later), which under normal conditions seems to be roughly in equilibrium with that in the fluid phase of the interstitium (24, 47). However, lymph also may not contain all of the osmotically active particles of the interstitium and there is always the possibility that the protein may not be in equilibrium with that in the fluid phase under abnormal conditions (1, 22).

The available data do suggest, however, that directional changes in the colloid osmotic pressure of the interstitium can conveniently be derived from lymph protein concentration (22, 24, 47) and that the normal absolute value is higher than generally appreciated (22, 24, 44–47). It has been suggested that under certain conditions this high value tends to defend against edema formation (7, 26, 27, 30, 46). Elevation of venous pressure, for example, would dilute the osmotically active particles in the interstitium, thereby increasing the colloid osmotic pressure gradient across the microvascular membrane and hence the force opposing filtration. Lymph protein concentration does vary inversely with lymph flow rate under resting conditions (24, 25, 48) and when venous pressure is elevated (47). In the dog forelimb, elevation of venous pressure first slightly increases lymph protein concentration (22), perhaps due to washout of protein from the free fluid phase of the interstitial space, but it then gradually falls below the control value (unpublished observation).

The protein and water apparently enter the lymphatic system via permeable terminal vessels. The terminal lymphatics in the mesentery of the cat and rabbit, for example (6, 49), originate as blind endothelial sacculations about 40–60 μm wide and 5–10 μm deep or as a network of delicate endothelial lined vessels 20–30 μm wide. The endothelial wall is extraordinarily thin, contains a few vesicles, and has no basement membrane. While there are no obvious discontinuities, the slits between the endothelial cells lack tight junctions (50). It is thought that the endothelium provides little more than mechanical support and that it is not a substantial barrier to the diffusion of water soluble materials up to and including plasma proteins. Contractile activity is inconsistent. Fibrils have been described which appear to anchor the endothelium to the tissue matrix (51). Like the true capillary, the terminal lymphatic has a low compliance and positive pressure (0–2 cm H_2O) (6, 9, 49).

Effluent channels are formed by the confluence of terminal vessels. Here the endothelial cells have close to tight junctions and a discontinuous to continuous basement membrane (50). Other features of these collecting lymphatics are one-way flap valves, compliant thick walls (at least 2–3 μm), muscle cells, anchoring filaments and contractile activity (49, 50). The latter undoubtedly results from a changing contractile activity of the smooth muscle cells but a contribution from the endothelial cells has not been ruled out (as in blood vessels, the endothelial cells appear to contain myofilaments (50). Movement of lymph down the length of the lymphatic vessels has been suggested to result from: 1) a gradient in pressure; 2) active contraction; and 3) passive contraction due to external compression (47). The valves facilitate unidirectional flow.

Lymph collected from the larger lymphatics contains considerable quantities of protein. That derived from skin and paw of the dog limb, for example, contains 2–3 g % of protein (22, 47). Both albumin and globulin are present, the concentration of the former being only slightly higher than the latter (the plasma albumin-globulin ratio is approximately one in the dog). The colloid osmotic pressure of lymph is, therefore, substantial, although considerably less than that of plasma, both because the protein concentrations are lower and because the extra osmotic effect of albumin is less at the concentrations seen in lymph (i.e., it exerts an osmotic effect more nearly in proportion to concentration).

Crystalloid Osmotic Pressure

Alterations in the concentrations of small molecules to which the microvascular membrane is very permeable can also directly influence transcapillary fluid flux. Thus the addition of crystalloids to plasma will cause transient water influx whereas addition to interstitial fluid will do the reverse (52). This transient fluid flux is to a large extent the direct result of the establishment of a crystalloid osmotic gradient across the membrane. It is transient because the gradient very quickly disappears. The magnitude of the fluid flux will depend upon the amount of the substance added and the ease with which it penetrates the membrane.

Under certain conditions this direct effect can be more sustained. A sustained direct effect requires that the addition be continuous and that there be an infinite sink into which the substance can disappear. Continuous addition to interstitial fluid in a local area meets these requirements. The gradient across the microvascular membrane will be sustained since the blood into which the substance diffuses is continuously replaced; consequently, plasma fluid continues to move into the interstitial space (this will in part be the indirect consequence of increased capillary hydrostatic pressure if the hyperosmolality produces arteriolar dilation).

During intra-arterial infusion of hypertonic solutions of sodium chloride, dextrose, and certain other small molecules, the dog forelimb (perfused at constant flow) continues to lose weight at a slow rate (53). The continued loss of weight indicates that fluid continues to move across the capillary membrane. This fluid flux probably results in part indirectly through effects on Starling's forces. The increase in interstitial fluid crystalloid osmotic pressure will withdraw water from the cells (a requirement for this cell dehydration is that the cell wall be relatively impermeable to the added agent). This in turn will dilute the colloid in the interstitial spaces and consequently increase the colloid osmotic pressure gradient across the microvascular membrane. Fluid will then move from the interstitial to the vascular space. It is also possible that the cell fluid raises interstitial fluid hydrostatic pressure (see section "Volume of Tissue Fluid"). This too would provide a force for moving interstitial fluid into the capillary. In contrast to adding the crystalloid to interstitial fluid, a rise in capillary hydrostatic pressure due to arteriolar dilation would tend to oppose the fluid flux into the capillary.

Surface Area

Microvascular surface area can change because of an alteration in the number of open channels or in the degree of distension of those already open. The former mechanism is probably the most important. Precapillary sphincters can add, or completely delete, a microvascular channel. On the other hand, only the venular end of the channel is measurably distensible. Change in surface area will not by itself influence fluid flux, i.e., the addition of a channel which is in balance with respect to Starling's and crystalloid forces will not influence the amount of water in the interstitial spaces. It will, however, have an important influence on fluid flux if the forces are out of balance.

INFLUENCE OF NATURALLY OCCURRING VASOACTIVE AGENTS ON NET FLUID TRANSFER

All substances which influence the contractile activity of vascular smooth muscle are potentially capable of altering net fluid flux across the capillary membrane through effects on capillary hydrostatic pressure. Some of these agents have a greater effect on fluid flux than others because, in addition to their effects on vascular smooth muscle, they also have effects on the membrane itself. Histamine and bradykinin, which have been implicated in urticaria, anaphylaxis, and other states, belong to the latter group. They will be considered first.

The evidence that the agent in fact alters fluid flux is first summarized. An explanation of the altered fluid flux is then attempted in the context of the physical determinants considered above. The capillary hydrostatic pressure gradient, colloid osmotic pressure gradient, and microvascular surface area are discussed in that order. Since effects on fluid flux may differ on local and systemic administration, the two routes of administration are considered separately.

Histamine

Local Administration

Skin and Skeletal Muscle   Several types of evidence demonstrate that locally administered histamine increases extravascular fluid volume in skin and skeletal muscle. Small to large doses of histamine (2–60 μg of base/min, i.a.) significantly increase organ weight, volume, and circumference (54–57). These changes are only partially attributable to increases in intravascular blood volume (57) since, especially with the larger doses, they greatly exceed those which can be produced by maximal vasodilation. In the canine forelimb, large doses of histamine increase weight by 10–20% in 30 min (57), and massive edema is readily detectable by visual and tactile examination. Lymph flow in skin and skeletal muscle is markedly increased by histamine (22, 36, 58), indicating that fluid filtration is elevated and that the edema does not result from decreased drainage via the lymphatics.

The increased net fluid filtration is due to both an increase in the transmural hydrostatic pressure gradient and a fall in the transmural colloid osmotic pressure gradient.

Histamine clearly increases the transmural hydrostatic pressure gradient owing to a marked increase in capillary hydrostatic pressure. Landis (59) demonstrated most convincingly by direct micropuncture pressure measurements that histamine increases capillary hydrostatic pressure well above normal plasma colloid osmotic pressure in skin. Additionally, small vein pressure (which represents a minimum for capillary hydrostatic pressure) is markedly elevated in skin and skeletal muscle from a control of 5–15 mm Hg to 20–40 mm Hg (54, 56, 57, 60). Hence, capillary hydrostatic pressure must be above these small vein pressures which already exceed normal plasma colloid osmotic pressure (approximately 20 mm Hg in dogs).

The mechanism of the rise in capillary hydrostatic pressure is attributable to a rise in capillary inflow subsequent to arteriolar vasodilation (54, 57, 60). Aortic and right atrial pressures are not affected by locally administered histamine except by high pharmacological doses. Even if these pressures are affected they change in a direction which would tend to lower capillary hydrostatic pressure. Likewise large artery and large vein resistances are not measurably affected by locally administered histamine in nonhypotensive doses which cause maximal edema formation (56, 57). Small vessel resistance, however, is markedly reduced (54, 57, 60).

Other evidence also indicates that the rise in small vein pressure and, by inference, capillary hydrostatic pressure, is flow-linked rather than the result of an increase in venous resistance. For example, histamine (5 µg of base/min) infused into the naturally perfused canine forelimb greatly increases small vein pressures in skin and skeletal muscle and limb weight (~25 g in 30 min) (57). If forelimb blood flow is held constant utilizing a blood pump, then small vein pressure is unchanged relative to control and the weight gain is largely abolished (~7 g in 30 min), presumably because capillary hydrostatic pressure is prevented from increasing (57, 60).

Clearly the increased transmural hydrostatic pressure gradient will tend to wane as fluid accumulates in the tissue and tissue pressure increases. This will limit filtration.

In addition to increasing the transmural hydrostatic pressure gradient, histamine also decreases the transmural plasma colloid osmotic pressure gradient owing to an increased interstitial fluid protein concentration subsequent to an enhanced rate of protein efflux from blood into interstitial fluid.

Many types of indirect evidence indicate that locally administered histamine increases the rate of macromolecule efflux from blood to tissue and consequently the interstitial fluid protein concentration. Histamine increases the deposition of carbon particles on the basement membrane of the venous microvessels (42, 61, 62) and the transport rate and concentration of dextrans in lymph (63, 64). More to the point, it has also been demonstrated that histamine greatly accelerates the passage of radioactive and dye-labeled protein (65–67), and, perhaps the best evidence, that the transport and concentration of protein in lymph increase, the latter approaching values found in plasma with large doses of this agent (22, 36, 63, 64, 68).

Other indirect evidence also supports an increase in interstitial fluid protein concentration. Histamine decreases the isogravimetric capillary hydrostatic pressure (69, 70). It also can increase extravascular fluid volume under conditions where capillary hydrostatic pressure cannot exceed normal plasma colloid osmotic pressure. In the dog forelimb perfused at constant flow with a blood pump, intra-arterial infusion of 60 μg of histamine base/min causes an increase in forelimb weight (50–120 g in 30 min), lymph flow, and lymph protein concentration, despite the fact that blood flow and small vein pressure remain unchanged relative to control (57). Furthermore, small artery pressure, which represents a maximum for capillary hydrostatic pressure, falls to 25 mm Hg or less in skin and skeletal muscle, a value itself near normal plasma colloid osmotic pressure. Capillary hydrostatic pressure must be less. Even more convincing, histamine still significantly increases weight while the forelimb is perfused at a pressure of 20 mm Hg (57).

The increased net protein efflux and interstitial fluid protein concentration clearly result from an increased microvascular "permeability" to plasma proteins. Histamine increases the capillary filtration coefficient (55, 69) and the permeability-surface area product for both albumin and globulin (72). The concentration of all plasma proteins, including the larger globulin species, increases in lymph (22, 72). The increase in globulin concentration is proportionately more than the increase in albumin concentration, resulting in a decrease in the albumin globulin ratio and a protein distribution close to that of plasma (22, 72).

The mechanism of the increased "permeability" is not clear but it seems to be to a large extent independent of the rise in capillary hydrostatic pressure. The albumin and globulin concentrations still rise in lymph under conditions designed to prevent the pressure increases (22). Microscopic observations show that the endothelial cells round up in such a manner as to open the intercellular clefts (40, 62, 73), and one investigator suggests that this results from contraction of actomyosin-like fibrils within the endothelial cells (73). Calculated pore size during administration of histamine has been reported to increase (71) or not change (74).

Still to be considered is whether the increased capillary hydrostatic pressure contributes to the increased protein "permeability." Is all the increase in microvascular "permeability" due to a direct pressure independent effect on microvascular membranes or is it in part indirect subsequent to passive stretching of clefts and fenestrae by the rise in capillary hydrostatic pressure? In the absence of histamine, many investigators have shown that conditions which likely increase capillary hydrostatic pressure also increase the size of the macromolecules that appear in lymph (35, 37, 38, 76, 77), leading to the concept of the "stretched pore phenomenon" (77). In the dog forelimb, mechanically increasing blood flow and small vein pressure for 30 min to levels seen during local administration of high doses of histamine slightly increases lymph total protein concentration due to a slight increase in albumin concentration; globulin concentration does not change (22). Similarly increasing venous pressure alone for 90 min slightly increases lymph total protein concentration in the first 10 min due to an increase in albumin concentration (as pointed out under "Interstitial Fluid Colloid Osmotic Pressure" this could represent washout from the free-fluid phase of the interstitial space). However, the total protein, albumin, and globulin concentrations then gradually fall so that after 90 min all three are below control values (75). The albumin-globulin ratio is increased at this time indicating a proportionately greater fall in globulin concentration than albumin concentration. These observations fail to provide evidence that increased pressure alone augments pore size in the dog forelimb and the changes do not resemble those seen during histamine infusion (increased albumin and globulin concentrations, the latter more than the former, resulting in a decrease in the ratio). However, the data do not rule out some stretching.

Elevation of venous pressure during histamine infusion also decreases albumin and globulin concentrations, but now proportionately, resulting in no further change in the ratio (75). Thus histamine makes the membrane relatively nonselective to plasma proteins and increased pressure does not further alter selectivity.

The concept of increased pore size has recently been challenged by Renkin and his co-workers (36, 63, 64, 72). They observed that, in the dog hindpaw, the effects of histamine on lymph flow, protein concentration, and protein transport (flow times concentration) are sustained and elevation of venous pressure during administration of histamine does not further increase protein transport. Since protein transport via fenestrae and intercellular clefts is thought to be sensitive to pressure whereas transport by vesicles is thought to be insensitive to pressure, they suggest that histamine accelerates protein efflux via increased vesicular transport. In the dog forelimb, however, the effects of histamine on lymph flow, protein concentration, and protein transport wane with time and elevation of venous pressure during histamine administration increases transport relative to that during administration of histamine alone (75). More work is obviously needed on this question.

Some evidence indicates that histamine also increases capillary surface area. This would increase net water efflux relative to that seen if capillary surface area were unaffected, assuming of course that Starling forces were already altered to favor increased net filtration. Histamine increases the capillary filtration coefficient (55, 69, 70) which, in part, may be attributable to an increase in capillary surface area. It has been shown that histamine increases the permeability-surface area product (72) and the extraction of rubidium (78, 79), also suggesting an increase in capillary surface area. The contribution of this increase in capillary surface area to histamine edema is difficult to assess, but it probably contributes much less importantly than the changes in transmural hydrostatic pressure gradient and the transmural plasma colloid osmotic pressure gradient, which are exceedingly large. For example, in the naturally perfused forelimb, high and low doses of histamine produce similar hemodynamic effects, i.e., total blood flow and small vein pressures increase to essentially the same level (57). Presumably, capillary surface area would be equally increased with both doses of histamine yet the weight gain is much greater with the large than with the small dose of histamine.

Both the rise in hydrostatic pressure gradient and the fall in colloid osmotic pressure gradient contribute significantly to histamine edema formation. With low doses (5 μg of base/min, i.a.) of locally administered histamine infused into the naturally perfused canine forelimb, weight increases by about 25 g in 30 min, and total blood flow, small vein pressure, and lymph flow are greatly increased (57). This same dose of histamine infused into forelimbs perfused at constant inflow still increases lymph flow greatly but small vein pressures are not altered relative to control and the weight gain is largely abolished (57). Hence, under conditions where microvascular pressure is prevented from rising, the edema with low doses of histamine is largely abolished. However, high doses (60 μg of base/min, i.a.) of histamine infused locally into the naturally perfused canine forelimb increase total blood flow and small vein pressures to only the same level as do the lower doses of histamine, yet maximal edema develops (weight increases 50–120 g in 10–30 min). If arterial inflow is now held constant, this same dose of histamine causes almost the same mean total weight gain in 30 min as when blood flow is allowed to vary naturally despite the fact that neither arterial inflow nor small vein pressures change relative to control. Thus, with high concentrations of histamine, the data strongly suggest that the fall in the transmural colloid osmotic pressure gradient exerts the dominant effect.

Other Vascular Beds   There are little data describing the actions of histamine on the storage of extravascular fluid in other vascular beds. In fact, the majority of these studies simply describe the effects of topical applications of histamine on small blood vessel diameter, microvascular permeability to large molecules and particles, and, in several instances, on venular pressure measured by direct micropuncture. Northover and Northover (80) found that in the rat mesentery topically applied histamine increases venular pressure (measured by direct micropuncture) and venular permeability to colloidal carbon particles. Buckley and Ryan (81) have also reported that histamine increases the permeability of mesentery venules and capillaries to macromolecules. The findings of Northover and Northover (80) and Buckley and Ryan (81) also suggest that increased venular permeability could not be accounted for by a rise in venular pressure. Hence, as in skin and skeletal muscle, the available data suggest that in the mesentery, at least, locally administered histamine increases microvascular pressure and microvascular permeability to plasma proteins.

Systemic Administration   It has been well established that intravenously administered histamine causes potent vasodilation at least initially, yet fails to promote edema formation in skin and skeletal muscle (56, 60, 82–84) as does locally administered histamine. Forelimb weight decreases with time (84). With large intravenous doses of histamine, the weight loss greatly exceeds that possible from reduction in vascular volume (84). Hence, with time, marked extravascular fluid reabsorption occurs despite the continued intravenous infusion of histamine (84). Aortic and right atrial pressures decrease and, with time, precapillary resistance increases (60, 84) subsequent to a decreased transmural vascular pressure and to a sympatho-adrenal discharge (60) owing to hypotension. Hence, capillary hydrostatic pressure probably decreases, promoting net fluid influx and a decrease in interstitial fluid volume. This is suggested by the marked sustained fall in small vein pressure in both skin and skeletal muscle (60,

84). Although venous resistance also increases with time (84), the rise in postcapillary resistance is not sufficient to counteract the fall in capillary hydrostatic pressure, for forelimb weight and, by inference, the interstitial fluid volume continue to decrease with time.

Surprisingly, the extravascular fluid reabsorption persists even though the calculated blood concentration of histamine is equal to that which causes massive edema formation when infused locally into a skin-skeletal muscle vascular bed. This finding, considered in the light of local studies which demonstrate an edemogenic action even in the face of low microvascular pressures, suggests that permeability is not influenced to the same extent during intravenous administration. Other major differences (i.e., besides the difference in capillary pressure) between the two situations are that the blood must first pass through the pulmonary circuit on intravenous infusion and that the hypotension elicits a sympathico-adrenal discharge. Clearly, additional experimentation is needed before this question can be resolved.

Bradykinin

Local Administration

Skin and Skeletal Muscle Small to large doses of bradykinin (0.8–10 μg of base/min, i.a.) significantly increase organ weight, volume, and circumference (56, 85–88). These changes are only partially attributable to increases in intravascular blood volume and largely represent a rise in extravascular fluid volume (edema) subsequent to increased net fluid filtration. In the canine forelimb, large doses of bradykinin increase weight 10–20% in 30 min (weight increase 50–140 g) and lymph flow up to 25-fold, and massive edema is readily detectable by visual and tactile examination (88).

The increased net fluid filtration appears to be due both to an increase in the transmural hydrostatic pressure gradient and a fall in the transmural colloid osmotic pressure gradient.

Direct measurements of capillary hydrostatic pressure during the local administration of bradykinin have not been made. However, small vein pressure and blood flow in both skin and skeletal muscle are markedly increased by bradykinin in the canine forelimb (56, 88). This suggests a large rise in capillary hydrostatic pressure, attributable to a rise in capillary inflow subsequent to arteriolar vasodilation (56, 88). Aortic and right atrial pressures are only minimally affected except with high pharmacological doses. Likewise, large artery and large vein resistances are at best minimally affected by doses which cause maximal edema formation (88), and, if anything, venous resistance falls. Small vessel resistance, however, is markedly reduced. Other evidence also indicates that the rise in small vein pressure and, inferentially, capillary hydrostatic pressure is linked to flow rather than to venous resistance (88). The rise in the transmural hydrostatic pressure gradient would fade away as tissue pressure rises.

In addition to increasing the transmural hydrostatic pressure gradient, bradykinin also decreases the transmural plasma colloid osmotic pressure gradient.

Like histamine, locally administered bradykinin increases the rate of macromolecule efflux (89–91) from blood to tissue and increases the concentration of protein in lymph (to values approaching those of plasma with large doses of this agent (88, 92, 93). Additionally, the magnitude of the rise in capillary filtration coefficient relative to that seen with exercise in skin-skeletal muscle preparations suggests an increase in protein permeability (55, 94). Measurements of forelimb weight and small vein pressure under conditions of natural inflow and constant inflow in response to small and large doses of this agent are also compatible with this interpretation (88).

The mechanism of the increased net protein efflux is still not completely clear but the prevailing view is that it results from an increase in microvascular pore size (95). Renkin and his co-workers (36, 63, 64, 72) have, however, recently threatened this concept. In any event, the increased protein permeability, as in the case of histamine, can occur independent of the rise in capillary hydrostatic pressure. Under conditions where capillary hydrostatic pressure probably is unchanged relative to control, or perhaps even falls, bradykinin still greatly increases lymph flow and lymph protein concentration, and causes massive edema (88). It is possible, however, that the increased capillary hydrostatic pressure contributes to the increased rate of protein efflux, i.e., protein transport.

Capillary surface area is also thought to be increased by bradykinin (55, 94, 95). However, as with histamine, it seems likely that the edema is largely influenced by changes in the transmural hydrostatic pressure gradient and the transmural plasma colloid pressure gradient rather than by the increase in capillary surface area. With the high doses, changes in permeability seem to dominate. The blood flow, blood pressure, and resistance responses to both low and high doses of bradykinin are similar (88), suggesting a similar increase in microvascular pressure and surface area in both situations, yet the weight gain is far greater with the high dose.

As with histamine, both the rise in capillary hydrostatic pressure and the fall in the transmural colloid osmotic gradient contribute significantly to bradykinin edema. With low doses of bradykinin, the weight gain is largely abolished if the rise in microvascular pressure is prevented by keeping blood flow constant at control levels with a pump (88). With large doses, the total weight gain in 30 min is similar, whether the forelimb is perfused naturally or whether flow is kept constant at control levels with the aid of a blood pump, suggesting that increase in permeability is the dominant factor with high doses of this agent (88). Nonetheless, the rise in capillary hydrostatic pressure is still important with large doses of bradykinin resulting in still more marked increases in lymph flow and lymph protein concentration, and the rate of weight gain is initially accelerated relative to the rate of weight gain when flow is kept constant at control levels (88).

Other Vascular Beds   There are little definitive data on the effects of locally administered bradykinin on the storage of extravascular fluid in other vascular beds. As with histamine, the majority of studies focus their attention on changes in vessel caliber, microvascular permeability to large molecules and colloidal carbon particles, and in some instances on venular pressure measured by direct micropuncture, largely in the rat mesentery.   Locally administered bradykinin increases flow through arterioles and opens precapillary sphincters to increase flow through a larger number of capillaries (96, 97), increases venular pressure (80), and increases venular permeability to colloidal carbon particles (80) in the rat mesentery. Hence, as in skin and skeletal muscle, it must be concluded that locally administered bradykinin increases microvascular pressure and microvascular permeability to plasma proteins.

Systemic Administration   There are little data on the effects of intravenously administered bradykinin on extravascular fluid volume and fluid filtration. The studies of Daugherty et al. (56) suggest that intravenously administered bradykinin fails to increase small vein pressure or to increase forelimb weight, suggesting that its classic edemogenic action is route dependent. Thus, like histamine, intravenously administered bradykinin fails to promote edema formation in skin and skeletal muscle, and, in fact may promote extravascular fluid reabsorption owing to a fall in capillary hydrostatic pressure.

Acetylcholine

Locally administered acetylcholine does not produce the classic signs of edema, yet still significantly increases weight (98) and lymph flow (22, 98) when infused into the naturally perfused, innervated canine forelimb. This weight gain ($\simeq$ 25 g in 30 min) is due, in part, to a rise in intravascular blood volume subsequent to vasodilation and, to a larger extent, to a rise in extravascular fluid volume subsequent to an increased net fluid filtration owing to a rise in capillary hydrostatic pressure (98). Small vein pressure is markedly increased in both skin and skeletal muscle (22, 56, 98). The rise in microvascular pressure is clearly flow-linked rather than the result of an increase in venous resistance. If forelimb blood flow is held constant at control levels with a blood pump, small vein pressure fails to increase relative to control and the weight gain is largely abolished (56). It is interesting to note that this same weight gain ($\simeq$ 25 g in 30 min) can be produced by mechanically increasing forelimb blood flow and small vein pressure to levels produced by acetylcholine (98).

There is no evidence that acetylcholine affects microvascular permeability to plasma proteins. Acetylcholine fails to increase lymph protein concentration (22, 98) in the canine forelimb. This is supported by other data. The capillary filtration coefficient in skeletal muscle is increased by acetylcholine about as much as by exercise (55) but far less than the increase produced by histamine and bradykinin (55).

Numerous other workers have also demonstrated that acetylcholine increases net fluid filtration and the capillary filtration coefficient (99, 100). Since this agent clearly fails to alter microvascular permeability to plasma proteins, the increase in capillary filtration coefficient must represent an increase in the capillary surface area available for exchange. Rubidium-86 extraction is also increased by locally administered acetylcholine in the isolated canine forelimb perfused at controlled flow, further evidence of an increased capillary surface area (101).

Serotonin

Serotonin (15 or 150 μg of base/min) infused locally into the naturally perfused innervated canine forelimb fails to promote an increase in limb weight relative to control (102), although lymph flow is substantially increased indicating an increased net fluid filtration (102). The rise in interstitial fluid volume relative to that caused by histamine and bradykinin is exceedingly small, as forelimb weight remains essentially unchanged. Since lymph protein concentration is unaltered (102), the increased fluid efflux cannot be accounted for by a fall in the transmural plasma colloid osmotic pressure gradient. The marked rise in small vein pressure, especially in skin, suggests that the fluid filtration increases subsequent to a rise in microvascular pressure. The rise in capillary hydrostatic pressure is attributable to intense large vein constriction in skin and to small vessel vasodilation (presumably arteriolar) in skeletal muscle (56, 102). The increased skin small vein pressure is clearly linked to the increased venous resistance as skin blood flow is markedly reduced. In skeletal muscle, the rise in microvascular pressure is clearly linked to flow rather than to venous resistance. Large vein resistance in skeletal muscle is unaltered or decreases, whereas small vessel resistance is markedly reduced, reflecting largely arteriolar vasodilation. Despite the marked reduction in total forelimb blood flow, skeletal muscle flow is either unchanged (low dose) or increases (high dose) relative to control. These data are consistent with findings in man. Several investigators have also failed to find evidence for substantial serotonin edema formation in the forearm and hand in man (103, 104).

In the rat, serotonin clearly increases microvascular permeability to plasma proteins. In fact, it is more potent than histamine in increasing venular permeability to macromolecules (dye-labeled proteins and carbon particles) in mesentery, ileum, cremaster muscle, and skin (42, 62, 80, 105–108). Hence, it must be concluded that the permeability-increasing effect of serotonin is species dependent. Interestingly, in the rabbit ear, serotonin also fails to alter permeability (109). In the cat intestine, the capillary filtration coefficient rises (110, 111) but increased microvascular permeability need not be invoked to explain the observation.

Serotonin is most unique in that it exerts varied effects on both series and parallel coupled vascular circuits (56, 102). Serotonin intensely constricts large arteries and large veins in skin (even more than small vessels, proportionately) whereas, if anything, it causes small vessel vasodilation in skeletal muscle (56, 102). However, serotonin uniformly constricts large artery strips from both skin and skeletal muscle in vitro (112). This suggests that the effect of serotonin in skeletal muscle may be indirect.

Norepinephrine and Epinephrine

Local Administration Norepinephrine fails to alter isogravimetric capillary pressure in the canine hindlimb, indicating that this agent does not affect microvascular permeability to plasma proteins (113). Hence, local infusions of this agent, in concentrations which do not affect systemic pressure, could only alter fluid fluxes via changes in microvascular pressure subsequent to alterations in the pre- to post-capillary resistance ratio. An increase in this ratio will decrease microvascular pressure promoting fluid reabsorption, whereas a decrease in this ratio will increase microvascular pressure promoting fluid filtration. Depending on the concentration, vascular bed (histological/anatomical differences of blood vessels), effectiveness of local regulation, and duration of infusion, either filtration, reabsorption, or no change in fluid fluxes might be predicted. Unfortunately, there are virtually no studies in which both weight/volume and vascular pressures-resistances have been simultaneously measured over prolonged periods of time in naturally perfused organs during local infusion of a wide dose range of this agent. The available data, while sparse and fragmentary, still, however, support this suggestion. In the naturally perfused denervated forelimb (54) and naturally perfused innervated ileum (114), norepinephrine either increases (occasionally resulting in frank edema), fails to alter, or decreases weight. The weight increase is associated with increases in small vein pressure suggesting an increased skin microvascular pressure and net fluid filtration. The weight decrease is associated with decreases in small vein pressure. From these studies, however, it is not clear if forelimb interstitial fluid volume may be increased in skin, skeletal muscle, or both. There is a lack of data concerning the effects of epinephrine infused locally for prolonged periods of time on fluid fluxes in naturally perfused organs. Presumably, effects similar to norepinephrine would be predicted except possibly in skeletal muscle, owing to the alleged vasodilation in precapillary vessels with low concentrations of this agent.

There is little agreement on the effects of norepinephrine and epinephrine on capillary surface area. In the human hand and calf (natural flow) (115) and isolated canine hindlimb (constant flow or constant perfusion pressure) (113), norepinephrine is reported not to alter the capillary filtration coefficient (no change in surface area), whereas in cat hindlimb skeletal muscle (constant perfusion pressure) (116), norepinephrine is reported to increase this coefficient (increase in surface area). In canine forelimbs and hindlimbs perfused at constant flow, both epinephrine (117, 118) and norepinephrine (116, 118) are reported to increase ^{86}Rb uptake indicating an increased surface area. In the canine gracilis muscle (constant flow), epinephrine and norepinephrine are reported to decrease ^{86}Rb uptake (119) and, in human skeletal muscle (natural flow), norepinephrine decreases the permeability surface area product (120) suggesting a decreased surface area. The reasons for these discrepancies are not understood but are probably related to the same factors listed above which cause variability in fluid flux.

Systemic Administration The alleged net fluid filtration and plasma volume depletion late in hemorrhagic and endotoxin shock are frequently attributed to the marked increase in circulating catecholamines. Indeed, early investigators, utilizing dilution techniques to measure blood volume, concluded that large intravenous doses of epinephrine and norepinephrine, infused continuously over prolonged periods of time, cause a progressive marked reduction in plasma volume (121, 122). In the dog, hematocrit increased and plasma volume decreased, suggesting a greatly increased fluid filtration. In fact, there was visual evidence of fluid loss from the vascular system into the intestinal lumen, intestinal wall, and pericardial sac. Fluid loss via these routes appears to be relatively small, however, and could not account for the reported large reductions in plasma volume. Hence, it was usually suggested that skeletal muscle was the likely sink for this fluid loss. The time course of the hematocrit increases and plasma volume decreases is, however, not consistent with extensive fluid loss by filtration. The rise in hematocrit appears maximal early, remaining relatively steady thereafter, and can be prevented by prior splenectomy (123). This suggests that the hematocrit increase is due mainly to the release of erythrocytes from the spleen. If marked progressive fluid loss by filtration occurred, a progressive rise in this variable would be predicted. The calculated decrease in plasma volume does not necessarily mean loss of fluid to extravascular spaces. In states associated with intense vasoconstriction, plasma volume measurements may be misleading owing to inadequate mixing of tag with blood (pooling), particularly if blood is sampled too soon. More recently, other investigators, utilizing similar dilution techniques to measure blood volume, have failed to find evidence for a large consistent plasma volume loss (124, 125), and still others have shown that the calculated decreases in plasma volume can be prevented by prior splenectomy (123). These conflicting observations and the suggestion that the alleged fluid loss occurs in skeletal muscle prompted the following study to examine fluid fluxes in skin and skeletal muscle specifically, utilizing a gravimetric technique.

Large amounts of epinephrine (3 or 6 μg of base/min) or norepinephrine (1.5 or 3 μg of base/min), infused intravenously for 3 hours, decrease forelimb weight markedly initially and then more slowly with time (126). The initial weight loss represents a reduction in vascular volume and might be greater were it not for increased net fluid filtration subsequent to marked increases in microvascular pressures in both skin and skeletal muscle (small vein pressures increase greatly). The rise in microvascular pressure is due to a large increase in systemic blood pressure and to postcapillary constriction. This increased fluid efflux is transient because vascular pressures return to, or near, control within about 15 min. Weight then declines more slowly with time. This secondary weight loss cannot be completely accounted for by a further decrease in vascular volume and therefore must, in part, represent extravascular fluid reabsorption. During this time, systemic pressure and small vein pressures in both skin and skeletal muscle fall well below control, suggesting a reduction in microvascular pressure, and plasma osmolality increases; both decreased microvascular pressure and hyperosmolality would promote fluid influx. Interestingly this fluid influx persists despite the fact that postcapillary resistance appears to be better maintained than precapillary resistance, i.e., the ratio of pre- to postcapillary resistance falls. These data provide no evidence for sustained net fluid efflux into skin and skeletal muscle during prolonged intravenous infusions of either epinephrine or norepinephrine.

Angiotensin and Vasopressin

Angiotensin fails to alter isogravimetric capillary pressure (113) in the isolated canine hindlimb, indicating that this agent does not alter microvascular permeability to plasma proteins. Although data are not available, there is no reason to suppose that vasopressin alters microvascular permeability to proteins either. Hence, like norepinephrine, locally infused angiotensin and vasopressin would be expected to produce alterations in net fluid transfer only via change in microvascular pressure subsequent to alterations in the pre- to postcapillary resistance ratio. Since these agents, if anything, constrict precapillary vessels proportionately more than postcapillary vessels, extravascular fluid reabsorption would be expected. This prediction is consistent with the observations of Haddy et al. (54). They reported that angiotensin and vasopressin infused locally into naturally perfused, denervated canine forelimbs either fails to alter limb weight or decreases the rate of weight gain suggestive of reduced fluid filtration. There is a real need to study the effects of a wide dose range of these agents in various naturally perfused vascular beds for prolonged periods of time in order to gain a more complete understanding of the local actions of these agents on net fluid transfer.

INFLUENCE OF NERVOUS SYSTEM ON NET FLUID TRANSFER

This area was reviewed in 1968 by Mellander and Johansson (127) in connection with the control of resistance, exchange, and capacitance function in the peripheral circulation. The account here therefore is brief and reference is sometimes made to their review rather than to original papers. As will become apparent, except in one instance, long-term effects on fluid flux have not been studied.

Local removal of sympathetic adrenergic vasoconstrictor fiber activity results in an increase in interstitial fluid volume in dog forelimb. Thus, section of the nerves leading to the limb results in a gradual small weight gain which diminishes with time, becoming negligible in about 20 min (54). The site (skin or skeletal muscle) of the increased fluid volume has not been determined. The fluid efflux apparently results from an increase in capillary hydrostatic pressure (54) and surface area (128) subsequent to arteriolar dilation. Its spontaneous termination could be related to increase in interstitial fluid hydrostatic pressure, decrease in interstitial fluid colloid osmotic pressure, and/or disappearance of the dilation.

The effect of locally enhanced sympathetic adrenergic vasoconstrictor fiber activity on interstitial fluid volume in skin has not been studied. Its effect on resistances, however, has been investigated and apparently depends, at least in part, on the stimulation frequency (129–132). Brief, low frequency faradic stimulation of the nerves leading to the dog hindpaw appears to constrict arteries proportionately more than veins (129–131). This would be expected to produce transcapillary fluid influx. Intermediate frequency stimulation can do the reverse, however, and on stopping stimulation the veins seem to relax more slowly than the arteries (129–131, 133, 134). This would be expected to cause efflux of fluid from the capillary, especially on stopping stimulation. Paw volume, after an initial decrease, remains constant (129) or increases (131) and then, on stopping stimulation, overshoots the control value (129, 131). High frequency stimulation (above that found physiologically) essentially closes the larger arteries but on stopping stimulation they relax faster than the veins (130, 131, 134, 135). Paw weight or volume studies during prolonged stimulation at low and intermediate frequencies are needed. Studies should also be directed at the physiological relevance of the intermediate frequencies.

Venous constriction is also pronounced in the skin of the cat hindlimb (133), dog forelimb (136), and dog hind leg (137) but apparently not in the rabbit ear (133, 138). Perhaps venous smooth muscle or nerve density is less prominent in ear skin than in leg skin.

Skeletal muscle has been studied more thoroughly. Short-term electrical stimulation of regional adrenergic vasomotor fibers leading to this tissue produces, over a wide frequency range, increases in the pre- and postcapillary resistances, the former proportionately more than the latter (127, 136, 139). Consequently, capillary hydrostatic pressure falls, resulting in fluid reabsorption (127, 139). Capillary surface area decreases (1, 14–16, 127, 139–141) but then apparently returns to the control value (127, 139). Responses are similar in canine subcutaneous adipose tissue except here capillary surface area appears to increase (142). Veins in skeletal muscle contain less smooth muscle and norepinephrine than veins in skin (143–145). This may account for the fact that the pre- to postcapillary resistance ratio always increases in skeletal muscle but usually decreases in skin with intermediate frequency stimulation.

Responses to short term electrical stimulation differ in intestine. In cat jejunum-ileum preparations, the increase in precapillary resistance is transient (autoregulatory escape) while the increase in postcapillary resistance and the decrease in capillary surface area (mucosa) appear to be sustained during stimulation (146, 147). During the brief peak constrictor response there appears to be an increase in the ratio of pre- to postcapillary resistance and fluid reabsorption as occurs in skeletal muscle, but in the steady state there is no significant net transcapillary fluid movement. Responses appear to be similar in the cat colon (148).

Reflex augmentation of vasoconstrictor fiber activity, for brief time periods, produced by 1) inactivation of the arterial baroreceptors; 2) stimulation of the chemoreceptors; or 3) stimulation of the somatic pressor afferent nerves (systemic arterial pressure held constant), causes net reabsorption of tissue fluid into the circulation from skeletal muscle in cat hindlimbs (149, 150). Reabsorption has also been demonstrated during 1) and 2) from skin in cat hindlimbs (149). Conversely, reflex inhibition of vasoconstrictor fiber activity produced by stimulation of the 1) arterial baroreceptors or 2) somatic depressor afferent nerves causes net filtration in skeletal muscle (149, 150). Filtration has also been demonstrated in skin during stimulation of arterial baroreceptors (149). On the other hand, no discernible net fluid transfer occurs across the capillary membranes in cat intestine with any of these maneuvers (149, 150). The fluid movement in skin and skeletal muscle undoubtedly results from changes in capillary hydrostatic pressure subsequent to resetting of the pre- to postcapillary resistance ratio, due to dominant changes in the precapillary vessels as described above.

A number of other studies also indicate that reflex activation and deactivation of the vasoconstrictor nerves by altering the pressure in the carotid sinus have little effect on veins in skin and skeletal muscle (151–155). Venous responses are seen only with very large changes in sinus pressure (151, 156). The effect of the resetting of the resistance ratio on capillary hydrostatic pressure, and hence fluid transfer, will augment that due to the change in arterial pressure (which elicits the resetting). These effects are important under a number of circumstances, hemorrhage, for example (149, 153, 157, 158). On the other hand, the effect on fluid transfer of the ratio resetting elicited by chemoreceptor stimulation might be obscured by the effect of the rise in arterial pressure (159).

Reflex activation of the sympathetic adrenergic vasoconstrictor system in man by other means, namely, a deep breath, ice to the forehead, or leg exercise, also has little effect on muscle veins but does influence skin veins (160).

The sympathetic adrenergic vasoconstrictor system can also be activated and inhibited by central stimulation but effects on fluid flux across capillary membranes have not been well studied. It is clear, however, that in cat hindquarters and intestine both arteries and veins as well as blood pressure can be influenced from a variety of central areas (hypothalamus, midbrain reticular formation, and medulla) via this system (161–163). The same seems to be the case for rat mesentery and skeletal muscle, although here venular responses seem to be absent or weak (164). It would therefore be surprising if fluid flux was not influenced, except perhaps in intestine (162).

Activation of the sympathetic cholinergic vasodilator system in the cat by peripheral nerve stimulation produces filtration in skeletal muscle (165). This apparently results from a rise in capillary hydrostatic pressure due to arteriolar dilation. Surface area appears to be unaffected (127, 166). The response is transient unless the frequency of stimulation is progressively increased (165). The system can also be activated in the cat by hypothalamic stimulation (165) but effects on fluid flux are not clear (165).

INFLUENCE OF SEVERAL TYPES OF STRESS ON NET FLUID TRANSFER

Exercise

Small Muscle Mass Vascular responses and transcapillary fluid movement during local skeletal muscle contraction were summarized in 1934 by Landis (167). He concluded that functional activity of skeletal muscle produces hyperemia, increased capillary pressure, and increased net efflux of water. Furthermore, he suggested that the increased water efflux that accompanies muscle activity is too large to be the sole result of the rise in capillary pressure and that osmotically active substances produced in the course of tissue metabolism may play a role. Basically these statements still hold today. However, the availability of additional quantitative data now allows a more exact definition of the mechanisms involved in the transcapillary fluid fluxes observed during exercise. It is now evident that the increased movement of fluid out of skeletal muscle capillaries during and for varying times following activation of skeletal muscle is due to at least three separate but interrelated mechanisms. These are: 1) a rise in capillary hydrostatic pressure subsequent to a fall in precapillary resistance out of proportion to a fall in postcapillary resistance; 2) an increase in microvascular surface area subsequent to recruitment of previously closed capillaries and distention of the venous ends of previously perfused capillaries; and 3) initially, at least, to an increase in interstitial crystalloid osmotic pressure subsequent to release of metabolic products from the activated skeletal muscle cells. Justification and/or experimental data to support each of these mechanisms is provided in the following discussion.

Casual inspection of a muscle just after a period of exercise reveals it to be shorter, larger in circumference, and firmer to the touch than an unexercised contralateral muscle (168). More definitive evidence for increased transcapillary fluid movement from plasma to extracellular space during muscle exercise is provided by a variety of physiological indices. Muscle weight increases and the specific gravity of muscle tissue decreases during exercise (168). Arteriovenous hemoglobin determinations reveal water loss from blood passing through active muscle (168). Muscle volume, measured plethysmographically, increases during exercise (169–176). The increase in volume is much greater than can be accounted for by calculated changes in intravascular volume. Moreover, volume varies as a function of the severity of muscle activity (169–171, 176) (this need not be the case for vascular volume). Baker and Davis (172), using plethysmographic and tracer techniques simultaneously, recently reported that extravascular volume decreases with the onset of exercise and then increases above the pre-exercise volume as exercise continues. They suggest that tissue pressure markedly increases during muscle contraction and forces extravascular fluid into the vascular system. However, as exercise continues, tissue osmotic and capillary hydrostatic forces override this effect and water moves into the muscle tissue. Lymph flow from skeletal muscle increases as the muscle goes from the resting to the active state (171, 177–180), as does tissue pressure (175, 176, 181). Finally, there is an increase in the interstitial and intracellular water content of exercising muscle (182).

The rate of transfer of fluid from intravascular to extravascular compartments during heavy local muscle exercise is reported to be as high as 1.49 ml/min/100 g of tissue (176) and during short-term exercise (15 min) total fluid accumulation in active muscle may reach 15 ml/100 g of active tissue (171, 176). Lymphatic drainage may have been obstructed in these studies (this would tend to magnify the volume to an unknown extent in an isolated exercising muscle). Fluid accumulation in active muscle reaches a maximum between 15 and 30 min after initiation of heavy exercise (171). Thus, initially, translocation of fluid from blood to tissue exceeds the removal capacity of the muscle lymphatic system. However, during long-term exercise the rate of fluid transfer out of the capillaries is reduced (see later) such that lymphatic drainage and net filtration appear to be nearly balanced.

In keeping with the Starling-Landis hypothesis, the rather remarkable increase in fluid transfer from blood to tissue associated with activation of skeletal muscle could result from 1) a decrease in the transcapillary colloid osmotic pressure gradient; 2) an increase in the transcapillary hydrostatic pressure gradient; and/or 3) a combination of 1) and 2). As pointed out under "Surface Area" an increase in the microvascular transfer area would affect the magnitude of the fluid shift, but could not in itself initiate the fluid efflux.

During exercise of a small muscle group, the only plausible way the transcapillary colloid osmotic pressure gradient could decrease would be through an increase in the permeability of the capillary membrane to protein. To date there is no good experimental evidence to support an increased membrane permeability to protein during muscle exercise. As stated earlier, lymph flow from contracting muscle increases but the protein concentration of this lymph either remains unchanged or falls slightly (178–180). The muscle lymph to plasma ratio for high molecular weight dextrans does not change during exercise (177) and there is minimal loss of labeled albumin from blood during its passage through contracting skeletal muscle (172). Thus, a fall in the transcapillary colloid osmotic pressure gradient does not appear to be involved in net fluid efflux observed during exercise. On the contrary, it seems reasonable to assume that with the large influx of water into the interstitial space the colloid osmotic pressure within this compartment actually falls. In fact, it has been suggested that an increase in the transcapillary colloid osmotic pressure gradient provides minimal protection against fluid accumulation (171).

On the other hand, there is good experimental evidence to support a rise in the transcapillary hydrostatic pressure gradient during exercise. Landis (183) showed by direct micropuncture of capillaries in the frog longissimus dorsi muscle that, during the hyperemic period following tetany, arterial and venous capillary pressures rise, on the average, by 5 and 6.5 cm H_2O, respectively. Kjellmer (169) and Lundvall (176) calculated that mean capillary hydrostatic pressure rises by 5–10 mm Hg during lower leg exercise in the cat. This is supported by the finding that muscle small vein pressure increases from 11–20 mm Hg in the dog (17). The magnitude of the capillary pressure rise appears to be related to the severity of the exercise. Between contractions and before large fluid accumulation in the muscle tissue there is little reason to suspect a change in interstitial pressure. Obviously, tissue pressure must rise (175) as volume accumulates in the interstitium and in the muscle cell (182). The rise in interstitial pressure will impede filtration (171, 175, 176).

Aortic and right atrial pressures do not change during exercise of a small muscle mass. Thus the rise in capillary pressure must result from a fall in pre- to postcapillary resistance ratio. The magnitude and possible mechanisms of these resistance changes have been treated in other reviews (10, 127, 184, 185).

Associated with the large decrease in vascular resistance is an increase in microvascular surface area. Krogh (186) showed that the number of blood-containing capillaries per unit of muscle is much greater in the active than in the resting state. Based on ^{86}Rb extraction experiments in isolated cat gastroc-nemius-plantaris and dog gracilis muscles perfused at constant flow, Renkin and Rosell (166) calculated that the capillary transport coefficient increases about 2-fold during maximum metabolic dilation (tetany). Numerous investigators (169, 171, 174, 176, 187) find that the capillary filtration coefficient increases 2- to 3-fold during skeletal muscle exercise. Since permeability is apparently unchanged, the rise in capillary filtration coefficient reflects an increase in microvascular surface area. Finally, there is an increase in vascular volume in skeletal muscle during exercise (172) which is also compatible with increased capillary density.

Kjellmer (169) concluded that the net outward movement of fluid during skeletal muscle exercise is due to an elevated capillary hydrostatic pressure and that the fluid efflux is augmented by an increase in capillary surface area. This conclusion was based largely on the finding that the increase in tissue volume is prevented when muscle blood flow is held constant. This averts the rise in capillary pressure (it appears there is little if any active response in veins of exercising muscle in the cat or dog (17, 169)). Subsequently, studies by Mellander et al. (173) and Lundvall (176) challenged the conept that increased capillary hydrostatic pressure and surface area are the only factors acting to promote fluid movement into exercising muscle tissue. These investigators measured the osmolality of venous blood leaving exercising cat skeletal muscle and found it to be from 4 to 40 mOsm/kg greater than arterial osmolality. There was a rough relationship between the increase in venous osmolality or total fluid accumulation and the severity of exercise. Furthermore, the venoarterial osmolar difference waned over a 15-min exercise period. They concluded that fluid accumulation in exercising muscle is mainly due to fluid osmosis across the capillary membranes caused by work-induced tissue hyperosmolality which primarily must involve the intracellular space and secondarily the interstitium. In fact, Lundvall suggested that, at all work intensities, more than 75% of the maximal transcapillary fluid transfer can be ascribed to fluid osmosis. Furthermore, he contended that the main factor limiting fluid transfer during prolonged work periods is a gradual decline in the transcapillary osmolar gradient, subsequent to decreased liberation of osmotic products by skeletal muscle cells.

Although most studies relating venous blood osmolality changes and muscle exercise have been performed in the cat, recent studies in man have yielded similar results (170). Changes in venous osmolality during exercise of the dog gracilis muscle are also directionally similar to those observed in the cat, but the absolute increase in venous osmolality is smaller and not as well maintained during the exercise period (188). Measurements of tissue fluid accumulation are lacking in gracilis.

Certain features of the tissue edema associated with muscle exercise are intriguing. Surprisingly, the magnitude of fluid accumulation reported in cat muscle is comparable to that produced in the dog foreleg by local intra-arterial infusions of histamine or bradykinin, agents known to greatly increase micro-vascular protein permeability (see Sections "Histamine" and "Bradykinin"), yet exercise apparently has little or no effect on the membrane. Assuming the measurements in isolated cat skeletal muscle apply to more physiological settings, the most logical explanation for this apparent discrepancy is that crystalloid osmotic forces are produced during exercise that act in conjunction with an increased microvascular pressure and surface area to sequester plasma water. However, before completely subscribing to this hypothesis, one must contend with Kjellmer's studies (169) showing that exercise edema can be prevented by holding muscle blood flow constant. Two possible explanations can be given for this discrepancy. First, as alluded to by Lundvall (176), under conditions of reduced blood flow velocity, large transcapillary crystalloid osmotic gradients would be less likely to develop and be maintained, and second, it is likely that capillary pressure falls rather than rises during exercise at constant flow. The former effect would tend to produce less transcapillary fluid efflux and the latter effect would actually promote absorption. Another interesting feature of exercise edema is the failure of lymph flow from the edematous tissue to increase to the degree expected. This may be related to the fact that much of the increased water is actually intracellular (182) and hence not available to the lymphatic system.

Large Muscle Mass   From the foregoing, one would expect exercise involving a large percentage of the total muscle mass to produce a drastic reduction in plasma volume due to fluid accumulation into the active muscle. This does not appear to be the case. Plasma volume does fall during short-term systemic exercise (170, 189–192). However, the reduction is at most in the order of 10–15% and in prolonged (1 hour or more) heavy exercise plasma volume may be unchanged or rise slightly (193, 194). This could mean that during such exercise 1) transcapillary fluid influx into active muscle is much less than during exercise of a small muscle mass; 2) lymphatic return from active muscle is facilitated; and/or 3) extravascular fluid from inactive tissue or organs is shifted into the vascular system. Two of these possibilities were systematically examined in a recent investigation (170). Changes in lower leg tissue volume were determined in healthy male subjects during exercise on a bicycle ergometer. Plasma volume and osmolality were also measured before and during exercise. During heavy exercise there was a significant increase in volume of the active muscle, averaging 4.5 ml/100 g of active tissue in 6–10 min. This is less than has been observed during exercise of a small muscle mass. Calculated total fluid loss into active muscle during the exercise period was about 1,100 ml but during this time plasma volume decreased by only 600 ml. This means that approximately 500 ml of extravascular fluid were transferred to the blood. Arterial plasma osmolality rose on the average by 22 mOsm/kg during exercise; a rise in osmolality of venous blood from the active muscle preceded the rise in arterial osmolality.

Thus, during short-term exercise of a large muscle mass, the direction, locality, and mechanisms of the transcapillary fluid flux may be summarized as follows. Volume accumulates in the active muscle tissue because of osmotic gradients subsequent to increased molecules created by elevated metabolism and as a result of increased filtration due to an elevated microvascular pressure. A contributory factor is an increased microvascular surface area. The smaller total fluid accumulation per unit of active muscle relative to exercise of a small muscle mass probably occurs because of 1) reduced transcapillary crystalloid osmotic gradients due to recirculation of blood with an increased osmolality; 2) a relatively smaller increase in microvascular transfer area (174); and 3) better lymphatic drainage. On the other hand, extravascular fluid is mobilized in nonactive tissue and transferred into the vasculature. The mechanisms responsible for this transfer are 1) hyperosmolality of the arterial plasma and 2) decreased microvascular pressure subsequent to the well documented increased sympathico-adrenal activity that accompanies exercise. Additional studies are needed to elucidate the mechanisms involved in the transcapillary fluid flux that accompanies long-term exercise.

Ischemia

A number of clinical reports indicate that edema can follow the relief of ischemia in an abnormal extremity of man. Arterial reconstruction for atherosclerosis of the lower limb is followed by warmth (relative to the normal extremity), lasting about 4 months. Edema is manifest in every extremity that displays warmth for over 7 days (195). Increased blood flow (measured plethysmographically) relative to the opposite limb, especially in the skin of the foot, is apparent and does not subside for about 4 weeks (196). Since the edema parallels the warmth in degree and duration, and disappears if the graft thromboses, and since the arteries distal to the occlusion exhibit pre-existing medial atrophy, Husni (195) suggests that the edema results from a high capillary hydrostatic pressure subsequent to the atrophy. In effect, this corresponds to perfusing the lung, which naturally has little arteriolar smooth muscle, at a pressure of 100 mm Hg. The pressure will, of course, rise higher when the reconstruction is complicated by phlebitis which obstructs the veins.

Edema is also reported following release of prolonged compression ("crush syndrome") (197). It is difficult to know whether the edema results from the ischemia, however, since the latter is frequently complicated by tissue and vascular damage.

Vascular occlusion with rubber bands leads, on release of occlusion, to edema in the rat hindlimb and rabbit ear. Application of rubber bands to the rat hindpaw for 2–60 min causes, on release of the bands, the rapid onset of redness and swelling (198). The latter is clearly apparent 60 and 180 min after release of the band and takes 2–4 days to disappear (198). The swelling is attenuated by prior administration of vasoconstrictor agents (198), suggesting that increased capillary hydrostatic pressure plays a role in its development.

Application of the rubber bands to the rat hindlimb for periods ranging between 30 min and 8 hours causes, on release, increased skin water content which is maximal with the 4.5-hour occlusion (199). An occasional skin venule labels with carbon after the 30 min occlusion; more extensive labeling is seen after a 2-hour occlusion (199). In muscle, water content is significantly elevated only after release of a 2.5-hour occlusion and progresses with occlusion duration up to 5.5 hours (199). Here venular labeling with carbon is never more than sporadic. Following release of a 2.5-hour occlusion, skin and muscle water contents are significantly elevated within 30 min and are maximal in about 2 hours (199). In the skin some venular labeling with carbon is evident between 30 and 90 min after release of the rubber band; muscle venules do not label (199). If vascular labeling with carbon is taken as a measure of vascular leakage, there is no comparison between the intense leakage of venules induced by local histamine injection and the minimal leakage of venules caused by occlusion. Thus, increased permeability to plasma proteins does not appear to play an important role in the edema seen in the rat hindlimb following occlusions up to 2.5 hours.

In the rabbit, application of a 1-cm-wide rubber strip to the base of the ear for 2, 6, and 18 hours causes, on release, an increased ear thickness proportional in magnitude to the duration of occlusion (197). Marked edema develops in ears occluded for 18–24 hours. Necrosis also develops, occasionally following an 18-hour occlusion, and always following a 24-hour occlusion (necrosis also invariably develops following an 18-hour occlusion if the ambient temperature is elevated). Following release of an 18-hour occlusion the thickness at first increases rapidly, doubling in the first 30 min. It then increases more slowly and is almost maximal at 2 hours. Edema fluid, obtained by a needle prick 1 hour after release, contains 5.2 g % of protein, mostly albumin. Following a 24-hour occlusion, an intravenous injection of Evans blue dye causes blueing of the ear (whether injected immediately or 1 hour following release). Thus, increased permeability to plasma proteins clearly plays a role in the edema seen in the rabbit ear following release of an 18-hour occlusion.

It is this type of data that has led to the assumption that ischemia of up to several hours duration leads to edema. As pointed out by Pochin (197), however, great care must be taken to avoid venous obstruction on removal of narrow rubber bands; there is always the possibility that the veins will not open as rapidly as the arteries. If this happens, the investigator studies venous obstruction in addition to ischemia.

Relevant to this question is the use of broad, air-inflated cuffs to produce ischemia. The pneumatic tourniquet is widely used as an adjunct to extremity surgery in man. Total occlusion is frequently maintained for 2 hours and the literature makes only vague reference to edema formation on release of occlusion (200, 201). This suggests that edema is not a serious problem. Ischemia is in fact produced. In the upper extremity, the p_{O_2} and pH of the venous blood distal to the tourniquet fall to 20 mm Hg and 7.19 units, respectively, by 1 hour, and to 4 mm Hg and 6.90 units by 2 hours (200). Corresponding values for p_{CO_2} are 62 and 104 mm Hg.

Furthermore, in the intact dog foreleg (no heparin), 2 hours of ischemia, produced by inflation of a broad pneumatic cuff above the elbow, does not produce edema on relief of occlusion (judged by weighing the limb with the shoulder joint as a fulcrum) (202). In the intact dog hindlimb, pneumatic cuff occlusion of the thigh for 2–4 hours is also without effect (judged by circumference) (203). These observations are similar to those of Lazarus-Barlow (204) in 1894 that relief of complete "anemia" (ischemia produced by tightly bandaging the limb to stop flow) of a dog limb for 1 hour does not lead to the development of edema (judged by changes in limb circumference). On the other hand, 5 hours of cuff occlusion of the thigh of the dog results in increased thigh circumference lasting a week which is partially prevented by hypothermia or heparin (203).

More definitive experiments have been conducted in the collateral-free dog forelimb attached to the body only by the major blood vessels and nerves (202). In this preparation, simple arterial occlusion produces complete ischemia (uninfluenced by bone flow), relief of ischemia is not complicated by venous spasm, and weight measurements are accurate to a fraction of a gram. Here relief of ischemia of moderate duration has little effect on water transport across microvascular membranes and the efflux that does occur appears more related to increased pressure on a greater surface than to increased permeability to plasma proteins. Relief of 2 hours of ischemia produces only a 1.7% increase in limb weight which disappears within 15 min and inspection and palpation fail to reveal signs of edema (202). By contrast, a 30-min intra-arterial infusion of histamine can produce an 18% increase in weight which does not disappear for long periods after stopping the infusion, and inspection and palpation reveal all of the classic signs of gross edema (see under "Histamine") (57). Furthermore, a large portion of the transient 1.7% increase in weight represents an increase in vascular volume since it is associated with markedly decreased vascular resistance and increased venous transmural pressure.

The slight fluid efflux that does occur apparently results largely from increased microvascular hydrostatic pressure and surface area. On relief of the ischemia, arterial inflow increases to high levels for a short period of time due to decreased arterial resistance (202), metabolically and myogenically induced (10, 127, 184, 185). Venous pressure is elevated during this same high flow period. When arterial inflow is not allowed to increase, venous pressure remains at the control level and limb weight does not rise (202). Thus, the weight increase can be abolished by preventing the natural consequence of the fall in arterial resistance, namely, increased flow and pressure. Studies in other preparations perfused at constant flow show that rubidium extraction (205) and capillary filtration coefficient (206) increase, suggesting increased surface area for exchange (as will be pointed out below, permeability does not change with short and intermediate periods of occlusion). This could result from opening of closed precapillary sphincters and passive distension of the venular end of already open capillaries by increased pressure (especially in the free-flow state).

The effect of ischemia on fluid exchange is small because the hemodynamic event is transient and the microvascular membrane remains relatively impermeable to plasma proteins. In the dog forelimb, lymph protein concentration increases essentially not at all (from 1.9 to 2.2 g %) following relief of 2 hours of ischemia (202). By contrast, it approaches that in plasma during infusion of histamine or bradykinin (see under "Histamine" and "Bradykinin" (22, 88). It has also been reported that calculated pore radius in the microvessels of the dog hindlimb is unchanged 20–40 min following relief of 30- and 60-min periods of ischemia and increases in only one-third of limbs subjected to 180 min of ischemia (206).

Thus, simple arterial occlusion of moderate duration does not lead to massive edema and the edema that does occur appears not to be related to an increase in the permeability of the capillary membrane to plasma proteins. Since ischemia produces hypoxia, it is necessary to reconcile this finding with the prevalent view that reduced oxygen tension increases the permeability of the capillary membrane to plasma proteins, thereby promoting edema formation in both systemic and pulmonary vascular beds. This belief stems to a large extent from studies by Landis in the frog mesentery (207) and Drinker in the dog lung (208). In 1928, Landis found that discrete capillary occlusion was sometimes followed by filtration distally (despite a fall in capillary hydrostatic pressure) when the mesentery was bathed with oxygen-free Ringer's solution. In other experiments, the effective colloid osmotic pressure across single capillaries was decreased 1 min after the relief of 3 min of ischemia of the entire intestine when the intestine was bathed with the same solution. The effect disappeared within 15 min and did not occur when the intestine was bathed with oxygenated Ringer's solution. The transient decrease in the effective colloid osmotic pressure indicates a transient increase in microvascular permeability to plasma proteins. This increased protein permeability due to oxygen lack, coupled with the higher capillary hydrostatic pressure due to reactive hyperemia, would promote net fluid filtration. Bathing the mesentery in Ringer's solution half-saturated with carbon dioxide produced no change in fluid movement and complete carbon dioxide saturation increased the rate of fluid filtration very slightly but the microvascular membrane remained impermeable to plasma proteins. Thus at oxygen tensions approaching zero, capillaries in the frog mesentery became more leaky to plasma proteins but carbon dioxide is without effect. The effects of lesser degrees of hypoxia were not reported and studies were not conducted on mammalian capillaries. In considering the relevance of these findings in a 1934 review, Landis stated that "... it is still doubtful whether the oxygen tension in functioning tissues ever becomes low enough to be of importance" (167).

The much quoted conclusion of Drinker, that hypoxemia promotes a development of pulmonary edema through an increase in the permeability of the lung capillaries, was apparently based upon one experiment on a dog subjected to increased airway resistance during inspiration plus hypoxia (208). In this animal, which was compared with another subjected to the same inspiratory resistance, breathing oxygen, there was at autopsy slightly more lung edema than in the dog given oxygen. However, it was subsequently shown that inspiratory resistance alone produces edema in some dogs but not in others and that the edema seen correlates positively with the pulmonary venous pressure during the last quarter hour of life (32, 209).

It is surprising how difficult it is to generate evidence for increased permeability in mammalian capillaries by more moderate, physiologically probable, degrees of hypoxia than used by Landis. Nairn (210) perfused the dog hindlimb at a constant rate with venous blood from the femoral vein of the opposite hindlimb (oxygen saturation $\sim$ 50%) for 6–7 hours and did not observe edema. Scott et al. (211) perfused dog forelimbs at constant flow with hypoxemic blood ($p_{O_2} \sim$ 10 mm Hg, pH unchanged) for 15 min and observed no changes in weight. Increased carbon dioxide tension is also without effect on the weight of this preparation (212) and there is no evidence that it increases capillary permeability to plasma proteins (213).

The failure to find an increase in fluid efflux from the capillary at constant flow does not necessarily mean that fluid efflux would not occur at natural flow. Local hypoxemia, like ischemia, produces arteriolar dilation. At natural flow (constant arterial pressure), this would produce increased capillary inflow and consequently increased capillary hydrostatic pressure if the veins did not dilate proportionately. There is in fact evidence that edema does occur under this condition (214), although definitive studies are lacking.

Shock

The early effects of hemorrhage on net transvascular fluid transfer are well understood. A variety of evidence indicates that hemorrhage rapidly promotes extravascular fluid reabsorption in proportion to the severity of the blood loss in bird, rat, cat, dog, man, and other primates (124, 215, 216). The plasma volume increases while the plasma protein concentration and hematocrit decrease (124) (the latter may not occur in the dog because its spleen has a large store of erythrocytes). Furthermore, the flow of lymph from the thoracic duct, after an initial increase, falls to low levels (124). The fluid is apparently reabsorbed to a large extent from skeletal muscle. This is supported by measurements of weight and volume of skin-skeletal muscle preparations in animals and man during hemorrhage and during pooling of blood in the lower extremities (149, 153, 157, 158, 215, 218–220). Weight and volume decrease with time due to reduction in vascular volume initially and then largely to net extravascular fluid reabsorption.

There is little reason to believe that the fluid reabsorption is initiated by a change in the colloid osmotic pressure gradient. The protein concentration in plasma falls while that in thoracic duct lymph does not change (124). Furthermore, there are no changes in the lymph to plasma ratios for administered dextrans in cervical and paw lymph (124). However, there is good reason to believe that the reabsorption is importantly related to a fall in the transmural hydrostatic pressure gradient owing to a reduction in capillary hydrostatic pressure. Aortic and right atrial pressures decrease and the precapillary vessels constrict out of proportion to the postcapillary vessels (149, 153, 157, 158, 215, 221). With mild hemorrhage the rise in postcapillary resistance is largely passive whereas with severe hemorrhage active venous constriction also occurs (153, 157, 158). Nonetheless, the pre- to postcapillary resistance ratio still initially increases. Steady state capillary surface area in skeletal muscle, as judged by the capillary filtration coefficient, is reported not to change in man (216) and rise in cat (217) and bird (215). Based on studies in the cat, it has been recently suggested that fluid also enters the vascular system osmotically from cells due to a rise in plasma osmolality subsequent to glucose release from the liver (218–220). In the dog, however, the rise in plasma osmolality is rather small (157, 158). Much work is needed, especially in man, to determine the relative contributions of the fall in capillary hydrostatic pressure and the rise in crystalloid osmotic pressure to the fluid influx.

The fluid that enters the vascular system during the early hypotensive period is relatively protein free (124). Thus the initial rapid influx of fluid will be tempered by a decrease in the colloid osmotic pressure gradient due to both a fall in plasma protein concentration and a rise in interstitial fluid colloid osmotic pressure. It will also be tempered by an increase in the transcapillary hydrostatic pressure gradient due to increased intravascular and decreased extravascular pressures.

There is less agreement on the direction of fluid transfer during the late oligemic and post-transfusion periods. Many investigators report a rise in hematocrit and falls in plasma and red cell volumes (measured by dilutional techniques) in the dog during these periods (124). This is frequently accompanied by bloody diarrhea and, at autopsy, swelling and hemorrhagic lesions are found in the wall of the small intestine. Furthermore, in a study in cat skeletal muscle, fluid reabsorption ceased as quickly as 20 min following hemorrhage and then was replaced by filtration (217). From studies such as these it has been concluded by many investigators that plasma volume loss by transcapillary filtration can occur in the late oligemic and post-transfusion periods and that this loss may contribute to irreversibility. In earlier times it was suggested that this alleged loss resulted from increased capillary permeability to plasma proteins due to ischemia and/or the release of naturally occurring edemogenic agents such as histamine or bradykinin. More recently it has been suggested to be the result of increased capillary hydrostatic pressure due to a fall in the pre- to postcapillary resistance ratio (maintained postcapillary constriction at a time precapillary resistance wanes owing to autoregulatory escape) (217, 222).

Other findings, however, are difficult to reconcile with the fluid filtration hypothesis. Some investigators using dilutional techniques fail to find evidence for plasma loss in the dog during the late oligemic and post-transfusion periods (124). Furthermore, in monkey, man, and splenectomized dog, measurements of plasma volume, plasma protein concentration, and hematocrit provide only marginal evidence for fluid loss from the vascular system in late oligemic shock (124). One laboratory finds that the volume of fluid loss via the intestinal route in dogs is rather small and that atropine prevents intestinal necrosis but fails to alter survival time or survival rates (223). Another laboratory finds that the enterectomized dog demonstrates the same hemodynamic pattern during irreversible hemorrhagic shock as the nonenterectomized dog (224).

Some of these discrepancies appear to be related to species variations. Bloody diarrhea, intestinal necrosis, and erythrocyte storage by the spleen are much more prominent in the dog than in monkey and man. Others are the result of differences in technique. Measurements of red cell and plasma volumes in hemorrhagic shock have been criticized because of the failure to obtain complete time-concentration curves, since it has been shown that the mixing of the injected test substance can be significantly delayed (124, 225, 226). Inadequate mixing would lead to the erroneous conclusion that volume is reduced when in actuality there may be little change in absolute volume.

There are other reasons to question the hypothesis. Complete ischemia for 2–5 hours has surprisingly little effect on microvascular permeability to plasma proteins in a canine skin and skeletal muscle preparation (see under "Ischemia"). Intravenous infusion of histamine at rates which lower blood pressure produces reabsorption rather than infiltration in this same preparation (see under "Histamine: Systemic Administration"). Furthermore, the flow of thoracic duct, paw, and cervical lymph decreases during sustained hemorrhagic hypotension (124, 227) and there is no change in the lymph to plasma ratio for administered dextrans in the paw and cervical regions (227).

These discrepancies stimulated studies in a canine forelimb preparation. Weight was measured for 4 hours after 25, 50, or 60% blood volume reduction (157, 158). Limb venous outflow pressure was permitted to fall as it does naturally in response to hemorrhage (this apparently was not the case in the study of cat skeletal muscle (217) quoted above. Also see Ref. 219). Forelimb weight decreased rapidly initially and then more slowly throughout the entire 4-hour period in all three groups. The slow weight losses were of sufficient magnitude to justify confidence that they did not represent changes in vascular volume. In no group was a rise in weight recorded during the late oligemic period, even though some of the animals in one group were clearly in irreversible shock (158). These data indicate that, in the canine forelimb, reabsorption of extravascular fluid continues in the late oligemic phase, and are consistent with the plasma volume measurements in monkeys, man, and splenectomized dogs.

In the forelimb studies (157, 158), arterial and venous pressures remained low and precapillary resistance appeared to remain high well into the late oligemic phase. Arterial plasma osmolality was also somewhat elevated during this period. All of these changes promote fluid reabsorption (124, 149, 153, 157, 158, 215, 217–220) and could account for the continued weight loss. The first three changes lower capillary hydrostatic pressure which in turn causes reabsorption of interstitial fluid. The increased osmolality withdraws fluid from the cellular compartment, making more available for reabsorption by the capillary. In some animals, there was evidence for a waning of the increased precapillary resistance in the late oligemic period. However, a terminal decrease in the pre- to postcapillary resistance ratio, had it occurred, would have been of little consequence with respect to fluid transfer. This ratio can have only minor effects on capillary hydrostatic pressure when arterial pressure is low (158).

It does, however, have an important effect when arterial pressure is normal and this is relevant to the post-transfusion period. Transfusion with blood raises arterial and venous pressures which in turn will raise capillary pressure. The increment in pressure will be greater with a low than with a high pre- to postcapillary resistance ratio. Thus one would expect more filtration on retransfusion if the pre- to postcapillary resistance ratio were low prior to transfusion and remained low post-transfusion. Filtration would be expected to cease if the resistance ratio is normalized or even if it remained low but arterial pressure again fell to a low level. These parameters have not been well studied in the post-transfusion period. Neither has the effect of blood transfusion on arterial plasma osmolality been investigated in this period.

Plasma volume, measured by dilutional techniques, is also often reported to decrease in dogs subjected to endotoxin shock (see references in Ref. 228). This has been attributed to increased microvascular permeability to plasma proteins and increased capillary hydrostatic pressure subsequent to decreased pre- to postcapillary resistance ratio and hepatic venous constriction (see references in Refs. 228 and 229). The gravimetric technique has been employed to assess changes in extravascular fluid volume in the canine forelimb during endotoxin shock (228–230). In this preparation, *Escherichia coli* endotoxin (2 or 5 mg/kg, i.v.) produces an initial rapid weight loss (0–10 min), attributable to reduced vascular volume, followed by a slow progressive weight loss (10–240 min), attributable to extravascular fluid reabsorption. Aortic and right atrial pressures remain low during the entire period and plasma osmolality rises slightly (228). Thus the fluid reabsorption may well result from decreased capillary hydrostatic pressure and osmotic withdrawal of fluid from the intracellular compartment. Evidence for fluid filtration is not observed in this skin-skeletal muscle preparation during the hypotensive period. Limb weight does rise, however, if 1,000 ml of cross-matched whole blood are administered over a 25-min period after 2 hours of hypotension, but it fails to reach the control level (230). This small rise in weight is associated with increased aortic and right atrial pressures and is essentially complete within 45 min. Weight then falls again, signaling reabsorption once more, as pressures decline. Thus even following transfusion there is little evidence for extensive fluid efflux into skin and skeletal muscle in canine endotoxin shock. Obviously, these studies do not necessarily apply to other vascular beds and species.

REFERENCES

1. Visscher, M.B., Haddy, F.J., and Stephens, G. (1956). The physiology and pharmacology of lung edema. Pharmacol. Rev. 8:389.
2. Staub, N.C. (1974). Pulmonary edema. Physiol. Rev. 54:678.
3. Starling, E.H. (1896). On the absorption of fluids from the connective tissue spaces. J. Physiol. (London) 19:312.
4. Hyman, C. (1944). Filtration across the vascular wall as a function of several physical factors. Amer. J. Physiol. 142:671.
5. Landis, E.M., and Pappenheimer, J.R. (1963). Exchange of substances through the capillary walls. Handbook of Physiology, Section 2: Circulation Vol. 2, p. 961. American Physiological Society, Washington, D.C.
6. Intaglietta, M., and Zweifach, B.W. (1974). Microcirculatory basis of fluid exchange. Adv. Biol. Med. Phys. 15:11.
7. Wiederhielm, C.A. (1974). Microcirculatory function in health and disease—the evolution of concepts and technology. Circulation 50: (Suppl. III), 3.

8. Guyton, A.C., Granger, H.J., and Taylor, A.E. (1971). Interstitial fluid pressure. Physiol. Rev. 51:527.
9. Wiederhielm, C.A. (1969). The interstitial space and lymphatic pressures in the bat wing. *In* A.P. Fishman and H.H. Heckt (eds.), Pulmonary Circulation and Interstitial Space. University of Chicago Press, Chicago.
10. Haddy, F.J., and Scott, J.B. (1968). Metabolically linked vasoactive chemicals in local regulation of blood flow. Physiol. Rev. 48:688.
11. Haddy, F.J. (1962). XIII. Renal circulation, D. Physiology. *In* D.L. Abramson (ed.), Blood Vessels and Lymphatics, p. 406, Academic Press, New York.
12. Deen, W.M., Robertson, C.R., and Brenner, B.M. (1974). Glomerular ultrafiltration. Fed. Proc. 33:14.
13. Renkin, E.M., and Robinson, R.R. (1974). Glomerular filtration. New Eng. J. Med. 290:785.
14. Haddy, F.J., and Scott, J.B. (1965). Role of transmural pressure in local regulation of blood flow through kidney. Amer. J. Physiol. 208:825.
15. Richardson, D.R., and Zweifach, B.W. (1970). Pressure relationships in the macro- and microcirculation of the mesentery. Microvasc. Res. 2:474.
16. Järhult, J., and Mellander, S. (1974). Autoregulation of capillary hydrostatic pressure in skeletal muscle during regional arterial hypo- and hyper-tension. Acta Physiol. Scand. 91:32.
17. Nagle, F.J., Scott, J.B., Swindall, B.T., and Haddy, F.J. (1968). Venous resistance in skeletal muscle and skin during local blood flow regulation. Amer. J. Physiol. 214:885.
18. Wiederhielm, C.A., Fox, J.R., Weston, B., and Heald, R. (1974). Microvascular vasomotion patterns in the bat wing. Fed. Proc. 33:337.
19. Gaskell, P., and Burton, A.C. (1953). Local postural vasomotor reflexes arising from the limb veins. Circ. Res. 1:27.
20. Haddy, F.J., and Gilbert, R.P. (1956). The relation of a venous-arteriolar reflex to transmural pressure and resistance in small and large systemic vessels. Circ. Res. 4:25.
21. Selkurt, E.E., and Johnson, P.C. (1958). Effect of acute elevation of portal venous pressure on mesenteric blood volume, interstitial fluid volume and hemodynamics. Circ. Res. 6:592.
22. Haddy, F.J., Scott, J.B., and Grega, G.J. (1972). Effects of histamine on lymph protein concentration and flow in the dog forelimb. Amer. J. Physiol. 223:1172.
23. Baez, S., Laidlow, Z., and Orkin, L.R. (1974). Localization of microvascular and microcirculatory responses to venous pressure elevation in the rat. Blood Vessels 2:260.
24. Garlick, D.G., and Renkin, E.M. (1970). Transport of large molecules from plasma to interstitial fluid and lymph in dogs. Amer. J. Physiol. 219:1595.
25. Renkin, E.M., and Garlick, D.G. (1970). Blood-lymph transport of macromolecules. Microvasc. Res. 2:392.
26. Johnson, P.C. (1965). Effect of venous pressure on mean capillary pressure and vascular resistance in the intestine. Circ. Res. 16:294.
27. Johnson, P.C., and Hanson, K.M. (1966). Capillary filtration in the small intestine of the dog. Circ. Res. 19:766.
28. Folkow, B., and Mellander, S. (1970). Measurements of capillary filtration coefficient and its use in studies of the control of capillary exchange. Capillary Permeability, Alfred Benzon Symposium II. Munksgaard, Copenhagen.
29. Gottschalk, C.W., and Mylle, M. (1956). Micropuncture studies of pressures in proximal tubules and peritubular capillaries of the rat kidney and their relation to ureteral and renal venous pressures. Amer. J. Physiol. 185:430.
30. Eliassen, E., Folkow, B., Hilton, S.M., Öberg, B., and Rippe, B. (1974). Pressure-volume characteristics of the interstitial fluid space in the skeletal muscle of the cat. Acta Physiol. Scand. 90:583.
31. Zweifach, B.W., and Intaglietta, M. (1971). Measurement of blood plasma colloid osmotic pressure. II. Comparative study of different species. Microvasc. Res. 3:83.
32. Haddy, F.J., Campbell, G.S., and Visscher, M.B. (1950). Pulmonary vascular pressures in relation to edema production by airway resistance and plethora in dogs. Amer. J. Physiol. 161:336.

33. Campbell, G.S., and Haddy, F.J. (1950). Improved design of apparatus for colloid osmotic pressure determination. J. Lab. Clin. Med. 35:117.

34. Rhodin, J.A.G. (1974). Histology, p. 354. Oxford University Press, New York.

35. Mayerson, H.S. (1963). The physiologic importance of lymph. Handbook of Physiology, Circulation, Section 2, Vol. 2, p. 1035. American Physiological Society, Washington, D.C.

36. Renkin, E.M., Carter, E.D., and Joyner, E.M. (1974). Mechanism of sustained action of histamine and bradykinin on transport of large molecules across capillary walls in the dog paw. Microvasc. Res. 7:49.

37. Arturson, G., Groth, T., and Grotte, G. (1972). The functional ultrastructure of the blood-lymph barrier. Computer analysis of data from dog heart-lymph experiments using theoretical models. Acta Physiol. Scand. 374 (suppl.):1.

38. Parving, H.H., Rossing, N., Nielsen, S.L., and Lassen, N.A. (1974). Increased transcapillary escape rate of albumin, IgG, and IgM after plasma volume expansion. Amer. J. Physiol. 227:245.

39. Baez, S., Lamport, H., and Baez, A. (1960). Pressure effects in living microscopic vessels. In A. Copley and G. Stainsby (eds.), Flow Properties of Blood, p. 122. Pergamon Press, New York.

40. Zweifach, B.W. (1964). Microscopic aspects of tissue injury. Ann. N. Y. Acad. Sci. 116:831.

41. Clough, G., Fraser, P.A., and Smoje, L.H. (1974). Compliance measurement in single capillaries of the cat mesentery. J. Physiol. 240:9P.

42. Majno, G., Gilmore, V., and Leventhal, M. (1967). On the mechanism of vascular leakage caused by histamine-type mediators. Circ. Res. 21:833.

43. Chvapil, M. (1967). Physiology of Connective Tissue, p. 179. Butterworths, London.

44. Haljamae, H., Linde, A., and Amundson, B. (1974). Comparative analysis of capsular fluid and interstitial fluid. Amer. J. Physiol. 227:1199.

45. Johnsen, H.M. (1974). Measurement of colloid osmotic pressure of interstitial fluid. Acta Physiol. Scand. 91:142.

46. Aukland, K., and Johnsen, H.M. (1974). Protein concentration and colloid osmotic pressure of rat skeletal muscle interstitial fluid. Acta Physiol. Scand. 91:354.

47. Taylor, A.E., Gibson, W.H., Granger, H.J., and Guyton, A.C. (1973). The interaction between intracapillary and tissue forces in the overall regulation of interstitial fluid volume. Lymphology, Vol. 6, p. 192. George Thieme Verlag, Stuttgart.

48. Joyner, W.L., Carter, R.D., and Renkin, E.M. (1973). Influence of lymph flow rate on concentrations of proteins and dextran in dog leg lymph. Lymphology, Vol. 6, p. 181, George Thieme Verlag, Stuttgart.

49. Zweifach, B.W. (1973). Micropuncture measurements in the terminal lymphatics. In J. Ditzel and D.H. Lewis (eds.), Bibliotheca Anatomica, No. 12, p. 361. Karger, Basel.

50. Todd, G.L., and Bernard, G.R. (1974). The cervical lymphatic ducts of the dog; structure and function of the endothelial lining. Microvasc. Res. 8:139.

51. Leak, L.V., and Burke, J.F. (1968). Ultrastructural studies on the lymphatic anchoring filaments. J. Cell Biol. 36:129.

52. Landis, E.M., and Sage, L.E. (1971). Fluid movement rates through walls of single capillaries exposed to hypertonic solutions. Amer. J. Physiol. 221:520.

53. Gazitùa, S., Scott, J.B., Swindall, B., and Haddy, F.J. (1971). Resistance responses to local changes in plasma osmolality in three vascular beds. Amer. J. Physiol. 220:384.

54. Haddy, F.J., Molnar, J.I., and Campbell, R.W. (1961). Effects of denervation and vasoactive agents on vascular pressures and weight of dog forelimb. Amer. J. Physiol. 201:631.

55. Kjellmer, I., and Odelram, H. (1965). The effects of some physiological vasodilators on the vascular bed of skeletal muscle. Acta Physiol. Scand. 63:94.

56. Daugherty, R.M., Scott, J.B., Emerson, T.E., and Haddy, F.J. (1968). Comparison of I.V. and I.A. infusion of vasoactive agents on dog forelimb blood flow. Amer. J. Physiol. 214:611.

57. Grega, G.J., Kline, R.L., Dobbins, D.E., and Haddy, F.J. (1972). Mechanism of edema formation by histamine administered locally into canine forelimbs. Amer. J. Physiol. 223:1165.

58. Carter, R.D., and Renkin, E.M. (1971). Influence of histamine on transport of macromolecules across the blood-lymph barrier. Physiologist 14:119.

59. Landis, E.M. (1930). Microinjection studies of capillary blood pressure in human skin. Heart 15:209.

60. Haddy, F.J. (1960). Effect of histamine on small and large vessel pressures in the dog foreleg. Amer. J. Physiol. 198:161.

61. Majno, G., and Palade, G.E. (1961). Studies on inflamation I. The effects of histamine and serotonin on vascular permeability: An electron microscopic study. J. Biophys. Biochem. Cytol. 11:571.

62. Majno, G., Palade, G.E., and Schoefl, G.I. (1961). Studies on inflamation II. The site of action of histamine and serotonin along the vascular tree: a topographic study. J. Biophys. Biochem. Cytol. 11:607.

63. Carter, R.D., Joyner, W.L., and Renkin, E.M. (1974). Effects of histamine and some other substances on molecular selectivity of the capillary wall to plasma proteins and dextran. Microvasc. Res. 7:31.

64. Joyner, W.L., Carter, R.D., Raizes, G.S., and Renkin, E.M. (1974). Influence of histamine and some other substances on blood-lymph transport of plasma protein and dextran in the dog paw. Microvasc. Res. 7:19.

65. Rowley, D.A. (1964). Venous constriction as the cause of increased vascular permeability produced by 5-hydroxytryptamine, histamine, bradykinin, and 48/80 in the rat. Brit. J. Exp. Pathol. 45:56.

66. Aschheim, E., and Zweifach, B.W. (1962). Quantative studies of protein and water shifts during inflamation. Amer. J. Physiol. 202:554.

67. Baumgarten, A., Melrose, G.J.H., and Vagg, W.J. (1970). Interactions between histamine and bradykinin assessed by continuous recording of increased vascular permeability. J. Physiol. 208:669.

68. Robinson, N.E., Jones, C.A., Scott, J.B., and Dabney, J.M. (1975). Effects of histamine and acetylcholine on equine digital lymph flow and composition. Proc. Soc. Exp. Biol. Med. 149:805.

69. Dietzel, W., Massion, W.H., and Hinshaw, L.B. (1969). The mechanism of histamine-induced transcapillary fluid movement. Pflügers Arch. 309:99.

70. Diana, J.W., and Kaiser, R.S. (1970). Pre- and postcapillary resistance during histamine infusion in isolated dog hindlimb. Amer. J. Physiol. 218:132.

71. McNamee, J.E., and Grodins, F.S. (1975). Effect of histamine on the microvasculature of the isolated dog gracilis muscle. Amer. J. Physiol. 229:119.

72. Renkin, E.M., and Carter, R.D. (1973). Influence of histamine on transport of fluid and plasma proteins in lymph. In A.G.B. Kovach, H.B. Stoner, and J.J. Spitzer (eds.), Advances in Experimental Medicine and Biology—Neurohumoral and Metabolic Aspects of Injury, Vol. 33, pp. 119–132. Plenum Press, New York.

73. Majno, G., and Shea, S.M. (1969). Endothelial contraction induced by histamine-type mediators: An electron microscopic study. J. Cell Biol. 42:647.

74. Diana, J.M., Long S.C., and Yao, H. (1972). Effect of histamine on equivalent pore radius in capillaries of isolated dog hindlimb. Microvasc. Res. 4:413.

75. Dobbins, D.E., Grega, G.J., Scott, J.B., and Haddy, F.J. (1975). The effect of mechanically increasing venous pressure in the canine forelimb on protein transport during the local administration of histamine and acetylcholine. Microvasc. Res. In press.

76. Landis, E.M., Jonas, L., Angevine, M., and Erb, W. (1932). The passage of fluid and protein through the human capillaries during venous congestion. J. Clin. Invest. 11:717.

77. Shirley, H.H., Wolfram, C.G., Wasserman, K., and Mayerson, H.S. (1957). Capillary permeability to macromolecules; stretched pore phenomenon. Amer. J. Physiol. 190:189.

78. Baker, C.H. (1966). Vascular volume changes following histamine release in the dog forelimb. Amer. J. Physiol. 211:661.

79. Baker, C.H., and Menninger, R.P. (1974). Histamine-induced peripheral volume and flow changes. Amer. J. Physiol. 226:731.

80. Northover, A.M., and Northover, B.J. (1970). The effect of vasoactive substances on rat mesenteric blood vessels. J. Pathol. 101:99.

81. Buckley, I.K., and Ryan, G.B. (1969). Increased vascular permeability: The effect of histamine and serotonin on rat mesenteric blood vessels in vivo. Amer. J. Pathol. 55:329.

82. Remington, J.W., and Baker, C.H. (1958). Blood and plasma specific gravity changes during acute alterations in hemodynamics in splenectomized dogs. Circ. Res. 6:146.

83. Deyrup, I.J. (1944). Circulatory changes following subcutaneous injection of histamine in dogs. Amer. J. Physiol. 142:158.

84. Grega, G.J., Dobbins, D.E., Parker, P.E., and Haddy, F.J. (1972). Effects of intravenous histamine on forelimb weight and vascular resistances. Amer. J. Physiol. 223:353.

85. Allwood, M.J., and Lewis, G.P. (1964). Bradykinin and forearm blood flow. J. Physiol. (London) 170:571.

86. Kontos, H.A., Magee, J.H., Shapiro, W., and Patterson, J.L. (1964). General and regional circulatory effects of synthetic bradykinin in man. Circ. Res. 14:351.

87. Diana, J.N., Colantino, R., and Haddy, F.J. (1967). Transcapillary fluid movement during vasopressin and bradykinin infusion. Amer. J. Physiol. 212:456.

88. Kline, R.L., Scott, J.B., Haddy, F.J., and Grega, G.J. (1973). Mechanisms of edema formation in canine forelimbs by locally administered bradykinin. Amer. J. Physiol. 225:105.

89. Konzett, H., and Sturmer, E. (1960). Biological activity of synthetic polypeptides with bradykinin-like properties. Brit. J. Pharmacol. 15:544.

90. Erdos, E.G., Wehler, J.R., and Levine, M.I. (1963). Blocking of the in vivo effects of bradykinin and kallidin with carboxypeptidase. Brit. J. Pharmacol. Exp. Ther. 142:327.

91. Berkinshaw-smith, E.M.I., Morgan, R.S., and Paylingwright, G. (1962). Permeability response of cerebral and cutaneous blood vessels to vasoactive agents. Nature 196:173.

92. Sturmer, E. (1966). The influence of intraarterial infusions of synthetic bradykinin on flow and composition of lymph in dogs. In E.G. Erdos, N. Buck, and F. Sicuteri (eds.), Hypotensive Peptides, p. 368. Springer, Berlin-Heidelberg-New York.

93. Lewis, G.P., and Winsey, N.J.P. (1970). The action of pharmacological active substances on the flow and composition of cat hindlimb lymph. Brit. J. Pharmacol. 40:446.

94. Eliassen, E., Folkow, B., and Hilton, S. (1973). Blood flow and capillary filtration capacities in salivary and pancreatic glands as compared with skeletal muscle. Acta Physiol. Scand. 87:11A.

95. Haddy, F.J., Emerson, T.E., Scott, J.B., and Daugherty, R.M. (1970). The effect of the kinins on the cardiovascular system. In E.G. Erdos (ed.), Handbook of Experimental Pharmacology, Vol. 25, Bradykinin, kallidin and kallikrein, p. 362. Springer, Berlin.

96. Hyman, C., and Paldino, R.T. (1966). Influence of intravascular and topically administered bradykinin on microcirculation of several tissues. In H. Harders (ed.), 4th European Conference on Microcirculation. S. Karger, Basel and New York.

97. Zweifach, B.W. (1966). Microcirculatory effects of polypeptides. In E.G. Erdos, N. Back, and F. Sicuteri (eds.), Hypotensive Peptides, p. 451. Springer, Berlin-Heidelberg-New York.

98. Kline, R.L., Sak, D.P., Haddy, F.J., and Grega, G.J. (1975). Pressure-dependent factors in edema formation in canine forelimbs. J. Pharmacol. Exp. Ther. 193:452.

99. Ablad, B., and Mellander, S. (1963). Comparative effects of hydralazine, sodium nitrite and acetylcholine on resistance and capacitance blood vessels and capillary filtration in skeletal muscle in the cat. Acta Physiol. Scand. 58:319.

100. Mellander, S. (1966). Comparative effects of acetylcholine, butylnorsynephrine (Vasculat), noradrenaline, and ethyl-adrianol (effontil) on resistance, capitance, and precapillary sphinctor vessels and capillary filtration in cat skeletal muscle. Angiologica 3:77.

101. Baker, C.H. (1965). Vascular volume changes resulting from vasodilation of the dog forelimb. Amer. J. Physiol. 209:60.

102. Merrill, G.F., Kline, R.L., Haddy, F.J., and Grega, G.J. (1974). Effects of locally infused serotonin on canine forelimb weight and segmental vascular resistances. J. Pharmacol. Exp. Ther. 189:140.

103. Glover, W.E., Greenfield, A.D.M., Kidd, B.S.L., and Whelan, R.F. (1958). The reactions of the capacity blood vessels of the human hand and forearm to vasoactive substances infused intra-arterially. J. Physiol. (London) 140:113.

104. Roddie, I.C., Shepard, J.T., and Whelan, R.F. (1955). The action of 5-hydroxytryptamine on the blood vessels of the human hand and forearm. Brit. J. Pharmacol. 10:445.

105. Zweifach, B.W. (1964). Microcirculatory aspects of tissue injury. Ann. N.Y. Acad. Sci. 116:83.

106. Rowley, D.A., and Benditt, E.P. (1956). 5-Hydroxytryptamine and histamine as mediators of vascular injury produced by agents which damage mast cells in rats. Exp. Med. 103:399.

107. Gabbiani, G., Badonnel, M.C., and Majno, G. (1970). Intra-arterial injections of histamine, serotonin, or bradykinin: A topographic study of vascular leakage. Proc. Soc. Exp. Biol. Med. 135:447.

108. Talone, G., and Bonasera, L. (1970). Kinetics of increased vascular permeability induced in rat skin by serotonin. Experimentia 26:66.

109. Ebert, R.H., and Graham, R.C. (1966). Observations on the effects of histamine and serotonin in the rabbit ear chamber. Angiology 17:402.

110. Biber, B., Fara, J., and Lundgren, O. (1973). Vascular reactions in the small intestine during vasodilation. Acta Physiol. Scand. 89:449.

111. Biber, B., Fara, J., and Lundgren, O. (1973). Intestinal vascular responses to 5-hydroxytryptamine. Acta Physiol. Scand. 87:526.

112. Iturri, S., and Emerson, T.E. (1971). Responses of arterial vessel strips from skin and muscle to serotonin. Proc. Soc. Exp. Biol. Med. 137:416.

113. Kaiser, R.S., and Diana, J.N. (1974). Effect of angiotensin and norepinephrine on capillary pressure and filtration coefficient in isolated dog hindlimb. Microvasc. Res. 7:207.

114. Estensen, R.D., and Gilbert, R.P. (1961). Response of ileal segment weight to prolonged levarternol infusion. Amer. J. Physiol. 201:628.

115. Mellander, S., and Nordenfelt, I. (1970). Comparative effects of dihydroergotamine and noradrenaline on resistance, exchange and capacitance functions in the peripheral circulation. Clin. Sci. 39:183.

116. Järhult, J. (1971). Comparative effects of angiotensin and noradrenaline on resistance, capacitance, and precapillary sphinctor vessels in cat skeletal muscle. Acta Physiol. Scand. 81:315.

117. Baker, C.H. (1969). Epinephrine-induced vascular volume changes in dog forelimbs with controlled flow. Amer. J. Physiol. 216:368.

118. Gabel, L.O., Winbury, M.M., Rowe, H., and Grandy, R.P. (1964). The effect of several pharmacological agents upon ^{86}Rb uptake by the perfused dog hindlimb. Amer. J. Pharmacol. Exp. Ther. 146:117.

119. Szwed, J.J., and Friedman, J.J. (1975). Comparative effects of norepinephrine, epinephrine, angiotensin on pre- and postcapillary resistance vessels in dog skeletal muscle. Microvasc. Res. 9:206.

120. Appelgren, L., and Lewis, D.H. (1968). Capillary permeability surface area product (CPS) of Renkin in human skeletal muscle. Acta Med. Scand. 184:281.

121. Freeman, N.E., Freedman, H., and Miller, C.C. (1941). The production of shock by the prolonged continuous injection of adrenalin in unanesthetized dogs. Amer. J. Physiol. 131:545.

122. Lillehei, R.C., Longerbeam, J.K., Bloch, J.H., and Manax, W.G. (1964). The nature of irreversible shock: Experimental and clinical observations. Ann. Surg. 160:682.

123. Weisse, A.B., Hilmi, I.K., and Regan, T.J. (1968). Circulatory effects of norepinephrine infusions in intact and splenectomized dogs. Amer. J. Physiol. 214:421.

124. Chien, S. (1967). Role of the sympathetic nervous system in hemorrhage. Physiol. Rev. 47:214.

125. Moss, A.J., Vittlands, I., and Shenk, E.A. (1966). Cardiovascular effects of sustained norephinephrine infusions. I. Hemodynamics. Circ. Res. 18:596.

126. Grega, G.J., and Haddy, F.J. (1971). Forelimb transcapillary fluid fluxes and vascular resistances in catecholamine shock. Amer. J. Physiol. 220:1448.

127. Mellander, S., and Johansson, B. (1968). Control of resistance, exchange, and capacitance functions in the peripheral circulation. Pharmacol. Rev. 20:117.

128. Baker, C.H. (1968). Effects of denervation on dog forelimb blood flow resistance and vascular volume. Amer. J. Physiol. 214:1304.

129. Kelley, W.D., and Visscher, M.B. (1956). Effect of sympathetic nerve stimulation on cutaneous small vein and small artery pressures, blood flow and hindpaw volume in the dog. Amer. J. Physiol. 185:453.

130. Davis, D.L., and Hamilton, W.F. (1959). Small vessel responses of the dog paw. Amer. J. Physiol. 196:1316.

131. Davis, D.L. (1963). Effect of sympathetic stimulation on dog paw volume. Amer. J. Physiol. 205:989.

132. Zimmerman, B.G. (1966). Separation of responses of arteries and veins to sympathetic stimulation. Circ. Res. 18:429.

133. Lee, J.S., and Visscher, M.B. (1957). Microscopic studies of skin blood vessels in relation to sympathetic nerve stimulation. Amer. J. Physiol. 190:37.

134. Hammond, M.C., Davis, D.L., and Dow, P. (1969). Rate of development of constrictor responses of the dog paw vasculature. Amer. J. Physiol. 216:414.

135. Davis, D.L. (1964). Sympathetic stimulation and small artery constriction. Amer. J. Physiol. 206:262.

136. Abboud, F.M., and Eckstein, J.W. (1965). Effects of norepinephrine and nerve stimulation on segmental vascular resistance in the perfused foreleg of the dog. Hypertension, Vol. 13, P. 49. American Heart Association, New York.

137. Browse, N.L., Lorenz, R.R., and Shepherd, J.T. (1966). Response of capacity and resistance vessels of the dog's limb to sympathetic nerve stimulation. Amer. J. Physiol. 210:95.

138. Davis, D.L., and Hamilton, W.F. (1959). Small vessel responses of the rabbit ear. Amer. J. Physiol. 196:1312.

139. Mellander, S. (1960). Comparative studies on the adrenergic neuro-hormonal control of resistance and capacitance blood vessels in the cat. Acta Physiol. Scand. 50 (Suppl. 176):3.

140. Renkin, E.M., and Rosell, S. (1962). The influence of sympathetic adrenergic vasoconstrictor nerves on transport of diffusible solutes from blood to tissues in skeletal muscle. Acta Physiol. Scand. 54:223.

141. Friedman, J.J. (1973). The influence of sympathetic adrenergic nerve stimulation on pre- and postcapillary resistance. Microvasc. Res. 6:297.

142. Oberg, B., and Rosell, S. (1967). Sympathetic control of consecutive vascular sections in canine subcutaneous adipose tissue. Acta Physiol. Scand. 71:47.

143. Shadle, O.W., Zukof, M., and Diana, J. (1958). Translocation of blood from the isolated dog's hindlimb during levarterenol infusion and sciatic nerve stimulation. Circ. Res. 6:326.

144. Mayer, H.E., Abboud, F.M., Ballard, D.R., and Eckstein, J.W. (1968). Catecholamines in arteries and veins of the foreleg of the dog. Circ. Res. 23:653.

145. Fuxe, K., and Sedvall, G. (1965). Distribution of adrenergic nerve fibers to blood vessels in skeletal muscle. Acta Physiol. Scand. 64:75.

146. Folkow, B., Lewis, D.H., Lundgren, O., Mellander, S., and Wallentin, I. (1964). The effect of graded vasoconstrictor fibre stimulation on intestinal resistance and capacitance vessels. Acta Physiol. Scand. 61:445.

147. Folkow, B., Lewis, D.H., Lundgren, O., Mellander, S., and Wallentin, I. (1964). The effect of the sympathetic vasoconstrictor fibers on the distribution of capillary blood flow in the intestine. Acta Physiol. Scand. 61:458.

148. Hultén, L. (1969). Extrinsic nervous control of colonic motility and blood flow. Acta Physiol. Scand. 77 (Suppl. 335):1.

149. Oberg, B. (1964). Effects of cardiovascular reflexes on net capillary fluid transfer. Acta Physiol. Scand. 62 (Suppl. 229):1.

150. Johansson, B., Lundgren, O., and Mellander, S. (1964). Reflex influence of "somatic pressor and depressor afferents" on resistance and capacitance vessels and on transcapillary fluid exchange. Acta Physiol. Scand. 62:280.

151. Browse, N.L., Donald, D.E., and Shepherd, J.T. (1966). Role of veins in the carotid sinus reflex. Amer. J. Physiol. 210:1424.

152. Epstein, S.E., Beiser, G.D., Stampfer, M., and Braunwald, E. (1968). Role of the venous system in baroreceptor-mediated reflexes in man. J. Clin. Invest. 47:139.

153. Haddy, F.J., Scott, J.B., and Molnar, J.I. (1965). Mechanism of volume replacement and vascular constriction following hemorrhage. Amer. J. Physiol. 208:169.

154. DiSalvo, J., Parker, P.E., Scott, J., and Haddy, F.J. (1971). Carotid baroreceptor influence on total and segmental resistances in skin and muscle vasculatures. Amer. J. Physiol. 220:1970.

155. Brender, D., and Webb-Peploe, M.M. (1969). Influence of carotid baroreceptors on different components of the vascular system. J. Physiol. 205:257.

156. Zingher, D., and Grodins, F.S. (1964). Effect of carotid baroreceptor stimulation upon the forelimb vascular bed of the dog. Circ. Res. 14:392.

157. Schwinghamer, J.M., Grega, G.J., and Haddy, F.J. (1970). Skin and muscle circulatory responses during prolonged hypovolemia. Amer. J. Physiol. 219:318.

158. Grega, G.J., Schwinghamer, J.M., and Haddy, F.J. (1971). Changes in forelimb weight and segmental vascular resistances following severe hemorrhage. Circ. Res. 29:691.

159. Parker, P.E., Dabney, J.M. Scott, J.B., and Haddy, F.J. (1975). Reflex vascular responses in kidney, ileum, and forelimb to carotid body stimulation. Amer. J. Physiol. 228:46.

160. Zelis, R., and Mason, D.T. (1969). Comparison of the reflex reactivity of skin and muscle veins in the human forearm. J. Clin. Invest. 48:1870.

161. Folkow, B., Langston, J., Obert, B., and Prerovsky, I. (1964). Reactions of the different series-coupled vascular sections upon stimulation of the hypothalamic sympatho-inhibitory area. Acta Physiol. Scand. 61:476.

162. Cobbold, A., Folkow, B., Lundgren, O., and Wallentin, I. (1964). Blood flow, capillary filtration coefficients and regional blood volume responses in the intestine of the cat during stimulation of the hypothalamic "defense" area. Acta Physiol. Scand. 61:467.

163. Baum, T. and Hasko, M.J., Jr. (1965). Response of resistance and capacitance vessels to central nervous system stimulation. Amer. J. Physiol. 209:236.

164. Gootman, P.M., Baez, S., and Feldman, M. (1973). Microcirculatory responses to central neural stimulation in the rat. Amer. J. Physiol. 225:1375.

165. Folkow, B., Mellander, S., and Oberg, B. (1961). The range of effect of the sympathetic vasodilator fibers with regard to consecutive sections of the muscle vessels. Acta Physiol. Scand. 53:7.

166. Renkin, E.M., and Rosell, S. (1962). Effects of different types of vasodilator mechanisms on vascular tonus and on transcapillary exchange of diffusible material in skeletal muscle. Acta Physiol. Scand. 54:241.

167. Landis, E.M. (1934). Capillary pressure and capillary permeability. Physiol. Rev. 14:404.

168. Barcroft, J., and Kato, T. (1916). Effects of functional activity in striated muscle and the submaxillary gland. Philos. Trans. R. Soc. London 207:149.

169. Kjellmer, I. (1964). The effect of exercise on the vascular bed of skeletal muscle. Acta Physiol. Scand. 62:18.

170. Lundvall, J., Mellander, S., Westling, H., and White, T. (1972). Fluid transfer between blood and tissues during exercise. Acta Physiol. Scand. 85:258.

171. Jacobsson, S., and Kjellmer, I. (1964). Accumulation of fluid in exercising skeletal muscle. Acta Physiol. Scand. 60:286.

172. Baker, C.H., and Davis, D.L. (1974). Isolated skeletal muscle blood flow and volume changes during contractile activity. Blood Vessels 11:32.

173. Mellander, S., Johansson, B., Gray, S., Jonsson, O., Lundvall J., and Ljung, B. (1967). The effects of hyperosmolarity on intact and isolated vascular smooth muscle. Possible role in exercise hyperemia. Angiologica 4:310.

174. Lundvall, J., Mellander, S. and Sparks, H. (1967). Myogenic response of resistance vessels and precapillary sphinctors in skeletal muscle during exercise. Acta Physiol. Scand. 70:257.

175. Kjellmer, I. (1964). An indirect method for estimating tissue pressure with special reference to tissue pressure in muscle during exercise. Acta Physiol. Scand. 62:31.

176. Lundvall, J. (1972). Tissue hyperosmolality as a mediator of vasodilation and transcapillary fluid flux in exercising skeletal muscle. Acta Physiol. Scand. 379 (Suppl.).

177. Arturson, G., and Kjellmer, I. (1964). Capillary permeability in skeletal muscle during rest and activity. Acta Physiol. Scand. 62:41.

178. Bach, D., and Lewis G.P. (1973). Lymph flow and lymph protein concentration in the skin and muscle of the rabbit hindlimb. J. Physiol. 235:16.
179. White, J.C., Field, M.E., and Drinker, C.K. (1933). On the protein content and normal flow of lymph from the foot of the dog. Amer. J. Physiol. 103:34.
180. Jacobsson, S., and Kjellmer, I. (1964). Flow and protein content of lymph in resting and exercising skeletal muscle. Acta Physiol. Scand. 60:278.
181. Wells, H.S., Youmans, J.B., and Miller, D.G., Jr. (1938). Tissue pressure (intracutaneous, subcutaneous, and intramuscular) as related to venous pressure, capillary filtration and other factors. J. Clin. Invest. 17:489.
182. Bergstrom, J., Guarnieri, G., and Hultman, E. (1971). Carbohydrate metabolism and electrolyte changes in human muscle tissue during heavy work. J. Appl. Physiol. 30:122.
183. Landis, E.M. (1931). Capillary pressure and hyperemia in muscle and skin of the frog. Amer. J. Physiol. 98:704.
184. Haddy, F.J., and Scott, J.B. (1975). Active hyperemia, reactive hyperemia and autoregulation of blood flow. Fed. Proc. 34:2006.
185. Johnson, P.C. (1975). The microcirculation and local and humoral control of the circulation. MTP International Review of Science, Series One, Vol. 1, P. 163.
186. Krogh, A. (1929). The Anatomy and Physiology of Capillaries. 2nd Ed. P. 30. Yale University Press, New Haven.
187. Folkow, B., and Halicka, H.D. (1968). A comparison between "red" and "white" muscle with respect to blood supply, capillary surface area and oxygen uptake during rest and exercise. Microvasc. Res. 1:1.
188. Scott, J.B., and Radawski, D. (1971). Role of hyperosmolarity in the genesis of active and reactive hyperemia. Circ. Res. 28 and 29 (Suppl. 1):1.
189. Hartley, H.L., Pernow, B., Haggenda, J., Lacour, J., deLattre, J., and Saltin, B. (1970). Central circulation during submaximal work preceded by heavy exercise. J. Appl. Physiol. 29:818.
190. Ebert, R.V., and Stead, E.A. (1941). Demonstration that in normal man no reserves of blood are mobilized by exercise, epinephrine and hemorrhage. Amer. J. Med. Sci. 201:655.
191. Kaltreider, N.L., and Meneely, G.R. (1940). The effect of exercise on the volume of the blood. J. Clin. Invest. 19:627.
192. Nylin, G. (1947). The effect of heavy muscular work on the volume of circulating red corpuscle in man. Amer. J. Physiol. 149:180.
193. Astrand, P., and Saltin, B. (1964). Plasma and red cell volume after prolonged severe exercise. J. Appl. Physiol. 19:829.
194. Saltin, B., and Stenberg, J. (1964). Circulatory response to prolonged severe exercise. J. Appl. Physiol. 19:833.
195. Husni, E.A. (1967). The edema of arterial reconstruction. Circulation 35 and 36 (Suppl. 1):169.
196. Gaskell, P. (1956). The rate of blood flow in the foot and calf before and after reconstruction by arterial grafting of an occluded main artery to the lower limb. Clin. Sci. 15: 259.
197. Pochin, E.E. (1942). Edema following ischemia in the rabbit ear. Clin. Sci. 4:341.
198. Brown, J.H., and Garrett, R.L. (1972). Inhibition of constriction-reduced reactive hyperemic edema. Proc. Soc. Exp. Biol. Med. 139:610.
199. Strock, P.E., and Majno, G. (1969). Vascular responses to experimental tourniquet ischemia. Surg. Gynecol. Obstet. 129:309.
200. Wilgis, E.F.S. (1971). Observations on the effects of tourniquet ischemia. J. Bone Joint Surg. 53-A:1343.
201. Solonen, K.A., and Hjelt, L. (1968). Morphological changes in striated muscle during ischemia. Acta Orthop. Scand. 39:13.
202. Miller, G.L., Kline, R.L., Haddy, F.J., and Grega, G.J. (1975). Effects of prolonged ischemia on canine forelimb weight, pressures, blood flows, and vascular resistances following relief of ischemia. Proc. Soc. Exp. Biol. Med. 149:581.
203. Paletta, F. X., Willman, V., and Ship, A.G. (1960). Prolonged tourniquet ischemia of extremities. J. Bone Joint Surg. 42-A:945.
204. Lazarus-Barlow, W.S. (1894). The pathology of edema which accompanies passive congestion. Philos. Trans. R. Soc. Lond. (Biol. Sci.) 185:779.

108   Haddy, Scott, and Grega

205. Friedman, J.J. (1971). 86Rb extraction as an indicator of capillary flow. Circ. Res. 28 and 29 (Suppl. 1): I-15.
206. Diana, J.N., and Laughlin, M.H. (1974). Effect of ischemia on capillary pressure and equivalent pore radius in capillaries of the isolated dog hindlimb. Circ. Res. 35:77.
207. Landis, E.M. (1928). Micro-injection studies of capillary permeability. III. The effect of lack of oxygen on the permeability of the capillary wall to fluid and to the plasma proteins. Amer. J. Physiol. 83:528.
208. Drinker, C.K. (1945). Pulmonary Edema and Inflammation, Harvard University Press, Cambridge.
209. Zinberg, S., Nudell, G., Kubicek, W.G., and Visscher, M.B. (1948). Observations on the effects on the lungs of respiratory air flow resistance in dogs with special reference to vagotomy. Amer. Heart J. 35:774.
210. Nairn, R.C. (1951). Oedema and capillary anoxia. J. Pathol. Bacteriol. 63:213.
211. Scott, J.B., Daugherty, R.M., Jr., and Haddy, F.J. (1967). Effect of severe local hypoxemia on transcapillary water movement in dog forelimb. Amer. J. Physiol. 212:847.
212. Radawski, D., Dabney, J.M., Daugherty, R.N., Jr., Haddy, F.J., and Scott, J.B. (1972). Local effects of CO_2 on vascular resistances and weight of dog forelimb. Amer. J. Physiol. 222:429.
213. Dietzel, W., Samuelson, S.A., Guenter, C.A., Massion, W.H., and Hinshaw L.B. (1969). The effect of changes in arterial p_{CO_2} on isogravimetric capillary pressure and vascular resistance. Proc. Soc. Exp. Biol. Med. 131:845.
214. Hendley, E.D., and Schiller, A.A. (1954). Change in capillary permeability during hypoxemic perfusion in rat hindlegs. Amer. J. Physiol. 179:216.
215. Dyoyosuaito, A.M., Folkow, B., and Kovach, A.G.B. (1968). The mechanisms behind the rapid blood volume restoration after hemorrhage in birds. Acta Physiol. Scand. 74:114.
216. Mellander, S., and Oberg, B. (1967). Transcapillary fluid absorption and other vascular reactions in the human forearm during reduction of the circulating blood volume. Acta Physiol. Scand. 71:37.
217. Lundgren, O., Lundvall, J., and Mellander, S. (1964). Range of sympathetic discharge and reflex vascular adjustments in skeletal muscle during hemorrhagic hypotension. Acta Physiol. Scand. 62:380.
218. Järhult, J., Lundvall, J., Mellander, S., and Tibblin, S. (1972). Osmolar control of plasma volume during hemorrhagic hypotension. Acta Physiol. Scand. 85:142.
219. Järhult, J. (1975). Osmolar control of the circulation in hemorrhagic hypotension. Acta Physiol. Scand. 423 (suppl.).
220. Järhult, J., (1973). Osmotic fluid transfer from tissue to blood during hemorrhagic hypotension. Acta Physiol. Scand. 89:213.
221. Haddy, F.J., Overbeck, H.W., and Daugherty, R.M., Jr. (1968). Peripheral vascular resistance. Annu. Rev. Med. 19:167.
222. Mellander, S., and Lewis, D.H. (1963). Effect of hemorrhagic shock on the reactivity of resistance and capacitance vessels and on capillary filtration transfer in cat skeletal muscle. Circ. Res. 23:105.
223. Smith, E.E., Crowell, J.W., Moran, C.J., and Smith, R.A. (1967). Intestinal fluid loss during irreversible shock. Surg. Gynecol. Obstet. 125:45.
224. Regan, T.J., LaForce, F.M., Teres, D., Block, J., and Hellems, H.K. (1965). Contribution of left ventricle and small bowel in irreversible hemorrhagic shock. Amer. J. Physiol. 208:938.
225. Noble, R.P., and Gregersen, M.I. (1946). Blood volume in clinical shock. I. Mixing time and disappearance rate of T-1824 in normal subjects and in patients in shock; determination of plasma volume in man from 10 minute samples. J. Clin. Invest. 25:158.
226. Shoemaker, W.C. (1962). Measurement of rapidly and slowly circulating red cell volumes in hemorrhagic shock. Amer. J. Physiol. 202:1179.
227. Arturson, G., and Thoren, L. (1964). Capillary permeability in hemorrhagic shock. Studies of the blood-lymph barrier with dextran as the test substance. Acta Chir. Scand. 129:345.

228. Weidner, W.J., Grega, G.J., and Haddy, F.J. (1971). Changes in forelimb weight and vascular resistances during endotoxin shock. Amer. J. Physiol. 221:1229.

229. Hinshaw, L.B., and Owen, S.E. (1971). Correlation of pooling and resistance changes in the canine forelimb in septic shock. J. Appl. Physiol. 30:331.

230. Weidner, W.J., Kline, R.L., Gersabeck, E.F., Haddy, F.J., and Grega, G.J. (1975). Effects of whole blood transfusion on forelimb weight and segmental vascular resistances in dogs previously injected with endotoxin. Circ. Shock 2:165.

International Review of Physiology
Cardiovascular Physiology II, Volume 9
Edited by Arthur C. Guyton and Allen W. Cowley
Copyright 1976   University Park Press   Baltimore

3
Reflex Control of Circulation by Heart and Lungs

G. MANCIA, R. R. LORENZ, AND J. T. SHEPHERD
Mayo Clinic and Mayo Foundation

INTRODUCTION

Recent reviews of the sensory receptors with afferent vagal fibers have discussed the types of receptors, the conduction velocities of their fibers, and the reflexes evoked when they are excited by natural stimuli and chemical substances (1, 2). The present review is limited to the receptors in the lungs, atria, and ventricles, with particular reference to their normal role in circulatory control. Reference is made to the function of receptors with myelinated and unmyelinated vagal afferents as well as to those with afferent sympathetic fibers. The interaction between reflexes originating in the heart and lungs and reflexes from the carotid baroreceptors and chemoreceptors is analyzed, and the reflex changes evoked by simulated pathological conditions are discussed.

CARDIOPULMONARY RECEPTORS

Receptors with Vagal Afferents

Pulmonary Receptors The receptors in the lungs have not been identified histologically. However, by use of electrophysiological techniques, three groups of receptors with vagal afferents have been described. For one group, the pulmonary stretch receptors, the natural stimulus is the normal inflation of the lungs. These receptors, located mainly in the respiratory bronchioles (3), are supplied with myelinated fibers and are characterized by a sustained discharge whose frequency is augmented in proportion to the depth of inspiration (1, 3, 5).

The second group, the irritant receptors located predominantly in the trachea and throughout the extrapulmonary bronchi, is supplied with myelinated fibers similar in size to the afferents of the stretch receptors. These receptors respond to mechanical or chemical irritation of the airways or to a sudden large lung inflation. Their activity consists of a brief high-frequency burst of firing with rapid adaptation (1, 6–8).

The third group consists of the juxtacapillary or "J" receptors that lie close to the pulmonary capillaries. These receptors are supplied mostly with unmyelinated fibers and normally are inactive but are stimulated by congestion of the lungs (9). Some vagal afferent fibers of small diameter have an irregular spontaneous discharge with no obvious relationship to the respiratory cycle but are stimulated by hyperinflation of the lungs (10).

Atrial Receptors   Atrial receptors with myelinated vagal afferent fibers are found predominantly at the junctions of the atrial walls with the intrapericardial portion of the systemic and pulmonary veins (Figure 1). They lie in the loose tissue of the endocardium and consist of complex unencapsulated endings (11, 12). It has been suggested that these receptors are of two types, one stimulated by atrial contraction and the other by changes in atrial volume (13). However, recent studies (14) indicate that these receptors have similar properties (Figure 2).

Receptors with unmyelinated vagal fibers are also located in the atria. These receptors have not been identified histologically, but they may be part of the

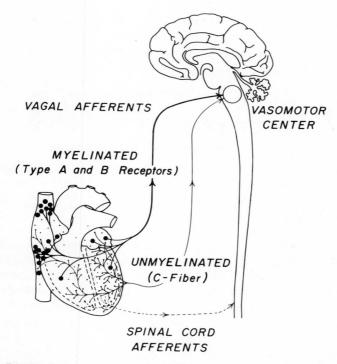

Figure 1. Distribution of cardiac receptors subserved by vagal afferents. Those connected to myelinated fibers have complex unencapsulated endings and are limited to discrete areas. Those connected to unmyelinated fibers are widespread and may be part of a system of fine nerve endings. In addition, afferent fibers originating from cardiac receptors run in cardiac sympathetic nerves to spinal cord. (From Shepherd, J.T., and Vanhoutte, P.M. (1975). Veins and Their Control. W.B. Saunders, London.)

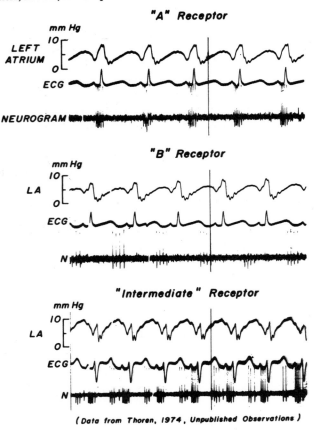

(Data from Thorén, 1974, Unpublished Observations)

Figure 2. Single-fiber recordings from left atrial receptors with myelinated vagal afferents. *Top,* receptor is firing during the *a* wave of atrial pressure pulse ("A" receptor). *Middle,* receptor is firing during *v* wave ("B" receptor). *Bottom,* receptor that is activated both during *a* wave and *v* wave ("intermediate" receptor). (Data from Thorén, P.N. (1974). Unpublished observations.)

system of fine nerve endings that are present throughout the subendocardial tissues (15, 16). Under normal conditions these receptors may be silent or have a sparse and irregular discharge, but they can be stimulated mechanically or by chemicals such as phenyldiguanide (17). Recent studies (18) have shown that these receptors can be activated by moderate increases in atrial pressure.

Ventricular Receptors   The ventricles give origin to myelinated fibers with a diameter similar to myelinated fibers originating from the atria. These fibers arise from receptors located in the subendocardium of both ventricles (19, 20), the parietal pericardium (21), and in or near the left main coronary artery (22). However, no complex unencapsulated sensory endings have been identified. These receptors usually discharge during ventricular systole and increase their firing as the ventricular pressure is increased. Those receptors located in or near a coronary artery discharge in phase with the coronary pressure pulse.

There also exists a diffuse and large population of receptors, predominantly in the left ventricle, with slowly conducting unmyelinated vagal fibers (20, 23, 24) (Figure 1). These receptors, like those in the atria, usually have only a sparse and irregular spontaneous discharge. They are activated by veratrum alkaloids and probably are the principal receptors responsible for the Bezold-Jarisch reflex (24). Recently it has been demonstrated that their firing rate is augmented by moderate increases in end-diastolic pressure and during coronary artery occlusion (24, 25).

Receptors with Sympathetic Afferents

Some of the cardiac receptors have afferent fibers that travel in the sympathetic nerves to the spinal cord. One group with myelinated afferents has a spontaneous discharge at normal intracardiac pressures. This discharge consists of single impulses in phase with atrial systole or diastole (atrial receptors) and ventricular systole (ventricular receptors) (26). Another group, also with myelinated afferents, discharges as the pressure in the coronary arteries is increased (27, 28) and has an increased rate of firing during the systolic bulging that accompanies coronary artery occlusion (29). A third group is supplied with unmyelinated fibers; these receptors do not have a spontaneous discharge but discharge irregularly during coronary occlusion (29). Like the receptors with vagal afferents, those with sympathetic afferents are strongly activated by veratrum alkaloids injected into the coronary circulation (26, 27).

REFLEXES FROM THE CARDIOPULMONARY REGION

Methodology

For analysis of the role of different reflexes that control the circulation, it is customary and proper to vary a single input to the central nervous system while eliminating or maintaining constant the inputs from other sensory zones. Thus, in the study of cardiopulmonary reflexes, the carotid sinus region usually is denervated or vascularly isolated and maintained at constant pressure (30), and the input from the aortic arch baroreceptors and chemoreceptors is eliminated.

The afferent fibers from the baroreceptors and chemoreceptors of the aortic arch and the root of the right subclavian artery form the aortic nerves that join the vagosympathetic trunks. In the rabbit, these aortic depressor nerves run separately from the vagosympathetic nerves along the length of the neck (31); in the cat, the rat, and especially the dog, they are often short and can only be identified as a separate nerve for a short distance below the nodose ganglion, particularly on the right side (32–36).

In 1931, Koch (32) described the aortic nerve in the dog. Since that time, many studies have shown that section of this nerve in the neck abolishes the reflex hypotension produced by distension of the vascularly isolated aortic arch, brachiocephalic trunk, and right subclavian artery (35–37), and also abolishes the hypertensive response to injection of cyanide (35, 37–42). Thus, this procedure acutely abrogates aortic baroreceptor and chemoreceptor reflexes.

This conclusion has been questioned by Ito and Scher (43, 44). In acute studies, these investigators found a decrease in aortic blood pressure in response to stimulation of the central segment of the sectioned "peripheral aortic nerves" after section of the cervical aortic nerves in one of three dogs and one of five cats. The hypotension was stated to be 10 mm Hg or less and was attributed solely to excitation of aortic depressor fibers. However, it should be appreciated that two of the nerves that comprise the "peripheral aortic nerve"—the dorsal and the ventromedial cervical cardiac nerves—supply filaments not only to the aortic arch but also to the pretracheal plexus, a primary area of nerve distribution for the heart (45). Afferent unmyelinated fibers originating in the heart have been shown to travel to the central nervous system via the cardiac and the aortic nerves; these fibers normally are silent or have a sparse irregular discharge and are not easy to identify by standard electrophysiological techniques. However, when stimulated, they are capable of exerting profound and varied effects on arterial blood pressure (46).

It should be emphasized that the evidence for loss of aortic baroreceptor and chemoreceptor reflexes after section of the cervical aortic nerves applies only to the acute situation. Infrequently, no aortic nerve can be identified (35, 36). When this occurs, the experiment has to be abandoned because the nerve probably lies within the vagal trunk.

Various approaches have been used to study the cardiopulmonary reflexes. These have included the administration of drugs to stimulate the receptors and electrical stimulation of afferent nerve fibers. Such studies have provided evidence of the presence of receptors in both the heart and the lungs, the ability of these receptors to cause cardiovascular changes, and the possible roles of myelinated and unmyelinated vagal afferents (47). Another approach is to interrupt the cervical vagal nerves by cutting them or cooling them. Usually, the nerve is cooled to $0°C$ and the absence of any further vascular changes when the nerve is sectioned during the cold block is taken as evidence that the cold-induced nerve block was complete.

After the vagi are sectioned at the diaphragm, the vascular changes reflect the tonic influence on the vasomotor center exerted by the spontaneous nerve traffic from the cardiopulmonary region prior to the interruption. Such studies reveal the combined influence from all of the receptors in the cardiopulmonary region—they do not permit discrimination among the separate influences of different receptor populations. Also, the procedure does not allow a complete study of the reflex control of the heart rate because it interrupts the vagal efferent fibers together with the afferents.

Reflexes from the Total Cardiopulmonary Region

In the anesthetized cat or dog with acute bilateral carotid sinus and aortic nerve section, interruption of the cervical vagi causes a sustained increase in systemic arterial pressure (Figure 3); the increase is unaffected by administration of atropine or by vagal section at the diaphragm (36, 40). These observations suggest that, like the carotid and aortic baroreceptors, the vagal fibers originating in the cardiopulmonary region exert a tonic inhibition of the vasomotor center.

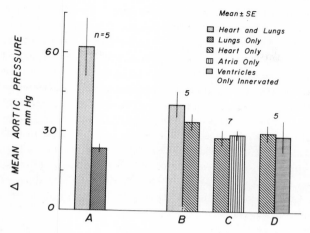

Figure 3. Data from various studies of changes in aortic blood pressure during vagal cold block in dogs with carotid sinus and aortic nerves cut; responses are shown as changes from control values. *A*, heart and lungs were in situ or only lungs were in situ; *B*, heart and left lung were in situ or only heart was in situ; *C*, lungs were removed and heart was in situ or only atria were in situ; *D*, lungs were removed and intact heart was in situ or heart was in situ but only ventricles were innervated. (From Mancia, G., and Donald, D.E. (1975). Demonstration that atria, ventricles, and lungs each are responsible for a tonic inhibition of the vasomotor center in the dog. Circ. Res. 36:310. By permission of American Heart Association.)

Several studies have confirmed and extended these observations in anesthetized cats, rabbits, and dogs (37, 48–50). Elimination of vagal afferents causes an increase in heart rate, constriction of the hindlimb and renal resistance vessels, constriction of the splanchnic capacitance vessels, and a decrease in volume of the limbs but no change in the tone of the saphenous veins (37, 48). Thus, the tonic inhibition is exerted on the central neurons controlling the sympathetic outflow to resistance and capacitance vessels, with the exception of the cutaneous veins. These veins, which have no consistent response to changes in activity of the carotid sinus mechanoreceptors (51, 52), are a specialized component of the capacitance system with a specific role in thermoregulation (53).

In anesthetized cats with their aortic nerves sectioned, Öberg and White (48) found that occlusion of the common carotid arteries decreased blood flow to a hindlimb and kidney perfused at constant pressure by 53 and 19%, respectively; when the vagi were blocked during the carotid occlusion, there was little further decrease in hindlimb flow but there was a decrease in renal flow that was much greater than that caused by the carotid occlusion. Similar experiments have been conducted in anesthetized dogs with their aortic nerves sectioned and a kidney and hindlimb perfused at constant pressure. When the pressure in the isolated carotid sinuses was abruptly decreased, so that baroreceptor activity changed from maximal to minimal, and vagal cold block was instituted, 83% of the decrease in hindlimb blood flow was due to withdrawal of carotid baroreceptor inhibition and 17% was due to withdrawal of the inhibition exerted by the cardiopulmonary receptors. The decrease in flow in the kidney was due equally to the removal of both receptor inputs. When the activity of the cardiopulmonary receptors was increased by augmenting the blood volume, the proportion of the decrease in flow due to vagal block was augmented (54) (Figure 4).

Two causes must be considered for this difference in effects on the kidney and hindlimb resistance vessels. The first is the different response of these two beds to direct electrical stimulation of their sympathetic nerves; in the dog, hindlimb blood flow decreases at much lower frequencies of stimulation than does renal blood flow. Thus, a given increase in sympathetic adrenergic outflow caused by a decrease in carotid baroreceptor inhibition will constrict the hindlimb more than the kidney resistance vessels; a further increase in outflow caused by vagal block will have less effect on the hindlimb than on the kidney because the former is now in the flatter and the latter in the steeper part of their stimulus-response curves (54).

Another cause is the different influences of the carotid baroreceptor and the cardiopulmonary receptor afferents on the pools of vasomotor neurons controlling the sympathetic activity to the hindlimb and kidney. In the cat, withdrawal of carotid baroreceptor restraint is followed by considerably lower firing rates in renal than in skeletal muscle vasoconstrictor fibers (55, 56); also, the afferent traffic from cardiac receptors is preferentially oriented toward those central neuron pools that control vasoconstrictor fiber discharge to the kidney (48, 57).

Blocking all traffic in the vagal afferents from the cardiopulmonary region demonstrates that receptors in this region exert a continuous inhibition on the vasomotor center; however, this provides no information on the size of the nerve fibers or on the receptors responsible. This inhibition could result from heart and lung receptors with myelinated vagal afferents, because these receptors have a high-frequency tonic discharge in phase with cardiac or respiratory events. However, this region also is richly endowed with receptors subserved by unmyelinated vagal afferents and, although their activity is usually sparse and irregular, it is possible that they could contribute to the tonic inhibition of the vasomotor center. This possibility was examined in the rabbit by cooling the cervical vagal nerves to different temperatures and examining the response of aortic blood pressure (58). These experiments were based on the evidence provided by Paintal (59) and Franz and Iggo (60) that cooling blocks the high-frequency nerve traffic in the myelinated fibers at about $7°C$ and the low-frequency nerve traffic in the unmyelinated fibers at about $2°C$.

It was found that vagal cooling to $6°C$, which blocked all myelinated afferent nerve traffic, caused an increase in aortic pressure; vagal cooling to $0°C$ caused a further substantial increase. The first pressure increase was difficult to interpret, but the second increase was due to blockade of nerve traffic in unmyelinated nerve afferents. Thus, receptors subserved by these unmyelinated afferents exert a tonic restraint on the vasomotor center. However, from such experiments, no conclusions can be reached as to the role of cardiopulmonary receptors with myelinated vagal afferents.

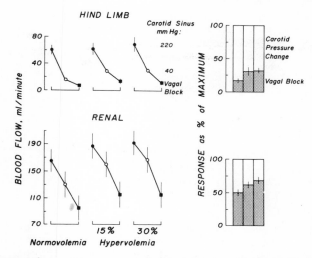

Figure 4. Changes (mean ± S.E.) in hindlimb and renal blood flow in response to decrease in carotid sinus pressure from 220 to 40 mm Hg and then to bilateral cervical vagal cold block at sinus pressure of 40 mm Hg in 13 closed-chest dogs with aortic nerves cut, carotid sinuses vascularly isolated and maintained at desired constant nonpulsatile pressure, and left hindlimb and left kidney perfused at constant pressure (120 mm Hg). The observations were made in sequence during normovolemia and during 15 and 30% increases in blood volume. Bar graphs show the mean (± S.E.) decrease in blood flow with decrease in carotid sinus pressure and with vagal block as percentage of mean decrease in flow caused by the combined procedures (maximum) during normovolemia and 15 and 30% hypervolemia. (From Mancia, G., Shepherd, J.T., and Donald, D.E. (1975). Role of cardiac, pulmonary, and carotid chemoreceptors in the control of hindlimb and renal circulation in dog. Circ. Res. In press. By permission of American Heart Association.)

Reflexes from the Lungs

Lung Inflation Reflex In dogs in which the input from arterial baroreceptors was controlled, inflation of the collapsed lungs caused dilatation of the muscle, skin, and mesenteric resistance vessels as a consequence of inhibition of the adrenergic outflow by vagal afferents from the lungs. The dilatation is proportional to the degree of lung inflation and has a low threshold; indeed, the vasodilatation appears at lung volumes less than the normal expiratory reserve volume. A decrease of this latter volume in closed-chest dogs caused an increase in aortic blood pressure, suggesting that this reflex is tonically active (61, 62).

In unanesthetized decerebrate cats after sino-aortic denervation the basal level of aortic pressure increased when the cervical vagi were sectioned. The contribution to this increase from both the "cardiac" and "pulmonary" vagal plexuses was demonstrated (after section of the right vago-aortic sympathetic trunk) by the stepwise increase in aortic pressure after two-stage sectioning of the left vagus above the emergence of the pulmonary plexus and, later, just above the cardiac fibers (40).

Further evidence of a tonic vasomotor inhibition from the lungs was obtained in sino-aortic denervated dogs (36) (Figure 3). To eliminate receptors other than those in the lungs, the dogs were maintained on extracorporeal circulation, the heart was removed, and the vagi were cut at the diaphragm. Vagal cooling caused a decrease in blood flow to a hindlimb and to a kidney (both perfused at constant pressure) and an increase in aortic blood pressure that was proportional to the tidal volume (54). These findings are difficult to reconcile with those of Hainsworth (42) who observed that systemic vasodilatation occurred only on hyperinflation; lung inflations at moderate pressure caused a small vasoconstriction, if anything. Hainsworth suggested that abnormal perfusion of the lungs, as used in other studies, might have altered the lung compliance and caused overdistension at smaller inflation pressures. However, this does not explain the fact that, with lungs only in situ, a tonic vasomotor inhibition was present during apnea with the lungs collapsed at atmospheric pressure (54).

The normal response to inflation of a normally perfused lung with moderate pressures is tachycardia (42, 63). Carotid chemoreceptor stimulation during controlled ventilation is accompanied by a slowing of the heart; when the ventilation is permitted to increase, the lung inflation reflex prevails and an increase in heart rate results (64, 65).

In man, a voluntary deep breath causes constriction of the resistance blood vessels in the fingers and of the cutaneous veins (66–68). It has been suggested that this is a spinal reflex mediated by receptors in the respiratory muscles (69, 70).

Other Reflexes from the Lungs   Receptors supplied by vagal afferents exist throughout the pulmonary vascular bed and stimulation of them by drugs (1, 71), hyperosmolarity (72, 73), or pressure (74, 75) causes systemic vasodilatation and bradycardia. Afferent vagal fibers arising predominantly at the bifurcation of the pulmonary trunk and along the main pulmonary arteries have a tonic pulsatile discharge at normal pulmonary arterial pressure (76, 77). Pulsatile distension (mean pressure between 20 and 60 mm Hg) of a vascularly isolated, innervated pulmonary artery caused systemic hypotension and sometimes bradycardia in 8 of 18 dogs; in 10 dogs there was little or no effect. The effects were augmented by carotid occlusion and abolished by vagal cooling (to 7–8°C) or section. When the pressure in the isolated vessel was increased to mean pressures greater than 80 mm Hg, systemic hypertension occurred; this response was abolished by vagotomy but not by cooling the vagus nerves to 7–8°C (78). It was concluded that the hypotension was induced reflexly by activation of the pulmonary arterial baroreceptors, but the mechanism of the hypertensive response was not determined.

Chemoreceptors with a blood supply from the pulmonary arteries are present immediately after birth but have never been found in adult animals (79–81). If present, they could cause reflex changes during exercise in proportion to the decrease in oxygen and increase in carbon dioxide tensions in the mixed venous blood. Duke et al. (82) claimed that perfusion of an isolated segment of a pulmonary artery with hypoxic hypercapnic blood caused increases in rate and depth of respiration and in efferent sympathetic traffic. However, other investigators have failed to demonstrate any ventilatory responses when the pulmonary artery was perfused with hypercapnic blood (83). The present consensus is that there are no chemoreceptors in the pulmonary arteries or the lungs.

The stretch receptors supplied by myelinated vagal afferents have a powerful tonic influence on the respiratory center. They adapt slowly and their activation begins at low lung distension and augments with it (1, 3–5); thus, they may be the major factor responsible for the vasomotor inhibition exerted by the lungs. Like the heart, the lungs are richly innervated with unmyelinated afferents that are activated in abnormal circumstances such as hyperinflation (10), congestion (9), and embolism (84). However, studies by Thorén et al. (58) indicate that the possibility of their participation in the tonic inhibition of the vasomotor center should not be excluded.

Reflexes from the Heart

Cardiac Reflex In 1866, Cyon and Ludwig (31), although wrong in their belief that the aortic depressor nerves arose from the heart, postulated that in rabbits vagally innervated cardiac receptors that normally are excited by an increase in intracardiac pressure have a depressor influence on the vasomotor center. About a century later, Aviado and Schmidt (38) found that increasing the pressure in the left and right sides of the heart in dogs independent of the systemic arterial blood pressure, with a pulmonary bypass and extracorporeal oxygenation to eliminate the reflexes from the lungs, caused a reflex bradycardia and a dilatation of the resistance vessels of the hindlimb. The response was abolished by vagotomy. In dogs with the beating heart hemodynamically isolated and the systemic circulation perfused extracorporeally at constant flow, Ross et al. (85) found that increasing left ventricular pressure from average values of 86/6 to 203/26 mm Hg and mean left atrial pressure from an average of 6 to an average of 36 cm H_2O, while maintaining right ventricular pressure constant, caused a 20–40% decrease in systemic vascular resistance; increasing right ventricular pressure, from an average value of 4/1 to 68/12 mm Hg, caused only minimal changes. This suggested that receptors in the left heart are more important than those in the right in modulating the systemic circulation.

To determine whether these receptors were in the left ventricle, Salisbury et al. (86) studied dogs, on cardiopulmonary bypass, in which the pulmonary artery was obstructed and the left atrium and the right ventricle were drained into the venous reservoir of the extracorporeal machine. When the left ventricular diastolic pressure was increased to 10–15 mm Hg by inflating a balloon in the outflow tract of the left ventricle, there were bradycardia and dilatation of the resistance vessels and systemic veins. These responses were dependent on the vagal afferent innervation. They used a degree of ventricular distension at which the mitral valve remained competent and the left atrial pressure did not change; hence, the reflex response was likely to have originated from left ventricular receptors.

Similar reflexes also can be evoked from the left atrium. When a vascularly isolated innervated pouch of the left atrium of dogs was distended, reflex systemic vasodilatation occurred in proportion to the pressure increase in the pouch. In these experiments, the aortic nerves were cut and the pressure in the isolated carotid sinuses was maintained at a low level so that the carotid mechanoreceptors were inactive. In similar circumstances, distension of the left pulmonary vein-atrial junctions by balloons, instead of using an isolated atrial pouch, caused similar reflex changes; the left lung was excluded from the circulation, and the blood flow to the left atrium through the right lung continued undisturbed during the balloon inflation (87).

In the cat, section of the vagal fibers of the cardiac plexus caused constriction of systemic resistance vessels (48). On the assumption that the majority of these fibers originate from the heart, these studies indicate that the receptors in the heart can exert a tonic inhibition of the vasomotor center. Further evidence for this comes from studies in dogs in which the only vagal input to the vasomotor center originated from the heart (36). The sino-aortic region was denervated and the vagi were cut at the diaphragm. With the venous return taken from the pulmonary arteries, oxygenated extracorporeally, and returned to the left atrium, the lungs were removed, leaving the intact working heart. Under these circumstances, cervical vagal cooling caused an increase in aortic blood pressure (Figure 3), due to systemic vasoconstriction, thus demonstrating that, at normal cardiac pressures the heart is a source of a continuous vasomotor inhibition. After removal of the heart, the vagal cooling was without effect.

Further analysis of this inhibition has been directed to its effects on different vascular beds and its origin from ventricular or atrial receptors. As to the former, section of the cardiac plexus in cats caused a large decrease in the blood flow to the kidney and only a small decrease in the blood flow to a hindlimb (both perfused at constant pressure), suggesting the removal of a greater inhibitory control of the former compared to the latter (48).

An assessment of the origin of the tonic inhibition originating from the heart has been performed in dog preparations with only the heart in situ (36). To study the ventricular receptors, the atria were denervated by dissecting them from the mediastinum, dividing and anastomosing the lower third of the atrial septum, and crushing the atrial walls circumferentially along the atrioventricular groove. After this procedure, which did not alter the intracardiac pressures, vagal cooling caused an increase in aortic blood pressure (Figure 3), due to systemic vasoconstriction; denervation of the ventricles by stripping the pulmonary trunk and the aortic root abolished the response. To study the atrial receptors, the heart was totally bypassed, the ventricles, major coronary vessels, and fossa ovalis were excised, and the atria were transformed into a common chamber by a plug-shaped cannula tied at the level of the atrioventricular groove. Perfusion of the atrial tissues with oxygenated blood was ensured through the cannulated sinus node artery; this blood drained into the atria from the coronary sinus and returned to the extracorporeal circuit via the plug-shaped cannula. The atria continued to beat vigorously and were maintained at normal pressure by changing the height at which the outflow cannula discharged above the extracorporeal circuit. Under these conditions, cervical vagal cooling caused an increase in aortic blood pressure (Figure 3). Removal of the atria abolished the response.

Thus, receptors in the atria and ventricles, with afferent vagal fibers, can exert a tonic vasomotor inhibition at pressures in the normal range. This inhibition is manifested when the input from the arterial baroreceptors is decreased or absent. The degree of inhibition in closed-chest animals, the role of myelinated and unmyelinated afferents, and the modification caused by general anesthesia require elucidation.

Bainbridge Reflex In 1915, Bainbridge (88) observed that large infusions of saline or blood caused an increase in heart rate in anesthetized dogs. He ascribed this to stimulation of receptors by the increased pressure in the right atrium. There has been a revival of interest in this reflex in recent years. To localize the stimulus to areas known to have numerous vagal afferent terminations, small balloons have been used to distend pulmonary vein-left atrial junctions, the left atrial appendage, and the superior vena cava-atrial junction (89–93). The distension caused a small but significant reflex decrease in vascular resistance in the hindlimb that was sustained and was preceded, in about half of the dogs studied, by a consistent large transient decrease (90). There was also a decrease in renal vascular resistance (92). The balloon distension invariably caused a reflex increase in heart rate with the afferent pathways in the vagal nerves and the efferent pathways in the sympathetic nerves (89).

Similar results were obtained by increasing the pressure in a pouch consisting of a part of the wall of the left atrium together with the left pulmonary vein-atrial junctions (94). The changes in sympathetic activity appeared to be confined to those fibers in the ansae subclaviae that innervate the sinoatrial node because there was no evidence of simultaneous changes in contractility of the ventricles (91).

Balloon distension of the pulmonary vein-left atrial junctions has been performed using larger balloons and with the aortic nerves cut and the carotid sinus pressure controlled to prevent any influence from the arterial baroreceptors (87). In these studies the distension was accompanied by reflex vascular changes which, although greater, resembled those reported by Carswell et al. (90). However, the heart rate accelerated only if the initial rate was less than 140–150 beats/min; if the initial rate was greater, a slowing occurred. The tachycardia was well sustained during the few minutes of distension, unlike the cardiac slowing which usually recovered toward the control values.

In anesthetized cats, electrical stimulation of the central stump of the right main cardiac nerve caused different vascular effects, depending on whether only the myelinated or both the myelinated and unmyelinated afferent fibers were stimulated (47). In the former case, there was an increase in heart rate and aortic blood pressure; in the latter, there were bradycardia and hypotension. Sleight and Widdicombe (23) described receptors in dogs, localized to small areas of the pulmonary veins at the hilus of the lungs, with slow-conducting fibers. These observations may explain the different effects on heart rate that have been reported with balloon distension. Thus, in those studies in which only tachycardia occurred, the myelinated fibers would be activated predominantly, whereas in those using larger balloons a mixed population of myelinated and unmyelinated fibers might have been involved (87).

Recently, Malliani et al. (95) observed that, in anesthetized cats, electrical stimulation of afferent sympathetic nerves with a cardiac origin caused a reflex acceleration in heart rate that largely persisted after section of the spinal cord. Thus, afferent information for the "Bainbridge reflex" also may travel in the sympathetic nerves and have a spinal integration.

Bezold-Jarisch Reflex   Extending earlier observations of von Bezold, Jarisch showed in 1937 that veratridine excites vagal receptors in the cardiopulmonary region and that this is accompanied by bradycardia and arterial hypotension, largely due to systemic vasodilatation (71). This drug activates various afferent vagal endings, both myelinated and unmyelinated, in the lungs, ventricles, and atria (1, 71). Because the stimulus provided by veratridine is artificial and activates almost all fibers indiscriminately, this reflex has limited physiological meaning. Nevertheless, it remains a useful method of investigation because receptors with little or no spontaneous discharge are excited, permitting them to be detected and studied.

Several other chemical substances excite heart and lung receptors and produce vagally mediated reflexes (1). With the possible exception of phenyldiguanide, which has been shown to stimulate the unmyelinated fibers more selectively, the limitations of these substances are the same as those described for veratridine.

Cardiac Chemoreceptor Reflex?   In the heart, no chemoreceptors sensitive to hypoxia and hypercapnia have been found whose stimulation causes the increase in ventilation, decrease in heart rate, and increase in systemic arterial blood pressure seen with stimulation of the arterial chemoreceptors (96, 97).

There is evidence from cats (80, 81) and dogs (79) that some aortic chemoreceptors receive a blood supply from the coronary circulation. In dogs, vessels have been traced in continuity from the left coronary artery to histologically verified aortic bodies, and chemoreceptor impulses were augmented when cyanide was injected directly into the left coronary artery (79). Eckstein et al. (98) have shown that the blood supply arose from the left anterior descendens artery in 55% of dogs and from the left common coronary artery in 16%. Injections of serotonin, cyanide, nicotine, and lobeline into the appropriate branch caused a reflex elevation of heart rate and arterial blood pressure, as did occlusion of the vessel. But whether this response was due to stimulation of aortic bodies or of coronary chemoreceptors is uncertain, since no studies were made after ablation of the aortic bodies.

Reflexes Mediated by Afferent Sympathetic Fibers

Peterson and Brown (99) studied the reflex influences of sympathetic cardiac afferent fibers, both myelinated and unmyelinated, by electrical stimulation of the central stump of the left inferior cardiac or pericoronary nerve of vagotomized cats. Stimulation of the myelinated fibers caused a slight increase in systemic blood pressure; stimulation of the unmyelinated fibers caused a much greater pressor response. These pressor responses were unaffected by section of the spinal cord at C-8 but were abolished either by section of the white rami 1 through 4 or by administration of an α-adrenergic blocking agent. These results suggested that cardiac sympathetic afferents, in contrast with cardiac vagal afferents, might exert an excitatory influence on those neurons that control the sympathetic outflow to the systemic blood vessels; this influence is predominantly transmitted by the unmyelinated fibers and operates through a spinal reflex. Stimulation of afferent sympathetic fibers also caused a reflex increase in myocardial contractility and in heart rate; hence, the excitatory influence is directed also to the sympathetic outflow to the heart (100). The stimulation not only increased the sympathetic firing to the heart but also decreased the vagal cardiac efferent firing (101). Thus, the afferent sympathetic system is capable of exerting ascending influences on the medullary centers.

Whether this excitatory reflex has any role in normal circulatory control is unknown. Lioy et al. (102) described pressor responses to mechanical stimulation of aortic sympathetic afferents by balloon distension of the thoracic aorta in vagotomized cats with spinal transection (C-1). They postulated the existence of tonically active spinal reflexes that operate through a positive feedback mechanism to set the excitability of the spinal vasomotor neurons to descending influences. This system thus would have a function similar to that of somatic spinal reflexes. Another possibility is that the pressor effects are part of the visceral response to pain. Sympathetic cardiac afferents, particularly of the fine myelinated or unmyelinated type, may be responsible for the pain of cardiac origin (28, 29, 103). It is known that an increase in blood pressure occurs on stimulation of the somatic unmyelinated afferents that mediate pain (104).

Afferent sympathetic fibers also originate in the lungs (105, 106). Pulmonary artery baroreceptors have been found in the extrapulmonary part of the pulmonary artery with afferent fibers in the left cardiac sympathetic nerve. It is likely that these are not active at the pressure normally prevailing in the pulmonary arterial circulation but only signal a transient increase in the pulmonary arterial pressure. Thus, they are excited by phasic mechanical events and show relatively rapid adaptation to sustained high pressure (107). Their behavior is different from that of the pulmonary arterial baroreceptors supplied by the vagus, in which the number of impulses per heartbeat and maximal frequency of discharges show a relative linear relationship to an increase in pressure (77).

CONTROL OF BLOOD VOLUME

The cardiopulmonary receptors are important in the regulation of blood volume through their effects on renal hemodynamics and on the secretion of hormones concerned with the regulation of salt and water balance.

Water Diuresis

The existence of receptors concerned with maintaining the constancy of blood volume was demonstrated in studies performed in both man and animals more than 20 years ago. Moderate increases or decreases in blood volume, with no concomitant alterations in osmotic pressure of the plasma, caused water diuresis or antidiuresis, respectively; these effects were attributed to changes in the plasma concentration of antidiuretic hormone (ADH), presumably through variations in its release from the hypophysis (108).

The volume receptors appear to be located within the thoracic cavity. Negative-pressure breathing, immersion of the body in water, and pressurization in an iron lung—all situations that lead to an increased central blood volume— cause a water diuresis accompanied by a decrease in plasma concentration of ADH. Positive-pressure breathing, orthostasis, constriction of the inferior vena cava, and positive acceleration, which pool blood in the peripheral veins and decrease central blood volume, cause an antidiuresis and an increase in plasma ADH (108).

To localize the volume receptors within the thoracic cavity, Henry et al. (109) occluded the pulmonary veins in anesthetized dogs and inflated a balloon to obstruct the mitral valve. Both of these procedures caused congestion of the lungs and increased the pressure in the right ventricle; however, only the mitral valve obstruction, which caused a marked increase in left atrial pressure, was accompanied by diuresis. This diuresis had a slow onset and was abolished by vagal cooling.

Baïsset and Montastruc (110) found that distension of a balloon in the left atrium of the anesthetized dog decreased the concentration of ADH in the plasma, a conclusion that had been drawn indirectly by Henry et al. (109).

When Johnson et al. (111) used smaller balloons to obtain moderate increases in left atrial pressure with a decrease in systemic arterial pressure of only a few millimeters of mercury, the plasma ADH concentration decreased linearly with an increase in left atrial pressure up to 7 cm H_2O. This indicated that a decrease in ADH concentration could occur with physiological increases in atrial pressure. Lydtin and Hamilton (112) and Zehr et al. (113) made similar observations in unanesthetized animals. Kappagoda et al. (114, 115) found that stretching the right atrial appendage or the superior vena cava atrial junction in anesthetized dogs, without obstructing venous return or altering right atrial pressure, resulted in an increase in urine flow similar to that obtained by distension of the left atrium. Crushing the base of the appendage or vagotomy abolished or largely decreased the response.

In recent studies using balloons to distend the left atrium (116) or the junction between the pulmonary vein and the left atrium (117) in dogs, a diuresis occurred in response to a blood-borne agent. However, the diuresis was not accompanied by a decrease in the antidiuretic activity of the plasma. It was concluded that the blood-borne agent was a diuretic substance and that stimulation of left atrial receptors is not accompanied by a decrease in the plasma ADH concentration (117). Further evidence that ADH is not responsible for the diuresis is the fact that it still occurred on stimulation of atrial receptors after destruction of the pituitary gland (118). In conscious dogs, decreases of mean atrial transmural pressure of 6–8 mm Hg did not significantly alter plasma ADH levels (119). Thus, the role of atrial receptors in the control of ADH has become open to debate.

In anesthetized vagotomized dogs (120) and cats (121), a decrease of carotid baroreceptor activity by occlusion of the common carotid arteries is accompanied by an increase in the plasma ADH concentration. This increase can be inhibited by stimulation of cardiac receptors that have unmyelinated fibers in the right vagus. Because these unmyelinated fibers arise in the ventricles as well as in the atria, this suggests that the receptors that exert an inhibitory influence on ADH release are not confined to the atria (122). An increase in the plasma ADH concentration occurs with the carotid sinuses isolated if the carotid baroreceptor activity is decreased by changing the carotid sinus pressure from pulsatile to nonpulsatile (123) or if the carotid body chemoreceptors are stimulated by their perfusion with hypoxic blood in vagotomized, open-chest, artificially ventilated dogs (124). Section of the aortic nerves of the anesthetized vagotomized rabbit also results in an increase in the plasma ADH concentration (125).

In dogs with chronic cardiac denervation, in which the pulmonary receptors and the arterial chemoreceptors and baroreceptors are intact, the diuretic response to acute intravascular volume expansion is attenuated (126) and returns to normal only when afferent reinnervation of the heart occurs (127). Thus, although several groups of receptors are involved in the inhibitory control of ADH, it is likely that those within the heart and supplied with vagal afferent fibers are the most important.

Sodium Excretion

In addition to an increase in water excretion, stimulation of the atrial receptors often causes an increase in sodium excretion by the kidney (112, 115–117, 128–131). Multiple efferent mechanisms seem to be responsible for this natriuretic effect. There is evidence for the existence of a humoral natriuretic agent in dogs. When an isolated kidney from a dog was perfused at constant pressure with blood from the femoral artery of another dog, distension of the left atrium of the second dog caused an increase in urine output and sodium excretion by the isolated kidney (116).

However, part of the natriuresis depends also on neurally induced increases in renal blood flow. The renal blood flow is continuously modulated by the activity of vagal afferents from the heart and also from the lungs (54). Stimulation of left atrial receptors causes a decrease in renal efferent sympathetic nerve traffic (132) and an increase in renal blood flow (92) and in glomerular filtration rate (128).

Either vagotomy or carotid sinus denervation attenuates significantly the diuretic response to intravascular expansion without influencing significantly the natriuretic response. This indicates that carotid sinus receptors and receptors that have afferent fibers in the vagus contribute substantially to the control of plasma volume by a mechanism that influences water excretion. By contrast, chronic cardiac denervation attenuates both the diuretic and the natriuretic response to acute isotonic (and isooncotic) intravascular expansion, indicating that cardiac afferent nerves may contribute to both salt and water homeostasis (126). Also, distension of the atrial appendages results in a diuresis, an increase in the rate of sodium excretion, and an increase in heart rate, and these responses are abolished or greatly decreased by crushing the bases of the appendages (114).

It has been suggested that those cardiac receptors innervated by vagal afferent fibers influence the circulating level of ADH and those innervated by sympathetic afferent fibers influence renal sodium excretion. However, posterior root section from C-8 to T-6, a procedure that interrupts sympathetic cardiac afferents, is followed by an unimpaired natriuresis with volume expansion, indicating that these sympathetic afferents are not essential for the response (133).

Renin Release

Vagal afferents from the cardiopulmonary region exert a tonic restraint on the release of renin by inhibiting sympathetic outflow to the kidney. This has been demonstrated in dogs with aortic nerves cut and carotid sinuses vascularly isolated (134). When the sinus pressure was set at a level below the threshold of the baroreceptors, bilateral vagal cold block caused an increase in aortic blood pressure, a decrease in renal blood flow, and a 3- to 5-fold increase in renin output. When the sinus pressure was maintained at the level of existing arterial pressure, there was only a small increase in arterial pressure and a small decrease in renal blood flow but a similar marked increase in renin release with vagal block (Figure 5).

These hemodynamic changes are consistent with previous observations showing that the vasomotor inhibition exerted tonically by the vagally innervated cardiopulmonary mechanoreceptors decreases as the inhibition from the carotid baroreceptors is increased. By contrast, the cardiopulmonary receptors seem to exert a relatively greater control of the renin secretion than of the circulation because the output of renin with vagal block was unaffected by the changes in carotid sinus pressure. The increase in renin release caused by vagal block was greater during the second minute than during the fifth minute of block. It may be that the prolonged vagal block caused an increase in arterial concentration of angiotensin II which in turn partially inhibited the release of renin (135). It is also possible that the sympathetic nerves have their major effect on the discharge of preformed renin.

The role of the arterial baroreceptors in the control of renin release is debatable (136–139). In a recent study, neither increasing nor decreasing carotid sinus pressure had any demonstrable effect on the arterial plasma renin activity in either vagotomized dogs or dogs with intact vagi (140).

After cardiac denervation in the dog, by surgical division and anastomosis of the heart caudal to the veno-arterial junctions, the increase in renin release in response to hemorrhage was decreased (141). Because this technique left most of the atrial receptors innervated and the ventricular receptors denervated, these

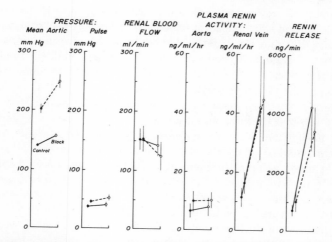

Figure 5. Effects of bilateral cervical vagal cold block at different carotid sinus pressures on aortic blood pressure, aortic pulse pressure, left renal blood flow, plasma renin activity in aorta and left renal vein, and renin release, shown as mean (± S.E.) of six dogs with aortic nerves cut. ——, sinus pressure maintained at level of mean aortic pressure before vagal block; – – –, sinus pressure maintained at 40 mm Hg. At low sinus pressure, vagal block caused greater increase in aortic pressure and greater decrease in renal blood flow, but the amount of renin released was not significantly different. (From Mancia, G., Romero, J.C., and Shepherd, J.T. (1975). Continuous inhibition of renin release in dogs by vagally innervated receptors in the cardiopulmonary region. Circ. Res. 36:529. By permission of American Heart Association.)

studies suggest that ventricular receptors are implicated in the control of renin release. In other studies, distension of the right atrium caused a decrease in renin secretion, suggesting that atrial receptors are involved (142).

It has not been established if the sympathetic nerves affect the renin release directly through their innervation of the juxtaglomerular apparatus or indirectly through changes in renal hemodynamics (143–147).

The release of renin through changes in activity of the cardiopulmonary receptors will affect the output of aldosterone and hence the tubular reabsorption of sodium (148).

INTERACTION OF REFLEXES FROM CARDIOPULMONARY REGION WITH CAROTID BAROREFLEX AND CHEMOREFLEX

There are many inputs to the central nervous system, and the reflex effects on the cardiovascular system are the result of the central integration of the information from the various afferents involved.

Cardiopulmonary Reflexes and Carotid Baroreflex

Inflation of the isolated lungs or balloon distension of the pulmonary vein-atrial junctions in dogs causes a decrease in systemic vascular resistance, and the magnitude of this decrease diminishes as the carotid sinus pressure increases

(87). In other studies in dogs with aortic nerves cut, the vagi were cooled at nonpulsatile carotid sinus pressures (vascularly isolated carotid sinuses) of about 50, 140, and 200 mm Hg to set the input from the carotid baroreceptors at minimal, intermediate, and maximal levels, respectively. With the sinus pressure at 50 mm Hg, vagal cooling caused an increase in arterial blood pressure, a constriction of the hindlimb resistance vessels, and, as expected, a much greater constriction of the renal resistance vessels (48, 54). However, with the sinus pressure at 140 mm Hg these responses were attenuated, and at a sinus pressure of 200 mm Hg there were no responses to vagal cooling (Figure 6). Similar results were obtained when the dogs were made hypervolemic (15 and 30% increases in blood volume) to augment the influence of the cardiopulmonary receptors. In this circumstance, the responses to vagal cooling were greater than those observed during normovolemia at low sinus pressure, but they decreased sharply and disappeared as the sinus pressure was increased.

Thus, the carotid sinus normally dominates in the interplay between carotid baroreflexes and cardiopulmonary reflexes. Only when the inhibitory influence of the carotid receptors is decreased does the cardiovascular inhibition exerted by the cardiopulmonary region become manifest. This is the case even in the renal circulation, which is strongly influenced by the cardiopulmonary inhibitory vasomotor reflex. This interplay between the carotid and cardiopulmonary vascular reflexes suggests that nerve traffic from the former area competitively inhibits that from the latter and that both act on the same neuronal pools at the vasomotor center.

Cardiopulmonary Reflexes and Carotid Chemoreflex

Stimulation of the carotid bodies causes slowing of the heart and systemic vasoconstriction whereas stimulation of the aortic bodies causes bradycardia or tachycardia and vasoconstriction (149, 150). These direct or primary reflex effects on the heart and systemic blood vessels are masked or overridden by secondary mechanisms evoked by the concomitant increase in ventilation, of which the most important is the vagal reflex that arises from the lungs when they are inflated and causes tachycardia and systemic vasodilatation (64, 65). There are naturally occurring situations, such as underwater diving, in which arterial hypoxia and hypercapnia are accompanied not by hyperventilation but by a decrease or cessation of respiratory movements. Under these conditions, the primary cardiac and vascular effects resulting from stimulation of the chemoreceptors by hypoxic hypercapnic blood are evident because of the absence of opposing secondary respiratory mechanisms, and bradycardia and systemic vasoconstriction occur (149). The cessation of respiratory movements during diving is caused by stimulation of trigeminal afferents originating from the nasal mucosa, leading to suppression of the chemoreceptor-respiratory reflex response and allowing the chemoreceptor circulatory reflex effect to be manifest (151).

In a study (152) in dogs with aortic nerves cut and ventilation controlled, the carotid chemoreceptors were stimulated by perfusion of the vascularly isolated carotid bifurcations with hypoxic hypercapnic blood while the input

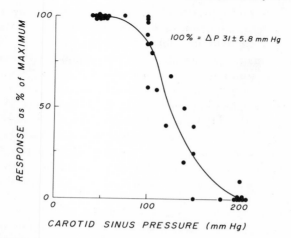

Figure 6. Interaction of cardiopulmonary and arterial mechanoreceptors, as shown by effect of blocking afferent activity of former by cooling vagi at different carotid sinus pressures (data from five dogs with aortic nerves cut). Note that maximal response to vagal cooling occurred when activity of arterial mechanoreceptors was minimal (carotid sinus pressure, 50 mm Hg). (From Shepherd, J.T. (1974). The cardiac catheter and the American Heart Association. Circulation 50:418. By permission of American Heart Association.)

from the carotid baroreceptors was kept constant at intermediate and maximal values. At each carotid baroreceptor input, the vagi were cooled before and during the chemoreceptor stimulation. The stimulation caused an increase in vascular resistance and augmented the vasomotor response to vagal cold block when the carotid sinus pressure was set at an intermediate level, but was without effect on the vascular system or on the response to vagal block when the sinus pressure was high.

Possible Central Interactions

The saphenous vein, which has no consistent response to changes in carotid baroreceptor activity, dilates with chemoreceptor stimulation when the carotid sinus pressure is intermediate but not when it is high. Thus, a maximal carotid baroreceptor input prevents the influence of the chemoreceptors on this vein without having a direct effect on the central neurons controlling the vein. This can be explained by a presynaptic inhibition by the carotid baroreceptor afferents of the afferent traffic from the chemoreceptors to the central neurons that control the saphenous vein (153).

This explanation might be applied more generally. If the carotid chemoreceptors do not influence the vasomotor center directly (other than the neuronal pools controlling the cutaneous veins) but exert their vascular effects by acting presynaptically in an agonist-antagonist fashion with the inhibitory traffic from the carotid baroreceptors, then one would not anticipate a chemoreceptor response when the baroreceptor inhibitory traffic was the dominant input. At the intermediate level of carotid sinus pressure, the presynaptic interaction of carotid baroreceptor and chemoreceptor inputs would be equivalent to a decrease in baroreceptor inhibition of the vasomotor center. This in turn would allow greater expression of the inhibition coming from the cardiopulmonary receptors. An alternative explanation is that the augmented vascular response to vagal block when carotid inhibition is withdrawn, or when the chemoreceptors are stimulated at an intermediate sinus pressure, simply is due to an increase in afferent vagal traffic induced by an alteration in cardiac performance.

Whatever the final explanation, it seems that the carotid sinus reflex dominates when the carotid sinus, cardiopulmonary, and carotid chemoreflexes interact. The cardiopulmonary receptors appear to regulate renin release in normal circumstances and to become effective in cardiovascular regulation only when the input from the carotid baroreceptors is decreased or the chemoreceptors are activated.

The present findings have certain implications in abnormal circumstances. For example, during hemorrhage, when the systemic arterial pressure and central venous pressure decrease, input from the carotid and cardiopulmonary receptors will be diminished, resulting in a strong constriction of the renal and muscle vascular beds. Renal blood flow decreased less in experimental cardiogenic shock, in which left atrial pressure was increased, than in hemorrhagic shock, in which left atrial pressure was decreased (154, 155).

REFLEXES FROM CARDIOPULMONARY REGION IN ABNORMAL CONDITIONS

Hemorrhage

During the 4 hours after bleeding of unanesthetized normal rabbits of 26% of their blood volume, the reflex vasoconstrictor effects were greatest in the kidney and next greatest in muscle, portal bed, and skin. After section of the carotid sinus and aortic nerves, reflex constrictor effects were absent in the portal, muscle, and skin beds, but significant vasoconstriction was still evident in the renal bed, although of smaller magnitude than in normal animals. These results suggest that the arterial baroreceptors are a major source of reflex activity after hemorrhage but that other reflexogenic zones contribute to the renal effects (156).

In anesthetized cats with aortic nerves cut and carotid sinuses and vagi intact, a 10% decrease in blood volume caused a decrease in blood pressure and constriction of renal and muscle resistance vessels. The responses were similar after hemorrhage with only the carotid baroreceptors intact and the vagi cooled. When the vagi were intact and the carotid sinus nerves were cut, the renal response to hemorrhage was maintained but the muscle response was largely attenuated and the hypotension was greater (157). In anesthetized dogs with aortic nerves cut, hemorrhage of 10% of the blood volume decreased the blood pressure by 13%. With only the carotid baroreflex operative, only the vagal reflex operative, and neither of these reflexes operative, the decreases in arterial pressure were 18, 24, and 42%, respectively. With either the carotid or the vagal baroreflex operative, hemorrhage caused similar degrees of renal and mesenteric vasoconstriction and of splenic contraction. In contrast, the constriction in the hindlimb was much less or absent in the latter circumstance (158).

Thus, the cardiopulmonary receptors subserved by vagal afferents participate in the circulatory adjustment to hemorrhage but their importance in control of blood pressure is less than that of the carotid baroreceptors, due to the dominant influence of the latter on the muscle resistance vessels. The reflex changes effected by the cardiopulmonary receptors involve mainly the renal circulation and the splanchnic resistance and capacitance vessels. Whether specific receptors or many receptors in the atria, ventricles, and lungs are involved remains to be determined. The similar tonic vasomotor inhibition exerted by the atrial and ventricular receptors at normal intracardiac pressures makes the latter more likely (36).

A rapid and severe hemorrhage in cats sometimes produces a reflex brady-cardia, similar to that seen in the vasovagal syncope in man. This slowing of the heart is associated with increased activity of cardiac receptors located mainly in the left ventricle and firing in unmyelinated vagal afferents. It is suggested that their stimulation results from vigorous contractions of a ventricle that is almost empty because of the rapid withdrawal of blood (159).

Hypercapnia

In anesthetized sino-aortic denervated rabbits with ventilation controlled, inter-ruption of vagal afferent traffic by cooling during normocapnia caused only a modest increase in hindlimb and renal vascular resistance, whereas during sys-temic hypercapnia (end-tidal carbon dioxide, 8–10%) these increases were poten-tiated but much more so in the kidney (160) (Figure 7). It appears that with hyper-capnia the vagal pulmonary receptors markedly increase their inhibitory control of the renal resistance vessels but increase their control of the muscle vessels only slightly. This serves to counteract the central excitatory action of

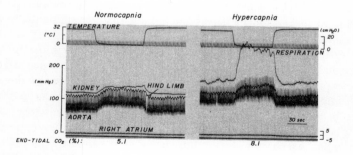

Figure 7. Effect of hypercapnia on circulatory changes caused by vagal cold block in a rabbit with sinus and aortic nerves cut. *Upper,* temperature of thermodes and intratracheal pressure. *Lower,* aortic blood pressure, kidney and limb perfusion pressures, and right atrial pressure. (From Ott, N.T., and Shepherd, J.T. (1973). Modifications of the aortic and vagal depressor reflexes by hypercapnia in the rabbit. Circ. Res. 33:160. By permission of American Heart Association.)

carbon dioxide which, for unknown reasons, seems more pronounced on the renal than on the muscle vasomotor neurons.

Similar effects were noted with vagal cooling in sino-aortic denervated dogs in a preparation with the ventilated lungs in situ and the heart removed (54). In addition, cardiac receptors with vagal afferents and aortic baroreceptors also exert a greater renal vasomotor inhibition during hypercapnia, thus helping to oppose the central excitatory effects of carbon dioxide (54, 161). The carotid baroreceptors act similarly because the reflex renal vasoconstriction induced by clamping of the common carotid arteries in cats is greater during hypoventilation (162). Thus, this phenomenon appears to be a general property of reflexes that inhibit the renal vasomotor neurons and not a specific function of receptors in the cardiopulmonary region.

This increased reflex inhibition of the renal resistance vessels would help to preserve the renal blood flow and to contribute to the restoration of acid-base balance in respiratory acidosis.

There is no information on the mechanism of the increase in afferent inhibition of the renal vasomotor neurons when these neurons are under the excitatory influence of carbon dioxide. In the studies on dogs with lungs in situ and heart removed, the ventilatory pressures were similar during normocapnia and hypercapnia so that there was no obvious mechanical cause for increased discharge from the receptor sites (54). Also, carbon dioxide is known to decrease the activity of pulmonary stretch receptors and of mechanoreceptors in general (163, 164). It seems likely that the increased afferent inhibition of the renal neurons is due at least in part to an interaction at the central level.

Reflexes during Myocardial Ischemia and Infarction

Experimental myocardial ischemia or necrosis produced by ligation of a major coronary artery or by intracoronary embolization or injection of necrotizing drugs causes arterial hypotension, due in part to reflex bradycardia and dilatation of resistance and capacitance vessels. These changes are more pronounced after section of the sino-aortic nerves and are abolished or substantially attenuated by vagotomy (165–168).

Whether this vagal depressor reflex plays an important role in the production of cardiogenic shock is a matter for speculation. Both the myocardial infarctions that are complicated by shock, and those that are not, often are of similar size and localization and are accompanied by similar decreases in cardiac output, the difference being that in the former group there is less compensatory vasoconstriction (169). This may be due to the inhibitory vagal reflex opposing the increased reflex constrictor drive from the arterial baroreceptors as a consequence of the decrease in systemic arterial blood pressure. Indeed, Kezdi et al. (170) have shown that, in experimental myocardial infarction in dogs, with a low cardiac output and shock there was the expected decrease in arterial baroreceptor activity but the postganglionic sympathetic nerve activity did not increase unless the vagal nerves were cut or blocked.

The vagal inhibitory reflex during myocardial infarction cannot be ascribed to any one particular type of receptor; these receptors are found not only in the left ventricle but also in the other cardiac chambers and are connected to myelinated and unmyelinated vagal afferents (25, 171–173). In the cat, transient occlusion of one coronary artery activates left ventricular receptors with unmyelinated vagal afferents, probably as a consequence of the mechanical bulging of the ischemic area of the left ventricular wall (25).

Although bradycardia and bradyarrythmia are frequent accompaniments of coronary occlusion, reflex tachycardia has also been reported in conscious dogs, due to increased sympathetic and decreased vagal activities (174).

In cats with spinal transection, activation of afferent sympathetic fibers by myocardial ischemia caused a reflex increase in efferent sympathetic traffic (175). Coronary artery occlusion in dogs caused excitation of both myelinated and unmyelinated afferent sympathetic nerve fibers whose receptor sites are in the left ventricular wall. The myelinated fibers discharged regularly with a close relationship to passive stretching or shortening of the bulging ischemic left ventricular wall. The unmyelinated fibers discharged irregularly and independently. The discharge frequently was more pronounced after release of occlusion than during occlusion (29). The role of this reflex and its interplay with the vagal reflex remain to be elucidated. When elicited by ischemia or infarction, it may increase ventricular contractility and thus oppose cardiac dilatation. However, this would induce an additional oxygen consumption by the heart and might facilitate arrythmias. The presence of the sympathetic nerves may be associated with a higher incidence of arrhythmias during coronary occlusion (176–179); the vagal nerves might offer some protection against them (180, 181).

Reflexes in Other Pathological Conditions

Several other pathological conditions may be associated with altered reflex control by the cardiopulmonary receptors. For example, during pulmonary edema or pneumonia, the activity of the "J" receptors in the affected pulmonary areas is increased (9, 182). Activation of these receptors has been shown to cause arterial hypotension and bradycardia (183) and to inhibit the skeletal muscle reflexes through connections at higher brain levels (184–186).

The polyuria associated with paroxysmal atrial tachycardia may be another example (187–189). It is conceivable that during the tachycardia stimulation of the atrial receptors by an increase in pulsation or circumference causes the diuretic response. Indeed, Kilburn (190) observed that pacing the heart at high rates induced polyuria before but not after vagotomy. However, it has been shown in dogs that rapid atrioventricular pacing produced no significant change in urine flow if the carotid and aortic baroreceptors and chemoreceptors were denervated (191). This also was the case in dogs with complete heart block and sino-aortic afferents intact when pacing was limited to the atria (39). Thus, elucidation of the relative importances of the atrial and the arterial afferents in this circumstance requires further study.

The role of cardiac and pulmonary receptors in conditions of chronic volume or pressure loads (congenital shunts, valvular diseases, or arterial and pulmonary hypertension) is unknown. Because of the increased stretch, these receptors may play a greater role than normally; however, these pathological conditions are associated with marked anatomical modifications of the structures where these receptors are located, so that a change in receptor sensitivity, or even destruction of them, cannot be excluded. In dogs with chronic mitral stenosis and extensive histological alterations in the left atrium, the increase in blood level of ADH in response to nonhypotensive hemorrhage was less than in normal dogs (192). In dogs in heart failure due to tricuspid insufficiency and pulmonary artery stenosis, the firing pattern of the atrial type B receptors (1) was impaired (193).

In man, small displacements of blood from the central circulation, caused by application of negative pressure to the lower body, are accompanied by a decrease in forearm and splanchnic blood flow. The vasoconstriction in these regions occurs without measurable change in arterial blood pressure, suggesting that reflexes from the cardiopulmonary region are responsible for the changes in blood flow (194–196).

In patients with aortic stenosis, the normal forearm vasoconstrictor response to leg exercise is inhibited or reversed, possibly because of activation of left ventricular mechanoreceptors. Such activation may contribute to syncope on exertion in such patients (197).

ACKNOWLEDGMENTS

The authors thank Drs. M.A. McGrath, P.N. Thorén, and R.H. Verhaeghe for reading the manuscript and Mrs. J.Y. Troxell and Mrs. P. Boccaccini for typing it.

REFERENCES

1. Paintal, A.S. (1973). Vagal sensory receptors and their reflex effects. Physiol. Rev. 53:159.
2. Coleridge, H., and Coleridge, J. (1972). Cardiovascular receptors. In C.B.B. Downman (ed.), Modern Trends in Physiology, pp. 245–267. Appleton-Century-Crofts, New York.
3. Armstrong, D.J., and Luck, J.C. (1974). Accessibility of pulmonary stretch receptors from the pulmonary and bronchial circulations. J. Appl. Physiol. 36:706.
4. Widdicombe, J.G. (1954). The site of pulmonary stretch receptors in the cat. J. Physiol. (Lond.) 125:336.
5. Miserocchi, G., Mortola, J., and Sant'Ambrogio, G. (1973). Localization of pulmonary stretch receptors in the airways of the dog. J. Physiol. (Lond.) 235:775.
6. Knowlton, G.C., and Larrabee, M.G. (1946). A unitary analysis of pulmonary volume receptors. Amer. J. Physiol. 147:100.
7. Widdicombe, J.G. (1954). Receptors in the trachea and bronchi of the cat. J. Physiol. (Lond.) 123:71.
8. Widdicombe, J.G., Kent, D.C., and Nadel, J.A. (1962). Mechanism of bronchoconstriction during inhalation of dust. J. Appl. Physiol. 17:613.
9. Paintal, A.S. (1969). Mechanism of stimulation of type J pulmonary receptors. J. Physiol. (Lond.) 203:511.
10. Coleridge, H.M., Coleridge, J.C.G., and Luck, J.C. (1965). Pulmonary afferent fibers

of small diameter stimulated by capsaicin and by hyperinflation of the lungs. J. Physiol. (Lond.) 179:248.

11. Nonidez, J.F. (1937). Identification of the receptor areas in the venae cavae and pulmonary veins which initiate reflex cardiac acceleration (Bainbridge's reflex). Amer. J. Anat. 61:203.

12. Coleridge, J.C.G., Hemingway, A., Holmes, R.L., and Linden, R.J. (1957). The location of atrial receptors in the dog: A physiological and histological study. J. Physiol. (Lond.) 136:174.

13. Paintal, A.S. (1953). A study of right and left atrial receptors. J. Physiol. (Lond.) 120:596.

14. Arndt, J.O., Brambring, P., Hindorf, K., and Röhnelt, M. (1974). Afferent discharge pattern of atrial mechanoreceptors in the cat during sinusoidal stretch of atrial strips in situ. J. Physiol. (Lond.) 240:33.

15. Johnston, B.D. (1968). Nerve endings in the human endocardium. Amer. J. Anat. 122:621.

16. Miller, M.R., and Kasahara, M. (1964). Studies on the nerve endings in the heart. Amer. J. Anat. 115:217.

17. Coleridge, H.M., Coleridge, J.C.G., Dangel, A., Kidd, C., Luck, J.C., and Sleight, P. (1973). Impulses in slowly conducting vagal fibers from afferent endings in the veins, atria, and arteries of dogs and cats. Circ. Res. 33:87.

18. Thorén, P. (1974). Characteristics of atrial receptors with non-medullated vagal afferents (abstract). Physiologist 17:344.

19. Paintal, A.S. (1955). A study of ventricular pressure receptors and their role in the Bezold reflex. Q. J. Exp. Physiol. 40:348.

20. Coleridge, H.M., Coleridge, J.C.G., and Kidd, C. (1964). Cardiac receptors in the dog, with particular reference to two types of afferent ending in the ventricular wall. J. Physiol. (Lond.) 174:323.

21. Sleight, P., and Widdicombe, J.G. (1965). Action potentials in afferent fibres from pericardial mechanoreceptors in the dog. J. Physiol. (Lond.) 181:259.

22. Brown, A.M. (1965). Mechanoreceptors in or near the coronary arteries. J. Physiol. (Lond.) 177:203.

23. Sleight, P., and Widdicombe, J.G. (1965). Action potentials in fibres from receptors in the epicardium and myocardium of the dog's left ventricle. J. Physiol. (Lond.) 181:235.

24. Öberg, B., and Thorén, P. (1972). Studies on left ventricular receptors, signalling in non-medullated vagal afferents. Acta Physiol. Scand. 85:145.

25. Thorén, P. (1972). Left ventricular receptors activated by severe asphyxia and by coronary artery occlusion. Acta Physiol. Scand. 85:455.

26. Malliani, A., Recordati, G., and Schwartz, P.J. (1973). Nervous activity of afferent cardiac sympathetic fibres with atrial and ventricular endings. J. Physiol. (Lond.) 229:457.

27. Brown, A.M., and Malliani, A. (1971). Spinal sympathetic reflexes initiated by coronary receptors. J. Physiol. (Lond.) 212:685.

28. Brown, A.M. (1967). Excitation of afferent cardiac sympathetic nerve fibres during myocardial ischaemia. J. Physiol. (Lond.) 190:35.

29. Uchida, Y., and Murao, S. (1974). Excitation of afferent cardiac sympathetic nerve fibers during coronary occlusion. Amer. J. Physiol. 226:1094.

30. Moissejeff, E. (1927). Zur Kenntnis des Carotissinusreflexes. Z. Gesamte Exp. Med. 53:696.

31. Cyon, E., and Ludwig, C. (1866). Die Reflexe eines der sensiblen Nerven des Herzens auf die motorischen der Blutgefässe. Arb. Physiol. Anst. 128.

32. Koch, E. (1931). Die reflektorische Selbststenerung des Kreislaufes. In B. Kisch (ed.), Ergebnisse der Kreisslaufforschung, Vol. 1, p. 234, Steinkopff, Dresden.

33. McCubbin, J.W., Masson, G.M.C., and Page, I.H. (1958). Aortic depressor nerves of the rat. Arch. Int. Pharmacodyn. Ther. 114:303.

34. Heymans, C., and Bouckaert, J.-J. (1933). Enervation des zones vasosensibles cardio-aortiques et sino-carotidiennes: techique influence sur les réflexes conditionnels. C. R. Soc. Biol. (Paris) 112:711.

35. Edis, A.J., and Shepherd, J.T. (1971). Selective denervation of aortic arch baroreceptors and chemoreceptors in dogs. J. Appl. Physiol. 30:294.

36. Mancia, G., and Donald, D.E. (1975). Demonstration that atria, ventricles, and lungs each are responsible for a tonic inhibition of the vasomotor center in the dog. Circ. Res. 36:310.
37. Mancia, G., Donald, D.E., and Shepherd, J.T. (1973). Inhibition of adrenergic outflow to peripheral blood vessels by vagal afferents from the cardiopulmonary region in the dog. Circ. Res. 33:713.
38. Aviado, D.M., Jr., and Schmidt, C.F. (1959). Cardiovascular and respiratory reflexes from the left side of the heart. Amer. J. Physiol. 196:726.
39. Goetz. K.L., and Bond, G.C. (1973). Reflex diuresis during tachycardia in the dog: Evaluation of the role of atrial and sinoaortic receptors. Circ. Res. 32:434.
40. Guazzi, M., Libretti, A., and Zanchetti, A. (1962). Tonic reflex regulation of the cat's blood pressure through vagal afferents from the cardiopulmonary region. Circ. Res. 11:7.
41. Krasney, J.A. (1971). Cardiovascular responses to cyanide in awake sinoaortic denervated dogs. Amer. J. Physiol. 220:1361.
42. Hainsworth, R. (1974). Circulatory responses from lung inflation in anesthetized dogs. Amer. J. Physiol. 226:247.
43. Ito, C.S., and Scher, A.M. (1973). Arterial baroreceptor fibers from the aortic region of the dog in the cervical vagus nerve. Circ. Res. 32:442.
44. Ito, C.S., and Scher, A.M. (1974). Reflexes from the aortic baroreceptor fibers in the cervical vagus of the cat and the dog. Circ. Res. 34:51.
45. Mizeres, N.J. (1955). The anatomy of the autonomic nervous system in the dog. Amer. J. Anat. 96:285.
46. Kulaev, B.S. (1963). Characteristics of afferent impulses evoked in cardiac nerves by chemical stimulation of epicardial receptors. Fed. Proc. 22:T749.
47. Öberg, B., and Thorén, P. (1973). Circulatory responses to stimulation of medullated and non-medullated afferents in the cardiac nerve in the cat. Acta Physiol. Scand. 87:121.
48. Öberg, B., and White, S. (1970). Circulatory effects of interruption and stimulation of cardiac vagal afferents. Acta Physiol. Scand. 80:383.
49. Pillsbury, H.R.C., III, Guazzi, M., and Freis, E.D. (1969). Vagal afferent depressor nerves in the rabbit. Amer. J. Physiol. 217:768.
50. Clement, D.L., Pelletier, C.L., and Shepherd, J.T. (1972). Role of vagal afferents in the control of renal sympathetic nerve activity in the rabbit. Circ. Res. 31:824.
51. Brender, D., and Webb-Peploe, M.M. (1969). Influence of carotid baroreceptors on different components of the vascular system. J. Physiol. (Lond.) 205:257.
52. Hainsworth, R., Karim, F., and Stoker, J.B. (1975). The influence of aortic baroreceptors on venous tone in the perfused hind limb of the dog. J. Physiol. (Lond.) 244:337.
53. Webb-Peploe, M.M., and Shepherd, J.T. (1968). Responses of the superficial limb veins of the dog to changes in temperature. Circ. Res. 22:737.
54. Mancia, G., Shepherd, J.T., and Donald, D.E. (1975). Role of cardiac, pulmonary, and carotid mechanoreceptors in the control of hindlimb and renal circulation in dogs. Circ. Res. 37:200.
55. Ninomiya, I., Nisimaru, N., and Irisawa, H. (1971). Sympathetic nerve activity to the spleen, kidney, and heart in response to baroceptor input. Amer. J. Physiol. 221:1346.
56. Kendrick, E., Öberg, B., and Wennergren, G. (1972). Vasoconstrictor fibre discharge to skeletal muscle, kidney, intestine and skin at varying levels of arterial baroreceptor activity in the cat. Acta Physiol. Scand. 85:464.
57. Öberg, B., and Thorén, P. (1973). Circulatory responses to stimulation of left ventricular receptors in the cat. Acta Physiol. Scand. 88:8.
58. Thorén, P.N., Mancia, G., and Shepherd, J.T. (1975). Vasomotor inhibition in rabbits by vagal nonmedullated fibers from cardiopulmonary area. Amer. J. Physiol. 229:1410.
59. Paintal, A.S. (1965). Block of conduction in mammalian myelinated nerve fibers by low temperatures. J. Physiol. (Lond.) 180:1.
60. Franz, D.N., and Iggo, A. (1968). Conduction failure in myelinated and nonmyelinated axons at low temperatures. J. Physiol. (Lond.) 199:319.
61. Daly, M. de B., Hazzledine, J.L., and Ungar, A. (1967). The reflex effects of

alterations in lung volume on systemic vascular resistance in the dog. J. Physiol. (Lond.) 188:331.

62. Daly, M. de B., and Robinson, B.H. (1968). An analysis of the reflex systemic vasodilator response elicited by lung inflation in the dog. J. Physiol. (Lond.) 195:387.

63. Anrep, G.V., Pascual, W., and Rössler, R. (1936). Respiratory variations of the heart rate. I. The reflex mechanism of the respiratory arrhythmia. Proc. R. Soc. Med. 119:191.

64. Daly, M. de B., and Scott, M.J. (1962). An analysis of the primary cardiovascular reflex effects of stimulation of the carotid body chemoreceptors in the dog. J. Physiol. (Lond.) 162:555.

65. Scott, M.J. (1966). The effects of hyperventilation on the reflex cardiac response from the carotid bodies in the cat. J. Physiol. (Lond.) 186:307.

66. Goetz, R.H. (1935). Der Fingerplethysmograph als Mittel zur Untersuchung der Regulationsmechanismen in peripheren Gefässgebieten. Pfluegers Arch. Ges. Physiol. 235:271.

67. Bolton, B., Carmichael, E.A., and Stürup, G. (1936). Vaso-constriction following deep inspiration. J. Physiol. (Lond.) 86:83.

68. Duggan, J.J., Love, V.L., and Lyons, R.H. (1953). A study of reflex venomotor reactions in man. Circulation 7:869.

69. Browse, N.L., and Hardwick, P.J. (1969). The deep breath-venoconstriction reflex. Clin. Sci. 37:125.

70. Gilliatt, R.W., Guttmann, L., and Witteridge, D. (1948). Inspiratory vaso-constriction in patients after spinal injuries. J. Physiol. (Lond.) 107:67.

71. Dawes, G.S., and Comroe, J.H., Jr. (1954). Chemoreflexes from the heart and lungs. Physiol. Res. 34:167.

72. Inglesby, T.V., Rainzer, A.E., Hanley, H.G., and Skinner, N.S., Jr. (1972). Cardiovascular reflexes induced by selectively altering pulmonary arterial osmolality. Amer. J. Physiol. 222:302.

73. Agarwal, J.B., Baile, E.M., and Palmer, W.H. (1969). Reflex systemic hypotension due to hypertonic solutions in pulmonary circulation. J. Appl. Physiol. 27:251.

74. Daly, I. de B., Lundány, G., Todd, A., and Verney, E.B. (1937). Sensory receptors in the pulmonary vascular bed. Q. J. Exp. Physiol. 27:123.

75. Aviado, D.M., Jr., Li, T.H., Kalow, W., Schmidt, C.F., Turnbull, G.L., Peskin, G.W., Hess, M.E., and Weiss, A.J. (1951). Respiratory and circulatory reflexes from the perfused heart and pulmonary circulation of the dog. Amer. J. Physiol. 165:261.

76. Coleridge, J.C.G., Kidd, C., and Sharp, J.A. (1961). The distribution, connexions, and histology of baroreceptors in the pulmonary artery, with some observations on the sensory innervation of the ductus arteriosus. J. Physiol. (Lond.) 156:591.

77. Coleridge, J.C.G., and Kidd, C. (1961). Relationship between pulmonary arterial pressure and impulse activity in pulmonary arterial baroreceptor fibres. J. Physiol. (Lond.) 158:197.

78. Coleridge, J.C.G., and Kidd, C. (1963). Reflex effects of stimulating baroreceptors in the pulmonary artery. J. Physiol. (Lond.) 166:197.

79. Coleridge, H.M., Coleridge, J.C.G., and Howe, A. (1970). Thoracic chemoreceptors in the dog: a histological and electrophysiological study of the location, innervation and blood supply of the aortic bodies. Circ. Res. 26:235.

80. Coleridge, H.M., Coleridge, J.C.G., and Howe, A. (1967). A search for pulmonary arterial chemoreceptors in the cat, with comparison of the blood supply of the aortic bodies in the new-born and adult animal. J. Physiol. (Lond.) 191:353.

81. Howe, A. (1956). The vasculature of the aortic bodies in the cat. J. Physiol. (Lond.) 134:311.

82. Duke, H.N., Green, J.H., Heffron, P.F., and Stubbens, V.W.J. (1963). Pulmonary chemoreceptors. Q. J. Exp. Physiol. 48:164.

83. Cropp, G.J.A., and Comroe, J.H., Jr. (1961). Role of mixed venous blood p_{CO_2} in respiratory control. J. Appl. Physiol. 16:1029.

84. Guz, A., and Trenchard, D.W. (1971). The role of non-myelinated vagal afferent fibres from the lungs in the genesis of tachypnoea in the rabbit. J. Physiol. (Lond.) 213:345.

85. Ross, J., Jr., Frahm, C.J., and Braunwald, E. (1961). The influence of intracardiac

baroreceptors on venous return, systemic vascular volume and peripheral resistance. J. Clin. Invest. 40:563.

86. Salisbury, P.F., Cross, C.E., and Rieben, P.A. (1960). Reflex effects of left ventricular distention. Circ. Res. 8:530.
87. Edis, A.J., Donald, D.E., and Shepherd, J.T. (1970). Cardiovascular reflexes from stretch of pulmonary vein-atrial junctions in the dog. Circ. Res. 27:1091.
88. Bainbridge, F.A. (1915). The influence of venous filling upon the rate of the heart. J. Physiol. (Lond.) 50:65.
89. Ledsome, J.R., and Linden, R.J. (1964). A reflex increase in heart rate from distension of the pulmonary vein-atrial junctions. J. Physiol. (Lond.) 170:456.
90. Carswell, F., Hainsworth, R., and Ledsome, J.R. (1970). The effects of distension of the pulmonary vein-atrial junctions upon peripheral vascular resistance. J. Physiol. (Lond.) 207:1.
91. Kappagoda, C.T., Linden, R.J., and Snow, H.M. (1972). A reflex increase in heart rate from distension of the junction between the superior vena cava and the right atrium. J. Physiol. (Lond.) 220:177.
92. Mason, J.M., and Ledsome, J.R. (1974). Effect of obstruction of the mitral orifice or distension of the pulmonary vein-atrial junctions on renal and hind-limb vascular resistance in the dog. Circ. Res. 35:24.
93. Burkhart, S.M., and Ledsome, J.R. (1974). The response to distension of the pulmonary vein-left atrial junctions in dogs with spinal section. J. Physiol. (Lond.) 237:685.
94. Ledsome, J.R., and Linden, R.J. (1967). The effect of distending a pouch of left atrium on the heart rate. J. Physiol. (Lond.) 193:121.
95. Malliani, A., Parks, M., Tuckett, R.P., and Brown, A.M. (1973). Reflex increases in heart rate elicited by stimulation of afferent cardiac sympathetic nerve fibers in the cat. Circ. Res. 32:9.
96. Brown, A.M. (1966). The depressor reflex arising from the left coronary artery of the cat. J. Physiol. (Lond.) 184:825.
97. Mark, A.L., Abboud, F.M., Heistad, D.D., Schmid, P.G., and Johannsen, U.J. (1974). Evidence against the presence of ventricular chemoreceptors activated by hypoxia and hypercapnia. Amer. J. Physiol. 227:178.
98. Eckstein, R.W., Shintani, F., Rowen, H.E., Jr., Shimomura, K., and Ohya, N. (1971). Identification of left coronary blood supply of aortic bodies in anesthetized dogs. J. Appl. Physiol. 30:488.
99. Peterson, D.F., and Brown, A.M. (1971). Pressor reflexes produced by stimulation of afferent fibers in the cardiac sympathetic nerves of the cat. Circ. Res. 28:605.
100. Malliani, A., Peterson, D.F., Bishop, V.S., and Brown, A.M. (1972). Spinal sympathetic cardiocardiac reflexes. Circ. Res. 30:158.
101. Schwartz, P.J., Pagani, M., Lombardi, F., Malliani, A., and Brown, A.M. (1973). A cardiocardiac sympathovagal reflex in the cat. Circ. Res. 32:215.
102. Lioy, F., Malliani, A., Pagani, M., Recordati, G., and Schwartz, P.J. (1974). Reflex hemodynamic responses initiated from the thoracic aorta. Circ. Res. 34:78.
103. Lindgren, I., and Olivecrona, H. (1947). Surgical treatment of angina pectoris. J. Neurosurg. 4:19.
104. Sato, A., and Schmidt, R.F. (1973). Somatosympathetic reflexes: Afferent fibers, central pathways, discharge characteristics. Physiol. Rev. 53:916.
105. Holmes, R., and Torrance, R.W. (1959). Afferent fibres of the stellate ganglion. Q. J. Exp. Physiol. 44:271.
106. Fegler, J. (1933). Recherches sur l'innervation sensitive antagoniste des voies respoiratoires inférieures. C. R. Soc. Biol. (Paris) 113:207.
107. Nishi, K., Sakanashi, M., and Takenaka, F. (1974). Afferent-fibres from pulmonary arterial baroreceptors in the left cardiac sympathetic-nerve of the cat. J. Physiol. (Lond.) 240:53.
108. Gauer, O.H., Henry, J.P., and Behn, C. (1970). The regulation of extracellular fluid volume. Annu. Rev. Physiol. 32:547.
109. Henry, J. P., Gauer, O.H., and Reeves, J.L. (1956). Evidence of the atrial location of receptors influencing urine flow. Circ. Res. 4:85.
110. Baïsset, A., and Montastruc, P. (1957). Polyurie par distension auriculaire chez le chien: rôle de l'hormone antidiurétique. J. Physiol. (Paris) 49:33.

111. Johnson, J.A., Moore, W.W., and Segar, W.E. (1969). Small changes in left atrial pressure and plasma antidiuretic hormone titers in dogs. J. Appl. Physiol. 217:210.
112. Lydtin, H., and Hamilton, W.F. (1964). Effect of acute changes in left atrial pressure on urine flow in unanesthetized dogs. Amer. J. Physiol. 207:530.
113. Zehr, J.E., Johnson, J.A., and Moore, W.W. (1969). Left atrial pressure, plasma osmolality, and ADH levels in the unanesthetized ewe. Amer. J. Physiol. 217:1672.
114. Kappagoda, C.T., Linden, R.J., and Snow, H.M. (1972). The effect of distending the atrial appendages on urine flow in the dog. J. Physiol. (Lond.) 227:233.
115. Kappagoda, C.T., Linden, R.J., and Snow, H.M. (1973). Effect of stimulating right atrial receptors on urine flow in the dog. J. Physiol. 235:493.
116. Carswell, F., Hainsworth, R., and Ledsome, J.R. (1970). The effects of left atrial distension upon urine flow from the isolated perfused kidney. Q. J. Exp. Physiol. 55:173.
117. Kappagoda, C.T., Linden, R.J., Snow, H.M., and Whitaker, E.M. (1974) Left atrial receptors and the antidiuretic hormone. J. Physiol. (Lond.) 237:663.
118. Kappagoda, C.T., Linden, R.J., Snow, H.M., and Whitaker, E.M. (1975). Effect of destruction of the posterior pituitary on the diuresis from left atrial receptors. J. Physiol. (Lond.) 244:757.
119. Goetz, K.L., Bond, G.C., Hermreck, A.S., and Trank, J.W. (1970). Plasma ADH levels following a decrease in mean atrial transmural pressure in dogs. Amer. J. Physiol. 219:1424.
120. Share, L., and Levy, M.N. (1962). Cardiovascular receptors and blood titer of antidiuretic hormone. Amer. J. Physiol. 203:425.
121. Clark, B.J., and Rocha E Silva. M., Jr. (1967). An afferent pathway for the selective release of vasopressin in response to carotid occlusion and haemorrhage in the cat. J. Physiol. (Lond.) 191:529.
122. Harris, M.C., and Spyer, K.M. (1973). Inhibition of ADH release by stimulation of afferent cardiac branches of the right vagus in cats. J. Physiol. (Lond.) 231:15P.
123. Share, L., and Levy, M.N. (1966). Carotid sinus pulse pressure, a determinant of plasma antidiuretic hormone concentration. Amer. J. Physiol. 211:721.
124. Share, L., and Levy, M.N. (1966). Effect of carotid chemoreceptor stimulation on plasma antidiuretic hormone titer. Amer. J. Physiol. 210:157.
125. Bond, G.C., and Trank, J.W. (1970). Effect of bilateral aortic nerve section on plasma ADH titer (abstract). Physiologist 13:152.
126. Gilmore, J.P., and Daggett, W.M. (1966). Response of the chronic cardiac denervated dog to acute volume expansion. Amer. J. Physiol. 210:509.
127. Gilmore, J. P., and Michaelis, L.L. (1968). Diuresis and natriuresis during volume expansion following cardiac reinnervation. Proc. Soc. Exp. Biol. Med. 128:645.
128. Arndt, J.O., Reineck, H., and Gauer, O.H. (1963). Ausscheidungsfunktion und Hämodynamik der Nieren bei Dehnung des liken Vorhofes am narkotisierten Hund. Pfluegers Arch. Ges. Physiol. 277:1.
129. Ledsome, J.R., and Mason, J.M. (1972). The effects of vasopressin on the diuretic response to left atrial distension. J. Physiol. (Lond.) 221:427.
130. Gillespie, D.J., Sandberg, R.L., and Koike, T.I. (1973). Dual effect of left atrial receptors on excretion of sodium and water in the dog. Amer. J. Physiol. 225:706.
131. Knox, F.G., Davis, B.B., and Berliner, R.W. (1967). Effect of chronic cardiac denervation on renal response to saline infusion. Amer. J. Physiol. 213:174.
132. Karim, F., Kidd, C., Malpus, C.M., and Penna, P.E. (1972). The effects of stimulation of the left atrial receptors on sympathetic efferent nerve activity. J. Physiol. (Lond.) 227:243.
133. McDonald, K.M., Rosenthal, A., Schrier, R.W., Galicich, J., and Lauler, D.P. (1970). Effect of interruption of neural pathways on renal response to volume expansion. Amer. J. Physiol. 218:510.
134. Mancia, G., Romero, J.C., and Shepherd, J.T. (1975). Continuous inhibition of renin release in dogs by vagally innervated receptors in the cardiopulmonary region. Circ. Res. 36:529.
135. Vander, A.J., and Geelhoed, G.W. (1969). Inhibition of renin secretion by angiotensin II. Proc. Soc. Exp. Biol. Med. 120:399.
136. Bunag, R.D., Page, I.H., and McCubbin, J.W. (1966). Neural stimulation of release of renin. Circ. Res. 19:851.

137. Skinner, S.L., McCubbin, J.W., and Page, I.H. (1964). Control of renin secretion. Circ. Res. 15:64.
138. Schmid, H.E., Jr. (1972). Renal autoregulation and renin release during changes in renal perfusion pressure. Amer. J. Physiol. 222:1132.
139. Hodge, R.L., Lowe, R.D., and Vane, J.R. (1966). Increased angiotensin formation in response to carotid occlusion in the dog. Nature 211:491.
140. Brennan, L.A., Henninger, A.L., Jochim, K.E., and Malvin, R.L. (1974). Relationship between carotid sinus pressure and plasma renin level. Amer. J. Physiol. 227:295.
141. Thames, M.D., Zubair-UL-Hassan, Brackett, N.C., Jr., Lower, R.R., and Kontos, H.A. (1971). Plasma renin responses to hemorrhage after cardiac autotransplantation. Amer. J. Physiol. 221:1115.
142. Brennan, L.A., Jr., Malvin, R.L., Jochim, K.E., and Roberts, D.E. (1971). Influence of right and left atrial receptors on plasma concentrations of ADH and renin. Amer. J. Physiol. 221:273.
143. Barajas, L., and Latta, H. (1967). Structure of the juxtaglomerular apparatus. Circ. Res. 21 (Suppl. 2): 15.
144. Wågermark, J., Ungerstedt, U., and Ljungqvist, A. (1968). Sympathetic innervation of the juxtaglomerular cells of the kidney. Circ. Res. 22:149.
145. Coote, J.H., Johns, E.J., Macleod, V.H., and Singer, B. (1972). Effect of renal nerve stimulation, renal blood flow and adrenergic blockage on plasma renin activity in the cat. J. Physiol. (Lond.) 226:15.
146. Johnson, J.A., Davis, J.O., and Witty, R.T. (1971). Effects of catecholamines and renal nerve stimulation on renin release in the non-filtering kidney. Circ. Res. 29:646.
147. La Grange, R.G., Sloop, C.H., and Schmid, H.E. (1973). Selective stimulation of renal nerves in the anesthetized dog: effect on renin release during controlled changes in renal hemodynamics. Circ. Res. 33:704.
148. Anderson, C.H., McCally, M., and Farrell, G.L. (1959). The effects of atrial stretch on aldosterone secretion. Endocrinology 64:202.
149. Daly, M. de B., and Ungar, A. (1966). Comparison of the reflex responses elicited by stimulation of the separately perfused carotid and aortic body chemoreceptors in the dog. J. Physiol. (Lond.) 182:379.
150. Angell-James, J.E., and Daly, M. de B. (1969). Cardiovascular responses in apnoeic asphyxia: role of arterial chemoreceptors and the modification of their effects by a pulmonary vagal inflation reflex. J. Physiol. (Lond.) 201:87.
151. Angell-James, J.E., and Daly, M. de B. (1973). The interaction of reflexes elicited by stimulation of carotid body chemoreceptors and receptors in the nasal mucosa affecting respiration and pulse interval in the dog. J. Physiol. (Lond.) 229:133.
152. Mancia, G., Shepherd, J.T., and Donald, D.E. Unpublished data.
153. Mancia, G. (1975). Influence of carotid baroreceptors on vascular responses to carotid chemoreceptor stimulation in dog. Circ. Res. 36:270.
154. Gorfinkel, H.J., Szidon, J.P., Hirsch, L.J., and Fishman, A.P. (1972). Renal performance in experimental cardiogenic shock. Amer. J. Physiol. 222:1260.
155. Hanley, H.G., Raizner, A.E., Inglesby, T.V., and Skinner, N.S., Jr. (1972). Response of the renal vascular bed to acute experimental coronary arterial occlusion. Amer. J. Cardiol. 29:803.
156. Chalmers, J.P., Korner, P.I., and White, S.W. (1967). Effects of haemorrhage on the distribution of the peripheral blood flow in the rabbit. J. Physiol. (Lond.) 192:561.
157. Öberg, B., and White, S. (1970). The role of vagal cardiac nerves and arterial baroreceptors in the circulatory adjustments to hemorrhage in the cat. Acta Physiol. Scand. 80:395.
158. Pelletier, C.L., Edis, A.J., and Shepherd, J.T. (1971). Circulatory reflex from vagal afferents in response to hemorrhage in the dog. Circ. Res. 29:626.
159. Öberg, B., and Thorén, P. (1971). Increased activity in left ventricular receptors during hemorrhage or occlusion of caval veins in the cat: A possible cause of the vaso-vagal reaction. Acta Physiol. Scand. 85:164.
160. Ott, N.T., Lorenz, R.R., and Shepherd, J.T. (1972). Modification of lung-inflation reflex in rabbits by hypercapnia. Amer. J. Physiol. 223:812.
161. Ott, N.T., and Shepherd, J.T. (1973). Modifications of the aortic and vagal depressor reflexes by hypercapnia in the rabbit. Circ. Res. 33:160.

162. Folkow, B., Johansson, B., and Lofving, B. (1961). Aspects of functional differentiation of the sympatho-adrenergic control of the cardiovascular system. Med. Exp. (Basel) 4:321.
163. Schoener, E.P., and Frankel, H.M. (1972). Effect of hyperthermia and Pa_{CO_2} on the slowly adapting pulmonary stretch receptor. Amer. J. Physiol. 222:68.
164. Estarillo, J., and Burger, R.E. (1973). Avian cardiac receptors: activity changes by blood pressure, carbon dioxide, and pH. Amer. J. Physiol. 225:1067.
165. Toubes, D.B., and Brody, M.J. (1970). Inhibition of reflex vasoconstriction after experimental coronary embolization in the dog. Circ. Res. 26:211.
166. Hanley, H.G., Costin, J.C., and Skinner, N.S., Jr. (1971). Differential reflex adjustments in cutaneous and muscle vascular beds during experimental coronary artery occlusion. Amer. J. Cardiol. 27:513.
167. Thorén, P. (1973). Evidence for a depressor reflex elicited from left ventricular receptors during occlusion of one coronary artery in the cat. Acta Physiol. Scand. 88:23.
168. Peterson, D.F., and Bishop, V.S. (1974). Reflex blood pressure control during acute myocardial ischemia in the conscious dog. Circ. Res. 34:226.
169. Agress, C.M., and Binder, M.J. (1957). Cardiogenic shock. Amer. Heart J. 54:458.
170. Kezdi, P., Kordenat, R.K., and Misra, S.N. (1974). Reflex inhibitory effects of vagal afferents in experimental myocardial infarction. Amer. J. Cardiol. 33:853.
171. Kolatat, T., Ascanio, G., Tallarida, R.J., and Oppenheimer, M.J. (1967). Action potentials in the sensory vagus at the time of coronary infarction. Amer. J. Physiol. 213:71.
172. Recordati, G., Schwartz, P.J., Pagani, M., Malliani, A., and Brown, A.M. (1971). Activation of cardiac vagal receptors during myocardial ischemia. Experientia 27:1423.
173. Zucker, I.H., and Gilmore, J.P. (1974). Atrial receptor discharge during acute coronary occlusion in the dog. Amer. J. Physiol. 227:360.
174. Peterson, D.F., Kaspar, R.L., and Bishop, V.S. (1973). Reflex tachycardia due to temporary coronary occlusion in the conscious dog. Circ. Res. 32:652.
175. Malliani, A., Schwartz, P.J., and Zanchetti, A. (1969). A sympathetic reflex elicited by experimental coronary occlusion. Amer. J. Physiol. 217:703.
176. Cox, W.V., and Robertson, H.F. (1936). The effect of stellate ganglionectomy on the cardiac function of intact dogs and its effect on the extent of myocardial infarction and on cardiac function following coronary artery occlusion. Amer. Heart J. 12:285.
177. Yodice, A. (1941). Sympathectomy and experimental occlusion of a coronary artery. Amer. Heart J. 22:545.
178. Harris, A.S., Estandia, A., and Tillotson, R.F. (1951). Ventricular ectopic rhythms and ventricular fibrillation following cardiac sympathectomy and coronary occlusion. Amer. J. Physiol. 165:505.
179. Ebert, P.A., Allgood, R.J., and Sabiston, D.C., Jr. (1968). The anti-arrhythmic effects of cardiac denervation. Ann. Surg. 168:728.
180. Corr, P.B., and Gillis, R.A. (1974). Role of the vagus nerves in the cardiovascular changes induced by coronary occlusion. Circulation 49:86.
181. Kent, K.M., Smith, E.R., Redwood, D.R., and Epstein, S.E. (1973). Electrical stability of acutely ischemic myocardium influences of heart-rate and vagal stimulation. Circulation 47:291.
182. Frankstein, S.I., and Sergeeva, Z.N. (1966). Tonic activity of lung receptors in normal and pathological states (letter to the editor). Nature 210:1054.
183. Paintal, A.S. (1955). Impulses in vagal afferent fibres from specific pulmonary deflation receptors. The response of these receptors to phenyl diguanide, potato starch, 5-hydroxytryptamine and nicotine, and their role in respiratory and cardiovascular reflexes. Q. J. Exp. Physiol. 40:89.
184. Deshpande, S.S., and Devanandan, M.S. (1970). Reflex inhibition of monosynaptic reflexes by stimulation of type J pulmonary endings. J. Physiol. (Lond.) 206:345.
185. Paintal, A.S. (1970). The mechanism of excitation of type J receptors, and the J reflex. In R. Porter (ed.), Breathing: Hering-Breuer Centenary Symposium, pp. 59–76. J. & A. Churchill, London.
186. Kalia, M. (1969). Cerebral pathways in reflex muscular inhibition from type J pulmonary receptors. J. Physiol. (Lond.) 204:92P.

187. Wood, P. (1963). Polyuria in paroxysmal tachycardia and paroxysmal atrial flutter and fibrillation. Brit. Heart J. 25:273.
188. Ghose, R.R., Joekes, A.M., and Kyriacou, E.H. (1965). Renal response to paroxysmal tachycardia. Brit. Heart J. 27:684.
189. Luria, M.H., Adelson, E.I., and Lochaya, S. (1966). Paroxysmal tachycardia with polyuria. Ann. Intern. Med. 65:461.
190. Kilburn, K.H. (1964). Fluid volume control and induced arrhythmias (abstract). Clin. Res. 12:186.
191. Goetz, K.L., and Bond, G.C. (1973). Reflex diuresis during tachycardia in the dog: Evaluation of the role of atrial and sinoaortic receptors. Circ. Res. 32:434.
192. Zehr, J.E., Hawe, A., Tsakiris, A.G., Rastelli, G.C., McGoon, D.C., and Segar, W.E. (1971). ADH levels following nonhypotensive hemorrhage in dogs with chronic mitral stenosis. Amer. J. Physiol. 221:312.
193. Greenberg, T.T., Richmond, W.H., Stocking, R.A., Gupta, P.D., Meehan, J.P., and Henry, J.P. (1973). Impaired atrial receptor responses in dogs with heart failure due to tricuspid insufficiency and pulmonary artery stenosis. Circ. Res. 32:424.
194. Zoller, R.P., Mark, A.L., Abboud, F.M., Schmidt, P.G., and Heistad, D.D. (1972). The role of low pressure baroreceptors in reflex vasoconstrictor responses in man. J. Clin. Invest. 51:2967.
195. Roddie, I.C., Shepherd, J.T., and Whelan, R.F. (1957). Reflex changes in vasoconstrictor tone in human skeletal muscle in response to stimulation of receptors in a low-pressure area of the intrathoracic vascular bed. J. Physiol. (Lond.) 139:369.
196. Johnson, J.M., Rowell, L.B., Niederberger, M., and Eisman, M.M. (1974). Human splanchnic and forearm vasoconstrictor responses to reductions of right atrial and aortic pressures. Circ. Res. 34:515.
197. Mark, A.L., Kioschos, J.M., Abboud, F.M., Heistad, D.D., and Schmid, P.G. (1973). Abnormal vascular responses to exercise in patients with aortic stenosis. J. Clin. Invest. 52:1138.

International Review of Physiology
Cardiovascular Physiology II, Volume 9
Edited by Arthur C. Guyton and Allen W. Cowley
Copyright 1976   University Park Press   Baltimore

4
Neurohormonal Control
of Plasma Volume

O. H. GAUER and J. P. HENRY
Free University of Berlin and University of Southern California, Los Angeles

DEFINITION OF VOLUME CONTROL

In 1951 the authors put forward the hypothesis that blood volume is regulated as an independent parameter through stretch receptors in the intrathoracic circulation (1). Since then numerous groups have contributed to this topic of the reflex mechanisms underlying volume control. In 1956 (2), 1963 (3), and again in 1970 (4) we reviewed the work. Our purpose is to update these older reviews, while restating the main lines of evidence that led to the initial hypothesis. In what follows the various chapters of the volume control story are presented as they developed and answers to a number of the problems raised by the hypothesis are attempted.

When speaking of blood volume control we do not imply the rigid homeostatic maintenance of a fixed volume, but the continuous adjustment of blood volume to the changing size of the vascular bed so that at all times an adequate "fullness of the blood stream" (5) is available to the left ventricle. Ernst H. Weber (6) was the first to recognize that this is the basic requirement for a functioning circulation. The final successful understanding of blood volume control mechanisms will comprise a knowledge of the control of erythrocyte volume, protein mass, and electrolyte fluid volume of the plasma. So far the regulation of blood cell volume and protein mass in response to a volume change has resisted efforts to assess them. However, the regulation of plasma fluid volume has proved easier, the reason being that as part of the extracellular fluid volume (ECFV) it is accessible to quantitative analysis for it is axiomatic that the body fluid volume results from the balance of fluid uptake and fluid loss through skin, lungs, and kidney.

The thirst mechanism and the excretory function of the kidney together regulate plasma volume by retaining or excreting ECFV. On the other hand, when the ECFV is held constant then the organism still has the option of increasing or decreasing plasma volume at the expense of the interstitial fluid volume by changing the filtration pressure in the microcirculation. But here the possibilities are limited because the interstitial space is an overflow basin or a reservoir of restricted capacity, which varies with the state of hydration.

MECHANISMS BY WHICH THE FILLING PRESSURE OF THE HEART CAN BE CONTROLLED

Figure 1 depicts the possible pathways for the control of blood volume. Let us follow the rectangular frame of the diagram and, starting in the right upper corner, describe the events following an enforced increase in blood volume. The increase in blood volume will involve an increase of plasma fluid volume and ECFV. At a given distensibility of the circulation the filling pressure of the heart and hence cardiac output will rise. As long as total peripheral resistance (TPR) remains constant, arterial blood pressure will increase and induce an increase in urine excretion. Provided fluid uptake remains constant, the ECFV will decrease. Thus blood volume will finally become normal again although with a raised hematocrit.

This scheme will certainly function when there are great changes of blood volume. However, there is now good evidence that under normal conditions the filling pressure of the heart is monitored through mechanoreceptors which control renal function (Figure 1 (II)) and influence the thirst mechanism (Figure 1 (I)) via the autonomic nervous system. Finally by affecting capillary filtration pressure this reflex shifts the dividing line between the interstitial fluid volume and the plasma volume to the left or to the right (Figure 1 (III)). Working through routes I, II, and III the mechanoreceptors responding to subtle changes in the cardiac filling pressure will induce corrections in plasma volume before the previously mentioned less sensitive mechanism that influences arterial pressure and urine flow via cardiac output has taken effect. Only when the volume gain is large enough to affect cardiac output, mean arterial pressure, or pulse pressure significantly will the arterial baroreceptors contribute to the above reflex control mechanism. This is indicated by the thin *dotted line* connecting the arterial pressure with receptors and the CNS.

Certain conditions must be fulfilled in order for these low pressure monitoring mechanisms to function:

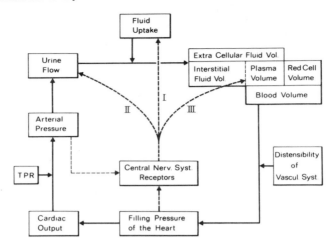

Figure 1. Pathways for the regulation of extracellular fluid volume and plasma volume by a mechanical feedback mechanism and by reflex mechanisms involving the autonomic nervous system. Both mechanisms can only function if the filling pressure of the heart is a well defined function of blood volume. Arrows I and II signify the major pathways by which extracellular fluid volume is controlled; arrow III indicates that with a constant extracellular fluid volume, plasma volume can still be changed by shifting the dividing line between plasma volume and interstitial fluid volume to the left or to the right (for details see text).

1. A well defined compliance relating blood volume and the filling pressure of the low pressure system must exist.
2. Receptors must be found in appropriate locations which should be capable of recording small changes of tension in the vascular walls of the low pressure system and discharge in appropriate areas of the CNS.
3. These areas must in turn influence efferent neurohormonal mechanisms controlling the three parameters vital for the control of plasma volume: thirst, renal function, and the redistribution of the ECFV.

The following discussion outlines the evidence that these criteria can be met.

Effective Compliance of Circulation

Inspection of Figure 1 shows that regardless of whether we subscribe to the classic theory of the mechanical feedback mechanism (7, 8) or to the hypothesis of reflex volume control involving receptors, the filling pressure of the low pressure system, which is closely related to the filling pressure of the heart, should be a clear-cut function of blood volume. When this problem first came to light, the need for an analysis of the distribution of the volume and distensibility of the various compartments of the circulation became apparent. It was found that from the point of view of volume control it would be advantageous to de-emphasize the anatomical distinction between a systemic and a lesser circulation and underscore the functional differences between the arterial system and the low pressure system.

Figure 2 accentuates the different size and distensibility of the two systems. The arterial system holds only 15% of the blood volume and is characterized by a very low distensibility. The high pressure in it is determined dynamically by cardiac output and peripheral flow resistance (TPR). The capacitance low pressure system preempts 85% of the blood volume and is highly distensible. It consists of the systemic veins, the pulmonary circulation, and the left heart in diastole. Normally the pressure is low and is basically determined by the interplay between a large capacity and the total blood volume.

When we put the animal in the horizontal position and arrest the heart a static pressure of approximately 9 cm H_2O will prevail and changes of blood

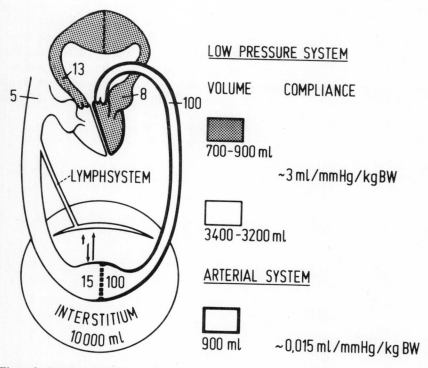

Figure 2. Distribution of blood volume and distensibility of the various compartments of the circulation. The arterial system holds approximately 15% of the estimated blood volume, and arterial pressure is entirely due to cardiac output and total peripheral resistance. The low pressure system consists of the systemic veins, the right heart, the pulmonary circulation, and the left ventricle in diastole. It holds 85% of the blood volume and the pressure is determined by the distensibility of the total vascular bed and the blood volume. The numbers in the various sections of the circulation indicate the mean pressures in mm Hg. On the right side the estimated volumes and compliances of the main compartments are indicated. The compliance of the arterial system is given at its working pressure of 100 mm Hg. The shaded area represents the filling reservoir of the left ventricle, studded with mechanoreceptors. (Gauer, O.H. (1972). Kreislauf des Blutes. *In* O.H. Gauer, K. Kramer, and R. Jung (eds.), Physiologie des Menschen, p. 234. (Urban and Schwarzenberg, Munich-Berlin-Vienna.)

volume will induce identical pressure changes in all compartments of the circulation. With the changing cardiac output of the active circulation and during orthostasis or the application of external pressures (i.e., pressure breathing or immersion in a bath), volume shifts will occur and the changes of pressure with hemorrhage and transfusion will no longer strictly parallel each other in the various compartments of the low pressure system. Thus the curve of response of pressure to changes of total blood volume is steeper in the left atrium than in the right (9).

Central venous pressure (CVP) was chosen as the reference parameter of the pressure-volume relationships of the total circulation, because this pressure plays a key role in the determination of stroke volume and represents the adequate stimulus for cardiac receptors during diastole (see below).

In a recent article by Echt et al. (10) the quotient Δblood volume/ΔCVP × kg BW was termed "Effective compliance of the circulation." The left atrial pressure might have been preferable because it more directly determines the performance of the main pump of the systemic circulation; however, it was not feasible to record it in the human subjects employed in this study. The results of this study as shown in Figure 3 demonstrate that the compliance of this compartment can be quite rigorously defined.

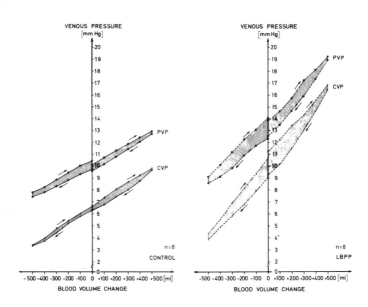

Figure 3. Changes in central venous pressure (*CVP*) and peripheral venous pressure (*PVP*) during a change in blood volume of ± 500 ml in a control group (*left*) and a group with lower body positive pressure (*LBPP*) (*right*); mean values from eight subjects. The experiments began with infusion; the sequence of events is indicated by the arrows; the full cycle was completed in 9 min. Note the change in compliance when the distension of the lower half of the vascular system is prevented by application of LBPP (see Echt et al. (10)).

It should be realized that the pressure-volume relationship defined as effective compliance not only reflects the elastic properties of the system but it may be modified by filtration processes in the microcirculation by redistribution of volume between circulatory compartments and possibly by "delayed compliance," i.e., the partial relaxation of the vessels that may occur after a pressure rise. In spite of these complicating factors, the effective compliance has remarkable constancy (Table 1). In normal supine man and in the dog we find the identical values of 2.3 ml/mm Hg/kg BW. In the upright posture, the dependent vascular bed is distended by the high hydrostatic pressure. At the same time the vessels in the upper half of the body are collapsed (Figure 4A). Because the pressure volume curve is nonlinear the gain in compliance of the cephalad portion of the circulation is not canceled by the loss of compliance in the tense dependent vascular bed. This explains the high effective compliance of 3.3 ml/mm Hg/kg BW in this posture. Confirming the suggestion of Glaser (cf. Echt et al. (10)) that during hemorrhage half of the blood volume lost comes from the intrathoracic pool, we find that the compliance of the intrathoracic compartment determined during LBPP represents about 50% of the compliance of the total circulation.

The question of a reflex modification of venomotor tone which supposedly results in a homeostatic control of "venous return" and prevents central venous pressure from falling following hemorrhage or orthostasis has been the subject of much debate (17–20). The bulk of evidence including direct measurements of peripheral venous tone indicates that as long as the blood loss does not exceed 10% of the estimated blood volume the peripheral capacitance vessels behave passively (17, 18). There may be changes in other parts of the capacitance system; however, the data show that they are clearly insufficient to stabilize the central venous pressure, which rises and falls predictably with such an asymptomatic nonhypotensive 10% change in blood volume. Thus continuous recording of CVP for 1 hour during and following a rapid moderate hemorrhage showed that after an initial overshoot, lasting 10–15 min, the pressure assumed a lowered level, which was maintained constant for the rest of the 50-min observation period (11).

Hemorrhage of 15–20% is a different matter and will lead to constriction of the splanchnic circulation. Such constriction was only one of the signs of circulatory emergency exhibited by the subjects of a recent study in which blood loss of this extent was employed (21). Others included pallor, emesis, dizziness, chest pain, and nausea. In spite of this evidence of visceral and peripheral venous constriction hepatic venous pressure did not return to normal.

Summary and Conclusions

1. Measurement of the effective compliance has shown that the distensibility of the circulation is constant as long as the changes of blood volume are moderate (< 10–15%). Thus the system effectively registers changes in blood volume as changes in vascular wall tension of the low pressure system.

Table 1. Comparison of effective vascular compliance and intrathoracic compliance measured in several different studies

Author	Reference	Compliance (ml/mm Hg kg^{-1} BW)	Animal
Effective vascular compliance			
Gauer et al.	11	2.7	Man, lateral decubitus position
Harlan et al.	12	2.3	Dog, arrested circulation (calculated data)
Ungewiss	13	2.3	Man recumbent
		3.5	Man essential hypertension
Shoukas and Sagawa	14	2.0–2.4	Dog constant perfusion of circulation
Echt et al.	10	2.3	Man recumbent
		1.7	Man norepinephrine infusion
Strey et al.	15	2.7	Man recumbent
Koubenec et al.	16	3.3	Man sitting upright in air
		2.4	Man sitting upright immersed to xiphoid
		1.9	Man sitting upright immersed to neck
Intrathoracic compliance			
Harlan et al.	12	1.0	Dog (calculated data)
Echt et al.	10	1.2	Man, lower body positive pressure
		0.9	Man, norepinephrine infusion and lower body positive pressure
Strey et al.	15	0.9	Man

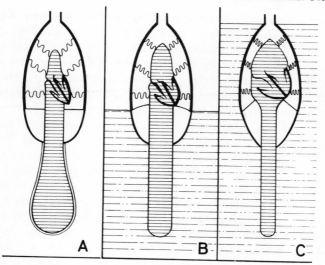

Figure 4. Effect of postural changes and immersion to the neck on the distribution of blood volume in the low pressure system. *A*, orthostasis; *B*, immersion to the diaphragm; *C*, immersion to the neck. Immersion to the diaphragm (*B*) corresponds to the condition in the supine posture. The difference of intrathoracic blood volume between *A* and *C* is approximately 700 ml (Gauer and Thron (17)).

2. The compliances of the extrathoracic and intrathoracic compartments are about equal. However, since the intrathoracic compartment has a considerably smaller volume than the peripheral capacitance vessels, its distensibility must be much greater. Receptors in the walls of the intrathoracic circulation should therefore be ideally suited to record changes in blood volume. For the benefit of the subsequent discussion it should be pointed out that the atria are the most distensible part of the intrathoracic vascular bed.

3. The persistence of a reduced CVP after moderate hemorrhage shows that the system does not restore CVP to normal by an acute constriction of the peripheral capacitance vessels. The evidence indicates that the normal filling state of the circulation is regained by the reflex control of volume, rather than by an increase in venous tone.

Afferent Limb for a Volume Control Reflex

In order to control either volume or pressure it must be measured by stretch receptors in the walls of the vessels. There are two possibilities. The first is that the receptors are uniformly distributed throughout the vascular bed and their input is integrated somewhere in the central nervous system. The second possibility is that an appropriate area monitors the whole system. In the case of the high pressure system the carotid and aortic baroreceptor areas point to the second choice. The results of the search for volume sensitive regions have pointed in the same direction.

Functional Logic of Stretch Receptors in Walls of Heart Chambers   While little is known about stretch receptors in the peripheral circulation, all four chambers of the heart are densely populated with various kinds of receptors (22–27). Neuroanatomists have been aware of their existence for many years and they have been held responsible for reflex changes of heart rate and peripheral flow resistance (23). In addition, due to their great distensibility, the atria appear ideally suited to signal the filling of the low pressure system in its strategically most important section. If this hypothesis is correct it should be possible to produce the diuresis of isotonic blood volume expansion not only by a blood transfusion but by any measure leading to increased filling and so to a stretching of the intrathoracic vascular compartments at the expense of the peripheral capacitance vessels. The most innocuous method of increasing intrathoracic blood volume is by the application of mild constant negative pressure breathing. This procedure induces an increase in urine flow in anesthetized dogs (28) as well as in man (29, 30). In most cases this is a water diuresis. This observation together with the slow onset of the effect and its persistence for several minutes after cessation of the stimulus suggested that a hormonal factor, probably ADH, might be involved. The problems surrounding the identification of the hormone are discussed later. The participation of ADH, the fluid regulating hormone par excellence in the reflex, would strongly support the idea that the observed diuresis is the expression of a true volume regulatory effort of the organism and is not simply due to a random spread of the effects on the renal circulation of stimulating a depressor reflex.

A review of the literature shows that all measures which induce a change of intrathoracic blood volume produce the expected effect on the excretory function of the kidney (Table 2).

The attempt to determine the volume-sensitive region more precisely was guided by the idea that the most distensible areas would be best suited to monitor small changes of intrathoracic blood volume. These are the atria. A cherry-sized inflatable balloon was put into the left atrial appendix under chloralose anesthesia and the chest was closed. Inflation of the balloon for 30–45 min produced clear-cut diuresis (31).

Autopsy and x-ray observation revealed that the balloon had moved out of the appendix into the atrium, causing a functional mitral stenosis. The pressures were elevated between the pulmonary valve and the mitral orifice and, although the balloon did not permit the hoped-for sharp delineation of the volume-sensitive area, it was remarkable that the diuresis occurred in spite of a considerable impairment of the pumping action of the heart leading to a reduction in cardiac output (31).

Procedures designed to induce a stepwise engorgement of the pulmonary circulation caused no effect other than a reduction in cardiac output, which if sufficiently severe was actually accompanied by an oliguria rather than the expected diuresis (31). Recently the Leeds group found they could induce a diuresis by inflating miniature balloons in the pulmonary vein left atrial junction or in the left and right atrial appendixes. These procedures allow an exact

Table 2. Measures changing intrathoracic blood volume[a]

1 Increase	2 Decrease
Blood transfusion	Blood loss
Constant negative pressure breathing NPB	Constant positive pressure breathing PPB
Transition to the horizontal posture; bed rest; weightlessness	Orthostasis; increased gravitational stress (human centrifuge)
Immersion in a waterbath of thermoneutral temperature	Venous tourniquets; vena cava ligation
Lower body positive pressure; LBPP	Lower body negative pressure; LBNP
Exposure to cold	Exposure to heat

[a]For references see Gauer et al. The procedures under column 1 induce diuresis and natriuresis, under column 2 oliguria.

location of the sensitive area with practically no impairment of the general circulation (32, 33). Their results confirm the earlier observations implicating stretch receptors in the atriovenous junctions in the response to volume change.

The water diuresis originating from a distension of the left atrium was termed by Homer Smith the Henry-Gauer reflex (34). This was perhaps unfortunate for it attached a specific significance to this heart chamber and to a reflex reduction in ADH secretion. But although impressed by the paramount importance of the filling pressure of the main power source of the circulation, the left ventricle, we have always considered that all chambers of the heart participate in volume control. Convinced of the all-important role of sodium in the control of volume, Homer Smith was reluctant to accept the evidence of a water diuresis as the expression of a true volume regulatory mechanism of the organism. As is shown later this requirement for control of sodium excretion in relation to cardiac receptors is now being met by the work of various investigators.

Receptors, Afferent Pathways, Central Representation

Receptors In 1937 Nonidez (23) described unencapsulated neural elements in the heart, which have great similarity to the complex mechanoreceptors found in other organs, e.g., in the skin. They are densely distributed over the intrapericardial part of the great veins and both atria. Nonidez suggested that they might be responsible for the Bainbridge reflex. They discharge into myelinated fibers in the vagus nerve. With the help of single fiber recordings, Paintal (26, 27) was able to distinguish between type A and type B receptors. The A receptors fire during atrial systole, while the B receptors become activated during the passive atrial stretch of ventricular diastole. They respond with great sensitivity to any change of transmural pressure and seem ideally suited to record the filling pressure of the heart. Arndt et al. (35) have recently shown that when exposed to the same sinusoidal stretch stimuli A and B receptors have identical response patterns. They concluded that their different behavior in situ is due to a variation in precise anatomic location and their exact arrangement in the tissues of the atria. In 1964 Coleridge et al. (36) described two types of afferent endings in the ventricular wall of the dog.

Besides the complex unencapsulated endings which predominate at the junctions of the atria and great veins, there is an endnet of fine unmyelinated fibers relatively uniformly distributed through the atria as well as through the ventricles (22, 25, 32, 33). Their exact function is not yet known but there is now evidence that the ventricular endnet activity is proportional to the degree of filling of the ventricles (37, 38). Preliminary information from Öberg's laboratory (39) indicates that sodium excretion is effected by stimulating the nerves to these receptors.

Afferent Pathways   The main afferent nerve carrying signals from cardiac receptors is the vagus. Cooling below 8°C interrupts the signal flow from atrial receptors and at the same time eliminates the diuresis that accompanies left atrial distension (40). More recent experiments showed that interruption of vagal pathways induces an increase in both the ADH (41) and renin (42) activity in the blood. This direct demonstration of the existence of a tonic inhibition of the water- and salt-retaining mechanisms by cardiac receptor activity is of great theoretical and practical importance for the concept of volume control (see p. 164).

This finding does not preclude the existence and importance of other afferent pathways conducting mechanoreceptor signals. This includes the well known painfibers of the heart in the anterior sympathetic roots of TH^1-Th^5 (43).

Central Representation   There is preliminary evidence of neural activity in the medulla oblongata synchronous with the heart beat. Some of this originates from cardiac receptors. Thus recordings in the nucleus of the tractus solitarius (44) together with the observation that mechanical stimulation of the various heart chambers leads to distinctive reflex changes of heart rate and peripheral vascular tone (33, 45) suggest that the afferent fibers connect with the reticular formation of the medulla oblongata.

Cutting the neural connections between the diencephalon and the midbrain eliminates the increased release of vasopressin in response to hemorrhage, while an osmotic stimulus remains effective (46). Section of the spinal cord at the level of C-1 also interferes with the hemorrhage-induced response of vasopressin, demonstrating that peripheral receptors are responsible.

Recently Menninger and Frazier (47) have investigated the central neurophysiological aspects of ADH release. Considering the difficulties, they obtained remarkable results. Recording with electrodes in the hypothalamus from a total of 162 single units in the supraoptic, the paraventricular, and the anterior and lateral nuclei of cats, they studied the response to two effective stimuli, i.e., they contrasted periods of left atrial stretch by a balloon raising the pressure to a physiological 6–12 cmH_2O with an intracarotid infusion of 1 ml of 1 M NaCl. This latter constitutes a vigorous osmotic stimulus. A third stimulus was used, i.e., a change of blood volume by approximately 10%. Under the conditions of this experiment it was, however, almost too weak to be effective. Their overall criterion for response was a 20% change from the control values; 16% of the cells qualified as responding to the atrial balloon and 70% to the osmotic stimulus. If, however, a 10% criterion had been used, 40% of the hypothalamic units would have been counted as responding to the mechanical stimulus.

All cells sensitive to atrial stretch were also sensitive to the osmotic stimulus. Some responded in the same and some in the opposite direction. The complex networks into which these cells are integrated have so many feedback loops that the direction of response cannot be predicted. Menninger and Frazier's successful documentation of a neural connection between cardiac receptors and hypothalamic units is compatible with the mechanoreceptor hypothesis but a number of questions remain unanswered. For example: Are the units which respond directly connected with the neurohypophysis? Are they related to the "osmoreceptor cells" of Verney or the osmoreceptor fibers originating in the liver (48)?

Summary The dense population of mechanoreceptors in the walls of the intrathoracic circulation, combined with the great distensibility of this part of the low-pressure system, provides theoretical grounds for expecting that stretch receptors involved in volume control would be found predominantly in this region. This implies that the diuresis of iso-osmotic expansion of the total blood volume should be triggered by intrathoracic receptors. The hypothesis that a local distension of the critical region by techniques for redistributing blood volume, i.e., NPB, change of posture, immersion, etc. should produce a diuresis indistinguishable from the diuresis of total blood volume expansion was confirmed.

By the use of invasive methods, it was possible to narrow down the volume-sensitive area to the heart and especially the left atrium.

The vagus nerve was identified as the main afferent pathway. Finally, it could be shown that mechanical stimulation of atrial receptors evoked activity in appropriately located single units of the hypothalamus.

TECHNIQUE OF WHOLE BODY IMMERSION

Despite the impressive effects of negative pressure breathing or inflation of a balloon in the left atrium on urine flow they were not accepted as proof of a true volume regulatory reflex mechanism for the following reasons (Smith) (34).

1. In individual cases the diuresis, which usually lasted for 30–45 min, declined before the stimulus was discontinued. It was argued that a reflex, which adapted so rapidly, could not lead to an effective long-term control of plasma volume.

2. Under the experimental conditions then being employed the diuresis was for the most part due to an increase in free water clearance. This neglected the doctrine that fluid volume is primarily determined by the control of sodium and that water follows passively.

At this stage it appeared indispensable to use volume stimuli of longer duration. In the search for a suitable technique whole body immersion in a bath of thermoneutral temperature was adopted. As in the case of negative pressure breathing the method produces an engorgement of the intrathoracic circulatory compartments. However, it is more effective and can be maintained for hours or even days, while moderate levels of negative pressure breathing become very disturbing after 1 hour.

In 1924, in the course of his elegant research on the physiological effects of baths of various temperatures, Bazett (49) found that immersion up to the neck in a bath of thermoneutral temperature may produce so profuse a diuresis that within 3–4 hours the hematocrit is significantly increased. This phenomenon remained without further intensive study for over 30 years, until the advent of the space age. Then in 1960 Graveline and Balke (50, 51) adopted immersion as a way to simulate the weightless condition. Immersion is a poor means to simulate the complex consequences of the weightless state with the exception of the cephalad redistribution of blood volume. Here the effects were similar except that the weightless state is even more potent (see below, p. 182). When a man walks into a bath of increasing depth, the amount of blood which is normally pooled in this dependent region (Figure 3A) will gradually be reduced as the intravascular hydrostatic pressure is balanced by the rising hydrostatic pressure on the body surface. With the water at the level of the diaphragm blood volume distribution is about equal to the distribution in the prone position (Figure 3B). As the water level rises above the diaphragm, the hydrostatic column between the water surface and the diaphragm will force blood volume into the intrathoracic space and the intrathoracic blood volume will increase at the expense of the extrathoracic volume. The blood pooling of orthostasis is now reversed (Figure 3C).

Hemodynamic Effects of Whole Body Immersion

Whole body immersion has developed into an invaluable tool for the exploration of volume control, and detailed studies have been made of its hemodynamic effects (52). Our data confirm and complement that of Lundgren's group (53). They show that the hydrostatic pressure in the bath causes the transfer of a considerable volume of blood into the thorax and in consequence there is an increase in CVP of approximately 12–18 mm Hg. Due to a compression of the thorax, the intrathoracic pressure measured as esophageal pressure rises from about 0.5 to 3.5 mm Hg. The resultant transmural filling pressure of the heart is increased by 8–13 mm Hg, and arterial pressure rises by the same amount (Table 3).

Because the physiological changes observed during immersion are thought to be initiated by a distension of the heart, especially of the atria, the changes in heart volume were recently determined by Lange et al. (54) using a biplane röntgenometric method. The individual changes of heart volume with the transition from standing upright in air to standing immersed varied widely between 120 and 320 ml. The mean value of 180 ± 62 ml is approximately 25% of the total blood volume shifted into the thorax (53). The volume change of the heart following a shift from the supine posture to immersion was 100 ± 69 ml. Assessment of the change in contour of the heart shadow indicates that the atria accommodate a lion's share of the extra blood taken up by the heart (Figure 5).

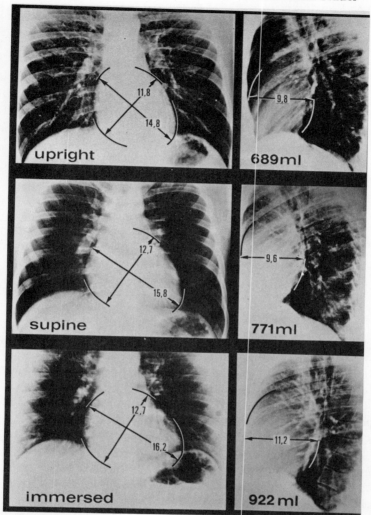

Figure 5. Effect of postural changes and immersion on heart volume; the volumes calculated from the diameters (cm) are given in the right-hand panels (see Lange et al. (54)).

Table 3 summarizes the primary and secondary effects of immersion of a standing subject on the circulation. It represents a compilation of Lundgren's (L) and our own group's (G) data.

The reflex effects on kidney function, filtration pressure in the microcirculation, and possibly the thirst mechanism, which together induce after due time a change in plasma volume, are discussed under "Response of Renal Function to Immersion" (p. 160).

Summary Immersion induces a very powerful distension of the heart, especially of the atria. A 30% increase of stroke volume and a rise of arterial

Table 3. Circulatory changes induced by whole body immersion[a]

Primary effects	
Central blood volume (L)[b]	+ 700 ml
Heart volume (G)[b]	+ 180 ml
Central venous pressure (G, L)	+ 12 to + 18 mm Hg
Intrathoracic pressure (G, L)	+ 4 to + 5 mm Hg
Transmural pressure (G, L)	+ 8 to + 13 mm Hg
Secondary effects	
Stroke volume (L)	+ 35%[c]
Cardiac output (L)	+ 32%[c]
Total peripheral resistance (L)	− 30%
Peripheral venous tone (G)	− 30%
Arterial pressure (L)	+ 10 mm Hg[d]

[a]Subject standing or sitting erect in air versus standing or sitting in water.

[b](G) Gauer et al.; (L) Lundgren et al.

[c]Heart rate was unchanged.

[d]The arteriovenous pressure gradient was not changed since CVP was increased by the same amount.

pressure by approximately 10 mm Hg are observed. A strong stimulation of stretch receptors in the circulation especially in the heart is to be expected, which will lead to reflex changes in cardiac performance, peripheral vascular tone, and body fluid volume control.

Response of Renal Function to Immersion

Identical Effects of Expansion of Intrathoracic Volume by Water Immersion and Massive Infusion of Saline In meticulously controlled experiments using a daily sodium uptake of 150 mEq, Epstein et al. (55) compared the effect on sodium and potassium excretion of whole body immersion of the upright seated subject with the effect of an expansion of the ECFV by a saline infusion of 2 liters administered in 2 hours to the same subject in the same position. The peak increment in sodium excretion ($V_{Na} \cdot V$) during immersion (177 ± 26 μEq/min) was indistinguishable from that of "seated saline" infusion (175 ± 29 μEq/min) (Figure 6). Similarly the kaliuretic response during immersion was identical with that induced by "seated saline." As was to be expected, saline infusion in the supine posture resulted in a peak $V_{Na} \cdot V$ which was significantly greater (313 ± 48 μEq/min). The data show that the volume stimulus of immersion is identical with that of an ECFV expansion induced by a 2-liter saline infusion in normal subjects assuming an identical seated posture.

These experiments elegantly and directly confirm that the decisive "volume stimulus" is a central hypervolemia. Referring to the attempts to assess the effects of a ECFV expansion by large saline transfusions on renal electrolyte

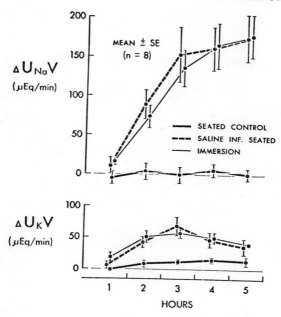

Figure 6. Comparison of the effects of immersion and acute saline infusion (2 liters/120 min) in the seated posture on the rates of sodium and potassium excretion in eight normal subjects. Data are expressed as the absolute changes from the preimmersion hour ($\Delta U_{Na}\dot{V}$ and $\Delta U_K\dot{V}$). Immersion resulted in significant increases in $\Delta U_{Na}\dot{V}$ (hours 2–5) and $\Delta U_K\dot{V}$ (hours 2–4), which were no different from the increases induced by saline infusion. (Epstein et al. (55)).

homeostasis and renin-aldosterone responsiveness in various clinical conditions, the authors suggest that whole body immersion may be the method of choice as an investigative tool for the study of the effects of volume expansion. This noninvasive technique has the advantage that it leaves the composition of the blood unaltered and avoids the complexities and hazards which accompany acute plasma expansion by massive saline infusions.

Influence of Degree of Hydration It has been suspected that the hydration régime plays a major role in the response of the kidney to volume stimuli. Therefore the following experiments were conducted (56) in two groups of seven and six subjects, respectively. For 3 days all received a standard diet containing 100 mEq of Na^+. Seven were well hydrated, the others were kept on the verge of thirst. To avoid the problem of the diurnal rhythm, the changes in water and mineral excretion with immersion were measured against the excretion at the same time of the preceding and following days. Figure 7 shows: (A) the diuresis is maintained throughout the period of immersion; in other words, the reflex does not adapt. (B) The response of the kidney depends on the degree of hydration. In the well hydrated subjects the diuresis of immersion is due to an increase in free water clearance. The average sodium excretion rose

162   Gauer and Henry

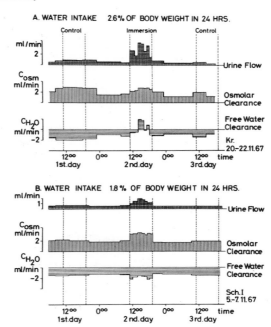

Figure 7. Urine flow, solute-obligated and solute-free water excretion in two subjects (Kr. and Sch. I) in two different states of hydration (A and B). Immersion is accompanied in both cases by an increase in urine flow which consists in an increase in free water clearance (C_{H_2O}) in the normally hydrated case (A) and in an increase of osmolar clearance (Co_{sm}) in the poorly hydrated one (B) (see Behn et al. (56)).

from a control of 118–180 mEq/hour/kg BW, or by 50%. In contrast the hydropenic subjects respond with an increase in the osmotic clearance and in this case sodium excretion rose from the control value of 60–152 mEq/hour/kg BW, or by 152%.

Redistribution of ECFV during Immersion

Comparison of the plasma volume reduction during immersion with the concurrent loss of total body fluid raised interesting questions regarding a reflex redistribution of ECFV between plasma volume and IFV. This is also probably initiated through cardiac mechanoreceptors. We found a decrease in plasma volume by approximately 10% in normally hydrated subjects and of 15% in slightly hydropenic subjects. The figures agree with those of previous workers.

Consideration of the interrelationships of body fluid compartments was greatly facilitated by the fact that during immersion the insensible water loss through the skin stops and fluid loss through expired air is insignificant because the subject's mouth is only 4–5 cm above a water surface of 34°C and so he is inhaling air fully saturated with water vapor. Therefore the only fluid loss during immersion is through the kidney. If we assume that urine volume is predominantly extracted from the total ECFV a comparison of ECFV reduction and

reduction in plasma volume is possible. Because the total plasma volume constitutes about one-fifth of the total ECFV one might expect the plasma volume
reduction to be about one-fifth of the urine volume excreted during the
immersion period. The remaining four-fifths would presumably be contributed
by the interstitial fluid volume. This reasoning fully applies to the findings in the
hydrated subjects. However, when dehydrated the plasma volume reduction is
greater than in the well hydrated subject while the urine volume excreted is
practically identical with that of the plasma volume. In some individuals, plasma
volume loss was even greater than the corresponding urine volume. The explanation would appear to be that in dehydration plasma volume control is predominantly achieved by an outward filtration of plasma fluid into the interstitial
space. As Öberg showed (57), cardiac distension is an effective way to reflexly
change the ratio of pre- to postcapillary resistance and hence filtration pressure.
This effect is mediated through the sympathetic nerves and moderated by
cardiac receptors; i.e., stimulation of cardiac receptors by injection of proto-
veratrine into the right atrium induced an increased fluid filtration from the
blood into the interstitial space.

Thirst and Intrathoracic Blood Volume

The control of fluid volume in the land animal depends on the voluntary
behavior of drinking in response to the urgent perception of thirst. The cellular
deficit of fluid as sensed by cells in the anterior diencephalon, the so-called
osmoreceptors, constitutes an important trigger. There is, however, as Fitzsimons
(58) points out, a second major stimulus. That is the state of the stretch
receptors in the low pressure part of the circulatory system. Recent papers show
that the renin-angiotensin system (loc.cit.) provides stimuli for extracellular or
hypovolemic thirst. Not only does it act locally in the hypothalamus to make
the animal drink, but it also provokes sodium and water retention by the kidney
through release of aldosterone.

The work of Fitzsimons was confirmed and extended by Stricker (59).
Reduction of intrathoracic blood volume by vena cava ligation as well as plasma
volume depletion by subcutaneous polyethyleneglycol (PG) injections both produce thirst in the rat. Thirst of vena cava ligation can be eliminated by
nephrectomy which removes the source of thirst-inducing angiotensin. Hyper-
oncotic thirst (injection of PG) cannot be prevented by this operation. To our
knowledge it is not clear whether this is a qualitative difference or whether
plasma volume depletion is simply a stronger stimulus for thirst induction using
pathways with a higher threshold. The strongest stimulus for drinking is a
combination of volume depletion and volume pooling in the periphery (59).

Whether the absence of thirst experienced by the astronaut (see p. 182),
whose heart chambers are overfilled, could be due to the direct effects of an
excessive barrage of nervous stimuli arriving in the hypothalamus from the
cardiac receptors or whether the renin-angiotensin mechanism intervenes is
unknown. In the same vein the thirst frequently complained of by the patient
who is developing congestive failure could be due to the opposite effect: recent

work has shown that the prolonged stretching of the heart chambers is associated with failure of the impulses coming from the stretch receptors (60). Fitzsimons points out that a deficiency in information from the cardiac receptors at the diencephalic level is interpreted as thirst.

It would be interesting to know the effects of immersion upon the thirst mechanism, but because of experimental difficulties no progress has yet been made in this area.(Compare *The Neuropsychology of Thirst: New Findings and Advances in Concepts* (1973), by A.N. Epstein, H.R. Kissileff, and E. Stellar (eds.), V.H. Winston and Sons, Washington, and articles by Fitzsimons, Stricker, and Epstein.)

Measured Changes of Plasma Volume during Immersion

As mentioned above, Bazett (49) found a striking increase in hemoglobin concentration after 3–4 hours of immersion. He viewed this as evidence of a reduced plasma volume. Later several groups independently showed that immersion of 6–8 hours' duration leads to a reduction in plasma volume of 8–15% (cf. Table 2 in Gauer et al. (4)). Their results demonstrate that, although the diuresis may wax and wane periodically, the reflex itself does not adapt quickly and a true correction of plasma volume is obtained.

Summary and Conclusions

From the evolutionary point of view it would appear to be advantageous if at a given fluid uptake the organism had available two different means of plasma volume control. The well hydrated subject can afford to simply excrete extracellular fluid; but the dehydrated subject can still achieve volume control while avoiding the irreversible loss of fluid through the kidney. He does this by translocating rapidly acquired plasma volume into the interstitial tissue spaces. In the "normally" hydrated subject both mechanisms seem to share in the control of plasma volume. In a series of "normally" hydrated subjects it was found that plasma volume reduction constituted about 40% of the total fluid loss (61), instead of the 20% to be expected in the well hydrated man.

We do not yet know what physiological switch determines the relative contribution of the above two mechanisms for plasma volume control—fluid excretion by the kidney or excess outward filtration in the microcirculation. It can only be said that with increasing dehydration the diuresis is attenuated in favor of increased filtration into the interstitial space. Even more obscure is the answer to the question: under what conditions does the organism use the thirst mechanism to change extracellular fluid volume in face of a constant urine flow?

NEUROHORMONAL EFFECTOR MECHANISMS OF VOLUME CONTROL

Figure 8 depicts the major factors regulating plasma volume in response to volume-conditioned stimuli. The main routes are the ADH mechanism for the control of water and the sympathetic renin-angiotensin aldosterone systems for

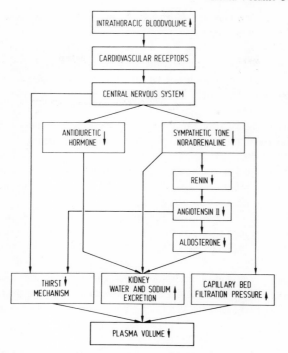

Figure 8. Major pathways for the reflex control of plasma volume following an expansion of the intrathoracic blood volume. An increase in intrathoracic blood volume will stimulate cardiovascular receptors thus conveying information to the central nervous system (CNS). The water and sodium regulatory mechanisms are activated via decreased ADH on one hand and sympathetic and renin-angiotensin control of kidney function on the other. Fluid uptake is diminished both due to action in the CNS and via decreased angiotensin effect on thirst. Restraint from fluid uptake and increased fluid excretion together decrease ECFV and hence plasma volume; furthermore, it is possible to affect plasma volume by changing the capillary filtration pressure via the sympathetic outflow. (For the sake of clarity several important points have been omitted, e.g., a possible diuretic and saliuretic hormonal factor, the interaction between the ADH and sympathetic, renin-aldosterone system as well as the possible effects of the various hormones on vascular tone.)

the control of sodium. The reaction of either hormone system to volume stimuli in noninvasive experiments like pressure breathing and particularly whole body immersion is presented first. These reactions are then viewed against the background of results obtained with invasive methods.

Control of Water by Antidiuretic
Hormonal Responses to Changes of Intrathoracic Blood Volume

In the search for a volume control mechanism the first clear connection between volume stimuli and a neuroendocrine response was the finding of a typical water diuresis. This has been described in the previous section. The evidence that it was an effect of vasopressin follows.

Evidence from Noninvasive Methods: Negative Pressure Breathing, Immersion, Orthostasis, Hemorrhage

Negative Pressure Breathing The diuresis induced by negative pressure breathing (NPB) or whole body immersion identified the triggering event of volume control as the engorgement of the intrathoracic circulation. When the subject was liberally hydrated his response was a typical water diuresis. The observation that in normal subjects the diuresis of NPB (30, 62) and immersion (63, 64) can be prevented by the infusion of exogenous vasopressin was strong support for the mechanoreceptor hypothesis that the diuresis is induced by a reflex reduction of the ADH activity in the blood.

Immersion More direct evidence of the role of ADH in volume control had to await the development of specific methods for the determination of very small concentrations of ADH in body fluids.

In 10 moderately dehydrated subjects that had been exposed to water deprivation for 14 hours prior to the experiment Epstein showed that during 5 hours of immersion the ADH excretion in the urine fell steadily to less than 50% of the control value (65). In the 1st hour of postimmersion recovery it rebounded to 190% of the preimmersion control (Figure 9). The plasma osmolarity remained unchanged throughout. It is noteworthy that this diuresis occurring simultaneously with the fall in ADH activity was entirely due to an increase in

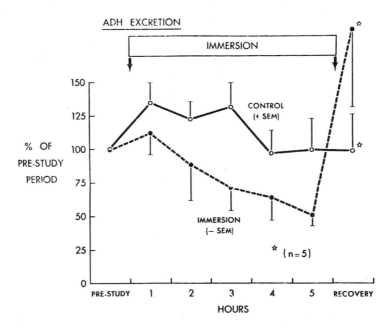

Figure 9. Effect of 5 hours of water immersion on urinary ADH excretion in the normal dehydrated subjects. Immersion was associated with a progressive decrease in ADH excretion. By the 5th hour of immersion, the ADH excretory rate was 50% of the prestudy hour. Recovery was associated with a prompt increase to 189% of the prestudy value (see Epstein et al. (65)).

osmolar clearance. Behn et al. (56) had also noticed that dehydrated subjects respond to immersion with an increase in osmolar clearance. On the other hand, the immersion diuresis of hydrated subjects is predominantly due to an increase in free water clearance. ADH was not measured in Behn's experiments. Better understanding of these events awaits the simultaneous measurements of plasma renin-angiotensin and ADH.

Orthostasis, Exposure to Heat, Cold, and Hemorrhage The intrathoracic blood volume is reduced by changing from the recumbent to the erect posture. At the same time the ADH concentration in the plasma rises by a factor of approximately 5 if the subject is suspended in such a way that blood can pool freely in the legs (66, 67). A similar increase is seen in "nonhypotensive" hemorrhage (68) or when a seated subject is exposed for 2 hours to an ambient temperature of 50°C (control temperature 26°C). A sudden change to a cold environment (13°C) leads to a significant reduction in ADH activity in the plasma. These changes occur within 15 min while serum sodium and osmolar concentrations remain constant throughout the experiments (66).

Severe hemorrhage is the most powerful known stimulus for ADH secretion, and plasma concentrations up to several hundred microunits per ml have been reported. There is ample evidence that at this point the arterial hypotension with its decreased stimulation of the aortic sinus and aortic arch receptors (69) is triggering the release of ADH together with the well known increase in sympathetic tone. The very high concentrations have repercussions on vascular tone (70).

Integration of Arterial and Cardiac Receptors in Control of ADH Secretion following Volume Loss Any manipulation of cardiac filling pressure usually leads to a change in cardiac output (CO). Thus during water immersion CO is increased by 35% (53) while orthostasis leads to a decrease of approximately 20–30% (17). Although mean arterial pressure may remain constant in either event, pulse pressure may change sufficiently to affect carotid sinus activity, thus influencing kidney function in the appropriate direction. It is therefore necessary to consider the effectiveness of the cardiac receptors as opposed to the arterial receptors in terms of volume control.

The observation of a high ADH activity after severe blood volume depletion stimulated Share and Levy (71) to investigate the role of carotid sinus (CS) and aortic arch receptors in the control of ADH secretion. Share's major contributions to the analysis of the problem have been discussed in extensive reviews (69, 72, 73).

Using the classic carotid sinus preparation Share and Levy (71) recorded the ADH titer in the blood in addition to the arterial blood pressure. They found a striking parallelism between the response of the two parameters: (*a*) clamping of both carotid arteries alone had little or no effect; (*b*) sectioning of the vagus nerves (elimination of aortic arch and cardiac receptors) significantly increased the ADH level. A subsequent carotid artery occlusion led to a further 2.5-fold increase in the ADH titer and the blood pressure rose, demonstrating the tonic inhibition exerted over the ADH secretion.

The effect of pulsating as opposed to continuous pressure in the CS on arterial blood pressure and the ADH in the blood was then investigated in the vagotomized dog (74). With the change to continuous pressure the ADH titer approximately doubled and the arterial blood pressure rose. Share and Levy comment that these changes of the ADH titer could only be elicited by very intense carotid sinus pulsation, i.e., 85 mm Hg. Reduction of pulse pressure from 39 to 9 mm Hg had little or no effect on arterial pressure and ADH titer (Share (69)). This agrees with the finding that the addition of a moderate pulsating component to the mean pressure adds only little to the function of the CS reflex in the dog (Kumada (75)). Thus the decisive parameter is the mean pressure in the CS while minor changes of pulse pressure such as those found in nonhypotensive hemorrhage and immersion appear to be less relevant.

A third series of experiments was performed by Share in dogs in which both CS were isolated, the vagus nerves were intact and a left atrial balloon was implanted (76). The arterial pressure was maintained constant to fix the excitation level of the aortic arch baroreceptors. Occlusion of the carotid arteries resulted in a marked increase in the blood titer of ADH. However, simultaneous distension of the left atrium not only prevented a rise of blood ADH but in six out of eight experiments actually reduced the level. Cutting the vagus nerves eliminated this reversing effect of atrial distension.

These three experiments demonstrate that although arterial pressure changes have a very strong influence on the ADH mechanism with volume loss, atrial receptors appear to be dominant in hypervolemia.

Evidence from Local Distension of the Left Atrium   The effect of inflation of a balloon in the left atrium on the antidiuretic activity in the blood was demonstrated by several groups using bioassay techniques (77–80).

However, the investigations of Johnson et al. (80) are of particular significance for two reasons. As in Share's experiment their extraction assay was specific for ADH. Furthermore, they used changes in left atrial transmural pressure which were well within the normal range of everyday life, yet an increase in this pressure by only 1–7 cm H_2O led to a significant decrease in the blood ADH titer (Figure 10). Such a change of atrial pressure is induced by a loss or gain of 10–20% total blood volume or a shift from the upright to the recumbent posture. This great sensitivity of the ADH mechanism to changes in left atrial distension is matched by the remarkable sensitivity of left atrial receptors to small changes in total blood volume (Henry and Pearce (40)).

Recently the reliability of this work has been seriously questioned by Kappagoda et al. (81). Close inspection of Johnson's (80) experiments shows that this criticism was not justified. His work involved a total of 45 balloon inflations in 16 dogs, 44 of which induced the fall in ADH and only one a rise. This is a remarkable record of precision from a system in which there are so may adverse influences such as level of anesthesia, surgical stress, and deterioration of the animal. The experimental design involved a comparison of the ADH concentration during a test distension with a mean of the control values before and after. Thus it permitted the specific testing of the effect of the experimental

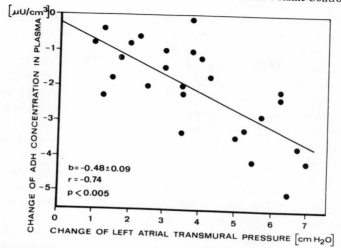

Figure 10. Change of left atrial transmural pressure and ADH concentration in the plasma; note the significant negative correlation (see Johnson et al. (80)).

forcing, i.e., increased left atrial pressure on the response, or plasma ADH. At the same time, it minimized uncontrollable adverse influences.

Although the mean difference in the concentration of ADH between the test and the control period was as small as 1.9 μU/ml this could easily be detected because the most sensitive range for the ADH assay was used, i.e., that between 6 and 4 μU/ml. The highly significant ($p<0.005$) negative linear relation between changes of LAP and changes in plasma ADH was noted at a point where the kidney has maximal concentration power. It is true that by working at the optimal range of the assay Johnson et al. (80) sacrificed the intensity of diuresis induced by the balloon, but by doing so they successfully attained a maximum of accuracy.

This work clearly demonstrates that left atrial distension is important in controlling the release of ADH. Where delicate experiments of this type are involved the burden of proof to the contrary has always rested on those who cannot confirm a result.

Share (82) has recently commented that the omission by Kappagoda et al. (81) of the procedure of extraction and concentration of the ADH and their use of large injection volumes of dog plasma into the test rat leave open the question whether their version of this bioassay technique was sufficiently sensitive.

While it had been clearly demonstrated that in the proper circumstances the diuresis induced by negative pressure breathing NPB or immersion can be prevented by a vasopressin infusion in normal man, Ledsome et al. (83) and Lydtin and Hamilton (84) in their animal experiments were unable to suppress the diuresis of left atrial distension by the intravenous administration of ADH. Recently Linden (32) has explained this anomaly by describing a new series of experiments in which Ledsome and Mason (85) arrive at the conclusion that the diuresis they observed was a consequence of the change from high to low infusion rates of vasopressin in an anesthetized animal. Linden (32) concludes that it is necessary to retain the hypothesis that the blood-borne agent affected by stimulation of left atrial receptors is most likely to be antidiuretic hormone. Ledsome and Mason also confirm the observations of Arndt et al. (86), commenting that in addition to a decrease in circulating levels of antidiuretic hormone the atrial distension also produces an increased osmolal clearance which is possibly due to a hemodynamic change in the kidney.

In a comprehensive paper, Kinney and DiScala (87) investigated in dogs the water diuresis of left atrial distension employing hydration, chronic and acute salt loading, deoxycorticosterone (DOCA) in excess, and distal tubular nephron blockade with diuretics. The diuresis was found in hydrated as well as salt-loaded animals and was independent of DOCA and presumed renin depletion. No significant saliuresis was documented. The diuresis could not be produced after distal tubular blockade and it was always stopped by exogenous vasopressin. The authors come to the conclusion that "antidiuretic hormone inhibition with distal tubular nephron water permeability changes appear to be the sole mechanism of the diuresis of left atrial distension in the dog." Their failure to observe an osmolal diuresis may have been due to the high level of hydration and excellent condition of the animals. As is discussed subsequently, left atrial distension may induce the above osmolal effect by its influence on the sympathetic outflow to the kidney.

Evidence from Nonhypotensive Hemorrhage   The evidence presented so far leaves no doubt that the intrathoracic receptors and especially the receptors in the left atrium are instrumental in the prevention of an overfilling of the bloodstream. It was difficult to demonstrate a similar role for the same receptors in the defense against a depletion of central blood volume. The problem stems from the fact that cardiac output and pulse pressure fall in the face of increasing blood loss. Under these circumstances the arterial baroreceptor mechanisms will be activated and they, too, can induce a reflex oliguria.

Henry et al. (68) investigated this problem with the method of "nonhypotensive hemorrhage." Dogs were bled in steps of 10% of the estimated blood volume, while the blood titer of ADH was determined and central venous and arterial pressures were recorded. With completion of the first step the ADH titer showed a large increase which was intensified with 20 and 30% bleeding. Mean arterial pressure remained unchanged for the first 20% bleeding while central venous pressure fell by 2–3 cm H_2O. Pulse pressure was mildly reduced. Pulse rate rose sharply with the first 10% bleeding. These results were corroborated by other experiments (Gupta et al. (88)) in which action potentials were recorded from high pressure and low pressure receptors during bleeding (Figure 11). The mean firing rate per unit time of the two sets of receptors closely followed the behavior of the pressures in the arterial and low pressure system. In the atrial receptors it fell to 50% with 10% hemorrhage and by 80% with 20% blood volume loss. The mean firing rate of the arterial receptors was very much less affected. It was concluded that the fall in atrial distension is the main cause of the considerable reflex rise in ADH titer during nonhypotensive hemorrhage.

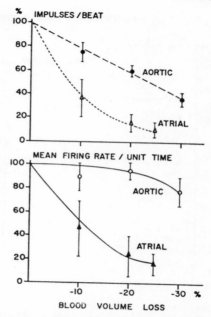

Figure 11. Comparison of the mean firing rate per second of six aortic (○) and six atrial (▲) fiber preparations during stepwise reduction of blood volume. The upper figure presents data as mean impulses per beat and the lower in terms of firing rate per second, i.e., with correction for the change in heart rate (vertical lines indicate the SD) (see Gupta et al. (88)).

By the use of continuous slow bleeding and a narrow spacing of the samples, Claybaugh and Share (89) obtained a high resolution picture of the events in the lower range of blood loss (Figure 12). When drawing blood at a rate of 0.28 ml/kg/min the plasma vasopressin titer had increased 75% over control at a cumulative hemorrhage of 4.2 ml/kg (about 5.2% blood volume). With the more rapid rate of blood withdrawal of 0.42 ml/kg/min the critical blood loss was only 2.1 ml/kg (2.6% blood loss). Arterial blood pressure and pulse pressure did not change significantly, but effective left atrial pressure (ELAP) fell continuously with the hemorrhage. Although the critical threshold for a significant ADH elevation is lower with the faster bleeding, the ADH secretion is volume and not rate dependent and parallels the fall in the effective left atrial pressure (ELAP) (Figure 12).

Renin activity was also measured simultaneously with the ADH concentration in the plasma. With the fast rate of bleeding, ADH and renin activities rose in parallel from a low threshold value. This might suggest that renin plays a role in ADH release and vice versa, thus confirming the observations of Bonjour and Malvin (90) and others (91). However, with slow hemorrhage this parallelism was not observed. These findings indicate that with moderate hemorrhage renin activity depends on the rate while the response of the ADH mechanism is determined by the cumulative loss of blood volume. Share has also recently studied the effects of hemorrhage in patients undergoing hemodialysis: with a

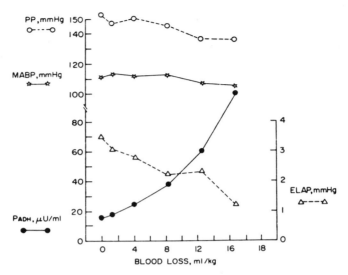

Figure 12. Effect of continuous, slow hemorrhage on mean arterial pressure (*MABP*), arterial pulse pressure (*PP*), effective left atrial pressure (*ELAP*), and the plasma ADH concentration (P_{ADH}) in the anesthetized dog; means of seven dogs. Arterial pressure was recorded from a femoral artery. P_{ADH} and ELAP were significantly different from initial levels when the cumulative blood loss was 4.2 ($p<0.05$), 8.4, 12.6, and 16.8 ml/kg ($p<0.01$). MABP was significantly changed ($p<0.05$) only when the cumulative blood loss was 16.8 ml/kg. PP did not change significantly from initial values (see Share (82)).

loss of only 6–10% of the blood volume plasma ADH concentration increased 70% (p = 0.05). There were no significant changes in systolic and diastolic blood pressure (82).

Summary It is readily apparent from experiments using atrial distension that the low pressure system receptors are strongly involved in the defense against circulatory congestion. It is much more difficult to separate the effects of the baroreceptors when working with diminished filling of the low-pressure system. However, studies with slow nonhypotensive hemorrhage demonstrate that the loss of no more than 6–10% of blood volume with a fall of no more than 2 cm H_2O in atrial pressure will induce a significant change in plasma ADH. Further evidence for a tonic inhibition of ADH release by cardiac receptors comes from experiments investigating the relationship between volume control and osmocontrol in conscious animals (see p. 178).

Control of Sodium by the Sympathetic System
and Renin Aldosterone Responses to Volume Changes

The control of sodium excretion is of critical importance in the control of blood volume and this section reviews the pathways connecting the salt-controlling hormone aldosterone with the cardiac receptor network.

Role of Various Cardiac Receptors Determining Na⁺ Excretion The work of
Muers and Sleight (92) and Öberg and Thorén (38) has demonstrated that the
ventricular receptors discharge into small nonmedullated C fibers in the vagi.
Most of these ventricular receptors arise from the left side and as Thorén has
shown (in press) their firing rate is sensitively proportional to varying levels of
diastolic pressure.

Öberg and Thorén (38) have contrasted their effect upon the skeletal
resistance vessels with that on the renal vascular bed. They found that the renal
bed is more sensitive to the ventricular receptors and the muscle bed to the
baroreceptors. In addition to reflex dilatation of the renal vessels by inhibition
of adrenergic vasoconstrictor activity the ventricular afferents lead to a reduction of heart rate due to vagal activation.

In related work using a balloon to stimulate the left atrial receptors Karim et
al. (93) showed that sympathetic activity in the nerves to the kidney decreased
while those to the heart increased and the activity in the nerves supplying muscle
was unaffected. These observations fit with those of Clement et al. (94) in which
in the rabbit nonhypotensive hemorrhage (10%) increased activity in a renal
nerve preparation by 30% and an equally modest infusion decreased it by 40%.
Interruption of the cervical vagi led to a 20% increase in activity. They concluded that low pressure intrathoracic receptors have appropriate control over
renal sympathetic nerve activity in response to changes in blood volume.

The preceding data throw light on the earlier studies of Bunag et al. (95)
demonstrating that nonhypotensive hemorrhage causes renal release of renin in
the anesthetized dog. When a nonhypotensive loss of central blood volume was
induced in man by assumption of the upright posture, Gordon et al. (96)
showed that renin excretion increased and they point to the simultaneous
increase of peripheral resistance and plasma norepinephrine as factors in maintaining the systemic arterial pressure.

Wagemark et al. (97) have demonstrated a sympathetic supply to the
juxtaglomerular cells and Davis (98) suggested that the renal nerves probably act
directly on them through a β-adrenergic receptor. In addition, as De Quattro and
Miura (99) point out, the renal nerves or circulating catecholamines released by
sympathetic stimulation or both together can influence glomerular filtration
rate. This in turn can determine renal tubular sodium load and the sodium
concentration at the macula densa.

The foregoing indicates that subendocardial receptor networks of the heart
chambers are connected via the sympathetic nervous system to the renin-
angiotensin-aldosterone mechanism controlling sodium excretion by the kidney.

*Use of Water Immersion to Study the Relation between Increase in Central
Blood Volume and Excretion of Sodium* The foregoing observations indicate that
the sympathetic stimulation which is elicited by activation of both the atrial and
the ventricular receptor networks can affect the renin-angiotensin-aldosterone
mechanism. Here the powerful technique for redistributing blood into the intrathoracic circulation by long-term water immersion has proved of critical importance. The effectiveness of the stimulus to the cardiac receptor network is shown

by the relatively enormous changes in central filling and pressure (Figure 5, Table 3). Since the arterial pressure rises by approximately 10 mm Hg, the baroreceptor mechanism must also be considered. However, Thron et al. (100) showed in normal man an increase of carotid sinus pressure is less effective in changing sympathetic tone than a decrease. It therefore appears justified to attribute much of the change of fluid and mineral balance to the cardiac receptor mechanism. These changes are often dramatic as in the case of a man reported by Eckert et al. (101, 102) who experienced a weight reduction of 5 kg with a salt loss of 40 g and a reduction of CVP of 7 cm H_2O all in 48 hours (Figure 13). This group also showed that the renin-angiotensin mechanism was involved (Korz et al. (103)) and they demonstrated a significant fall in renin-like activity while Behn et al. (56) established an increase in the Na/K quotient. More recently, Epstein et al. have used this technique in a series of impressive studies (55, 104–106).

They have demonstrated that the immersion-induced increase in central volume produced a profound suppression of plasma renin activity over a period of 4 hours and with it a like fall in plasma aldosterone. During a 60-min. recovery period the values returned toward normal (Figure 14) (106). They concluded that the renin-angiotensin axis was effectively mediating aldosterone secretion in response to central blood volume expansion.

Various Studies of the Effects of Left or Right Atrial Distension on Neuroendocrine Mechanisms Considering the great similarity of structure and of distribution of their nervous elements one might expect that by and large the two atria would have the same function in volume control. There is, however, one difference which may be important: an artificial mitral stenosis induced by a balloon will only affect receptors in the myocardium and at the atrial-venous junctions while a significant elevation of pressure in the right atrium will in addition have repercussions on the coronary circulation, because the coronary

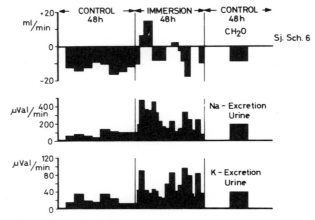

Figure 13. Influence of a 48-hour water immersion on free water clearance, Na^+ and K^+ excretion. The effect although operating in waves did not subside for 48 hours. During this time the subject lost 5 kg of BW and 40 g of salt (see Eckert et al. cited in Gauer (202)).

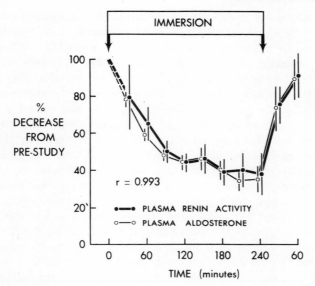

Figure 14. Comparison of the effects of immersion on plasma renin activity (PRA) and plasma aldosterone in subjects in balance on a 10-mEq Na diet. Data are expressed in terms of per cent change from the pre-immersion hour. The suppression of plasma aldosterone paralleled the suppression of PRA throughout the immersion period. Similarly, both PRA and plasma aldosterone recovered in parallel following cessation of immersion. (Epstein et al. (106)).

sinus empties into this side and congestion of the outflow may lead to the stimulation of nervous elements around the upstream coronary venous bed.

The preliminary observations of Lazzara et al. (107) on a diuresis initiated by coronary stretch receptors appear to be relevant here. They made measurements of the renal function of dogs before, during, and after occlusion of the coronary sinus with a balloon catheter. Urine flow and osmolar clearance increased two to three times; creatine clearance increased by 10–15%. Free water clearance decreased and renal artery flow was unchanged or slightly increased. The responses appeared to be absent in the denervated kidney and they attributed the effect to vagal afferents excited by distension of the coronary vein or capillaries.

In this same context and in analogy to their experiments in which the pulmonary vein-left atrial junctions were distended with miniature balloons (108), Kappagoda et al. (109) devised a balloon catheter which permitted local distension of the junction between the superior vena cava and the right atrium and so raised right atrial pressure without obstructing venous return. A moderate increase in urine flow and Na^+ excretion was observed. A mild but significant diuresis was also obtained by the distension of the right and left atrial appendages (110). Cooling of the right vagal nerve attenuated the response to distension of the right atrial appendage.

Further interesting studies exist: early experiments in which the right atrium was mechanically stimulated have furnished suggestive evidence that its receptors contribute to the regulation of aldosterone secretion (111). More recently Brennan et al. (78) distended balloons in the left and right atria of open chest animals. In the dogs in which left atrial pressure was increased they report that ADH concentration in the blood was lowered significantly while renin activity was not significantly suppressed. Right atrial extension caused a significant decrease in renin activity but had no significant effect on the ADH titer.

Further evidence for a special role of right atrial receptors in the control of adrenocorticosteroids comes from the experiments of Cryer and Gann (112). A small hemorrhage of 5 ml/kg in 3 min caused a significant increase in the secretion of cortisone, whereas hemorrhage combined with simultaneous balloon inflation in the right atrium led to a diminished response. Cryer and Gann's experiments indicate that right atrial receptors play a role in the control of adrenocorticosteroid secretion during small hemorrhage. So far this effect has not been directly demonstrated for the left atrial receptors. We must, however, remember that left atrial distension may lead to an increase in urine flow and sodium excretion which in all probability is independent of the ADH mechanism (86).

The studies of Arndt et al. (86) with left atrial balloon distension demonstrated an increase of PAH and inulin clearance and pointed to changes in glomerular filtration rate and renal plasma flow. A shift in renal blood flow distribution was also suspected. One or more of these events could be involved in the diuretic reflex. They postulated the action of a depressor reflex of the same type that affects the heart in response to activation of cardiac receptors. The previously mentioned discovery by Karim et al. (93) and Öberg and Thoren (38) and Clement et al. (94) of such a sympathetic reflex affecting the renal nerves constitutes a separate and important route, by which the excretion of salt as well as water can be affected independent of the ADH mechanism.

In this same connection the observations of Goetz et al. (113) on the excretion of sodium following manipulation of cardiac receptor areas are of interest. In order to investigate the effect of an isolated reduction of atrial transmural pressure without an alteration of the various circulatory parameters, they devised the method of the atrial pouch. The pericardium was incised circumferentially below the A-V groove. The cut edge of the atrial pericardium was then sutured to the A-V groove. With the help of an indwelling catheter in the pericardium, the pouch could be filled with saline several days after the animal had recovered from the operation. The ensuing simultaneous reduction of the wall tension of both atria resulted in a significant decrease in urine flow and sodium excretion (113) while the ADH concentration in the plasma remained unchanged (114). An explanation of this finding has not been definitively established.

Summary The existing evidence shows that the right atrial receptor zones contribute to the control of renin and corticosteroids. Although suggestive data are available the evidence is not yet sufficient to assign with certainty qualitatively different functions to the left and right atrial receptor zones. Furthermore, it is difficult to affect the right atrium without influencing coronary sinus drainage with its possible effects upon the ventricular subendocardial network.

INTEGRATION OF VOLUME AND OSMOCONTROL

If the osmotically active mass of the organism does not change during water deprivation the osmocontrol mechanism of Verney will restore the normal volume at the same time that it restores the normal osmotic pressure. This concept needs amplification by pointing out that the dehydration is also causing a fall in left atrial pressure and hence an activation of the volume control mechanism (115). Thus in the case of water deprivation volume control and osmocontrol both using the ADH mechanism will act in unison to restore volume and osmotic pressure.

There are, however, conditions in which osmocontrol and volume control compete. In the classic experiments of McCance the subjects were kept in a hot environment for 11 days and forced to sweat (116). While they were thus losing water *and* salt, only water was offered ad libitum. After a weight loss of about 2 kg they started to abandon the homeostatic control of osmotic pressure by retaining excess water. In discussing this experiment Leaf comments that the developing dilutional hyponatremia can be interpreted teleologically as a "lesser evil than circulatory collapse" (117). The great influence of volume control in the determination of the release of ADH may be seen from the following rough estimate: since plasma volume is approximately 7.5% of whole body water, blood volume reduction as the consequence of the fluid loss of 2–3 liters, experienced by McCance's subjects, was only 150–200 ml. In other words, they were even more sensitive to volume loss than the dogs in the following experiments.

Arndt (118) used trained dogs to provoke the competition of volume control and osmocontrol in the following elegant way. He showed that the infusion of water into a carotid loop in an amount which changed the local osmotic pressure in the hypothalamic region by about 2% would cause a water diuresis. If the animal was bled 8% of its blood volume during the first 10 min of water infusion, the diuresis was completely suppressed. Reinfusion of the shed blood resulted in a profuse diuresis. During these procedures heart rate and blood pressure did not change even to a minimal degree. These results are in full agreement with more recent investigations in the unanesthetized ewe in which the diuretic effect of a 1.2% reduction in plasma osmolality could be canceled out by a 10% hemorrhage (119).

It is interesting to discuss these findings in the light of recent work by Dunn et al. (120). By intraperitoneal (i.p.) injections of polyethyleneglycol and hypo- or hypertonic saline and combinations thereof they were able to establish thoroughly in the rat the effects of changes of blood volume and osmotic pressure on the antidiuretic activity in the blood. They found a linear relationship between osmolality and arginine-vasopressin (AVP) concentration in the blood. For a 7% increase in osmolality the AVP concentration rose from near zero to 20 pg of AVP/ml. The function relating percent blood volume loss with AVP concentration was exponential; with a change of blood volume of 7% the change of AVP was very small. Only when the blood loss exceeded 10–15% did the AVP concentration rise steeply to over 40 pg/ml.

It would, however, be rash to conclude from these data that the osmocontrol mechanism is more sensitive than the volume control mechanism. The two entities have different physical dimensions and are therefore not comparable. In fact a 7% change of osmolality would mean a gross interference with vital processes on the molecular level while a 7% change of blood volume is irrelevant to a healthy resting animal.

The point at issue is illustrated in Figure 15 using the data from Figure 6 in the work of Dunn et al. (120) which relates plasma osmolality with AVP concentration in normal animals and animals which have lost approximately 6.3 and 14.5%, respectively, of their blood volume. With increasing plasma volume depletion the function relating AVP and osmotic pressure becomes steeper and is shifted to the left, i.e., the osmotic threshold is lowered. This type of diagram is familiar to the physiologist investigating the combined effects of two stimuli on one parameter. A classic example is the family of response curves relating respiratory drive to changes in CO_2 concentrations in the inspired air at various alveolar O_2 tensions (121).

The data of Dunn et al. fully explain Arndt's findings: if we reduce the osmotic pressure of the normal animal from 300 to 295 mosm/kg (2%) the AVP concentration will fall from approximately 9 to 4 pg/ml and a diuresis will ensue (Figure 15, *arrow A*). However, when this osmotic stimulus is combined with a blood loss of 6.3%, the AVP concentration and hence urine flow will not change (Figure 15, *arrow B*). This means that a blood loss of 6.3% which in Arndt's experiments was hemodynamically imperceptible except for a small reduction in central venous pressure offsets the response of the hypothalamic system to a standard osmotic stimulus. This finding is evidence of the delicate central integration of volume control and osmocontrol and the great weight which volume control carries in the partnership.

Recent work (122) confirms that an increased blood volume will lower the osmotic reactivity of the hypothalamohypophysial antidiuretic system (Moses and Miller (123)).

PATHOPHYSIOLOGICAL CONSIDERATIONS OF VOLUME REGULATION

In the consideration of a volume-regulatory reflex the question comes up why the gradual onset of heart failure is so often associated with disturbance of fluid volume regulation as to be called congestive heart failure. The distension of the heart chambers in this condition has always raised questions as to the efficacy of the regulatory mechanism. In this section this and other puzzling pathophysiological states related to volume control are discussed.

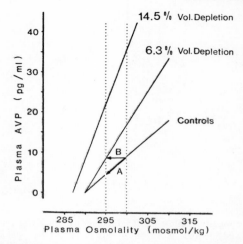

Figure 15. The effect of volume depletion on the sensitivity of the osmocontrol mechanism (adapted from Dunn et al. (120)). This diagram depicts the relationship between plasma osmolality and plasma AVP (arginin-vasopressin) under normal conditions and in two states of blood volume depletion. A 2% reduction of the osmotic pressure from 300 to 295 mosm will lower the AVP concentration in the control animal from approximately 8 pg to 4 pg/ml and a diuresis will ensue (A); however, if we deplete blood volume by only 6.3%, the AVP concentration will remain unchanged on simultaneous application of the same osmotic stimulus (B) and diuresis will not occur (see text).

Congestive Heart Failure

There is evidence that in heart failure the information coming from the various receptor networks in the subendocardium of the atria and the ventricles is altered. Thus Zehr et al. (124) have demonstrated that in chronic mitral stenosis in dogs there are extensive endocardial fibrosis and calcification which could involve the nerve nets and damage them. They also report that a small nonhypotensive hemorrhage induced a marked attenuation in the increase of blood ADH levels in the stenotic group as contrasted with the controls. This occurred in spite of a 2-fold greater decrease in left atrial pressure in the stenotic group and a relatively constant mean and pulsatile arterial pressure in both groups.

Greenberg et al. (60) have demonstrated, using chronic right-sided failure, that even when the frequency of atrial receptor discharge was considered in spikes per cycle it was very much less in the animals in failure despite huge superimposed atrial pressures ranging up to 30–40 cm H_2O induced by overtransfusion in both normal and experimental animals. As Figure 16 shows, the curves of atrial receptor response in spikes per cycle to the venous pressure increment differed sharply in the two cases.

One of Greenberg's dogs was a rare case suffering from chronic spontaneous heart failure with left- as well as right-sided lesion. This edematous and dyspnoeic animal showed extremely depressed atrial firing responses.

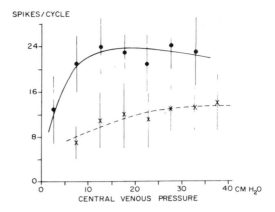

Figure 16. Effect of congestive failure on atrial receptor discharge in dogs (——, experimental group; —— ——, controls). Atrial receptor discharge in spikes per cycle was plotted against central venous pressure (CVP) grouped in increments of 5 cm H_2O. CVP was increased by progressive infusion. Note the low firing rates in congestive failure and the reduced response to pressure increase (see Greenberg et al. (60)).

This case taken with other evidence suggesting that both atria were distended in these experiments indicates that the depressed response of the animals in failure was not due to an experimental artifact by which the overtransfusion pressure failed to be transmitted to the left atrium. The sum total of evidence convinced the two groups that the sensitivity of cardiac receptor mechanisms as measured by ADH and electrical response to changes in wall tension can be impaired in congestive heart failure in dogs. They concluded that the impairment probably involved the subendocardial network and that it could result in a blunting of the response of the neuroendocrine mechanisms involved in the regulation of sodium and water metabolism.

Paroxysmal Tachycardia and Diuresis

Spontaneous paroxysmal atrial tachycardia results from a rapid succession of rhythmic impulses arising from an irritable ectopic focus in the atrium. The attacks occur suddenly and may proceed with few symptoms as long as the rate is not in excess of 200/min. Wood (125) has demonstrated that atrial mean pressure and pulse pressure may abruptly increase to double the resting state. At the lower heart rates there are usually no overt signs of heart failure but a water or a salt diuresis frequently occurs, suggesting that the various subendocardial receptor networks may be stimulated by the increased size and rate of heart chamber contraction.

In preliminary studies with dogs Kilburn (126) reports that electrical pacing of the normal left atrium to 200/min produced a diuresis in 30 min; but that the response failed if the cardiac output dropped by more than 50%. He found the diuresis was inhibited by infusion of ADH or vagal cooling. Mersch and Arndt (127) demonstrated that pacing in cats (210–390 beats/min) is accompanied by increased atrial circumference and pressure. Zucker and Gilmore (128) have shown in open chest dogs that periods of atrial fibrillation are accompanied by increased atrial pressure and receptor discharge, while Haynal and Matsch (129) report that the ADH level in the blood decreases in patients discharging abundant dilute urine during attacks. Taken together the evidence thus indicates that an increased cardiac receptor discharge is associated with the condition. Rosing et al. (130) have demonstrated the close tie with cardiac competence. At pacing rates of 200 to 250 beat/min in dogs there was no change in atrial pressure and no significant diuresis. But with higher heart rates (328/min) as long as aortic blood pressure was maintained, atrial pressure increased by an effective 8 cm H_2O and there was a water diuresis together with a natriuresis. With still higher heart rates the blood pressure falls and there is no significant natriuresis. Thus as Kilburn and Wood have both noted, if heart failure supervenes a diuresis or natriuresis can no longer be expected.

Goetz and Bond demonstrated (131) the expected failure of rapid atrial pacing to alter ventricular rate and arterial and central venous pressure in dogs with complete heart block. There was also no significant change in urine flow or sodium excretion. Pacing by no more than 169/min in normal control dogs elicited a diuresis and natriuresis. Seeking for a mechanism they found no response to rapid pacing in dogs with chronic sinoaortic denervation. They therefore concluded that the baroreceptors participate in eliciting the diuresis of tachycardia. On the other hand they also showed that atrioventricular pacing of dogs with complete heart block elicited an increase in urine flow and sodium excretion. Their conclusion was that with appropriate stimulation the cardiac as well as the sinoaortic receptor networks play a part in the natriuresis and diuresis of paroxysmal tachycardia.

Postcommissurotomy Hyponatremic Syndrome

Shu'ayb et al. (79) have studied the mechanism of the increase in antidiuretic hormone following surgical operations. They note that it is elevated after a variety of procedures. However, they confirm previous work showing that after mitral commissurotomy the ensuing hyponatremia and water retention are unaccountably pronounced and prolonged. Since there was a favorable response to alcohol it was suggested that a change in cardiac afferent activity due to a reduction in atrial pressure might be responsible. They considered the fact that after commissurotomy there are marked changes in the pressures within the heart chambers. They demonstrated that discontinuation of acute left atrial distension in dogs causes a significant short-term elevation of ADH. Shu'ayb et al. concluded (79) that the greater severity of the metabolic alteration after commissurotomy was due to the specific addition of an altered cardiac receptor traffic to the general ADH elevating effects of the trauma of surgery. Thus they provided a mechanism explaining the 1955 observations of Goodyer and Glenn (132) who were among the first of several to report unexpected difficulties with water balance following the newly introduced mitral operation (133). Goodyer and Glenn had postulated "an unusually prolonged antidiuretic stimulus of uncertain etiology as the responsible mechanism for water retention and dilutional hyponatremia."

Vagus Nerve Destruction and Inappropriate Secretion of ADH

Occasionally the vagus nerves are invaded by bronchogenic carcinoma, mediastinal tumors, or chronic inflammatory processes. In 1961 we suggested that the ensuing nerve damage might then lead to inappropriate ADH responses with dilutional hyponatremia by interfering with the passage of impulses from the cardiac receptors (134). Bronchogenic carcinoma is occasionally associated with the inappropriate secretion of ADH. However, it is now known that these tumors can synthetize ADH in vitro. This then must be considered the probable mechanism of the water intoxication that may be observed (135, 136).

However, the possibility remains that interference with the vagi may have implications in terms of volume regulation. A current preliminary report (137) describes two cases of carcinoma of the mediastinum who suffered from recently developed orthostatic hypotension. Studies were made to see if responses attributable to cardiac receptors were disturbed. When immersed in a water bath they failed to produce the usual increase in free water clearance. They also failed to show the fall in plasma ADH levels developed by normal controls.

Distortion of Volume Control during Adaptation to the Weightless Condition

Circumferential measurements show that the astronauts in a recent Skylab report quickly lose approximately 2 liters (!) of blood and extracellular fluid from the lower extremities upon entry into the weightless condition (138). This dramatic mobilization of fluid, which is greater than would be expected when the pressures involved are compared with those observed in postural changes at 1 G, leads to a severe engorgement of the circulation and tissues of the cephalad regions of the body; the face becomes puffy and the neck veins stand out as in congestive heart failure. By and large this is an exaggeration of the circulatory condition found in whole body immersion. An increased arterial blood flow in the extremities was determined plethysmographically (139). It is in all likelihood due to an increased cardiac output as a consequence of the increase in ventricular filling. Flow resistance was reduced and so was venous tone. This complies with the findings in immersion (Table 3).

The cardiac receptors which initiate the hemodynamic reflex changes of weightlessness and immersion are in all likelihood also responsible for the observed changes in fluid and electrolyte balance. There is a fluid loss of about 1 liter within a few days upon entry into the weightless state (140). This fluid loss is not due to excess excretion but to a drinking deficit of about 300 ml per day and the fluid is regained promptly by voluntary ingestion after return to earth gravity. Blood volume was reduced by an estimated 10%. In seven of the nine astronauts ADH excretion in the urine was reduced. There was a steady excess excretion of potassium and sodium, the latter in spite of a continuously elevated aldosterone excretion during the whole mission (142). The effect of a natriuretic factor has been discussed. There are many open questions and the difficulties may not only be due to the problems of experimentation during space flight but also to our insufficient knowledge of the basic physiological mechanisms involved even under normal conditions. We may, however, conclude that many of the observed changes are initiated by signals from the heart and that as in whole body immersion the organism by and large adapts to a new equilibrium of fluid balance after a few days.

FINAL SUMMARY

Recent evidence related to the hypothesis of volume control as an independent parameter has clearly demonstrated that sodium as well as water responds to volume stimuli. These responses are not locked together rigidly but depend on the condition of the animal. Thus hydration during immersion leads predominantly to a water and sodium diuresis but moderate dehydration induces an increase in osmotic clearance alone. With further increase in dehydration this osmotic diuresis yields to a movement of plasma fluid into the interstitial spaces. There is evidence that this movement is in response to changes in afferents from the heart. Thirst too appears to be activated by the stretch receptors monitoring the extracellular fluid compartment.

The key role of the ADH mechanism in volume control has been firmly established for many years. More recent investigations demonstrate that sympathetic tone and the renin-aldosterone mechanism react equally sensitively to volume stimuli. An increasing number of observations suggests that diuretic-saliuretic factors may also be involved (4, 105, 142). Extensive work shows that changes in blood volume greatly influence the sensitivity of the osmocontrol mechanisms. There is little doubt that the diagram which describes quantitatively the combined effects of changes of blood volume and osmolar pressure on the ADH activity which was elaborated by Dunn et al. for the rat (Figure 15) is also applicable to other mammals with only minor modifications.

Consideration of the interface between the high and low pressure system indicates that the altered secretion of vasopressin during volume increase is primarily due to changes in the activity of cardiac receptors. With decreasing volume the first changes also appear to be primarily due to the cardiac receptors but no clear cut line can be drawn between the two systems and the arterial baroreceptors assume increasing influence as volume loss progresses. In this way the baroreceptor activating mechanisms supporting systemic arterial pressure are subtly integrated with the cardiac receptor networks in the atria and ventricles to maintain an adequate cardiac filling and output.

With time the stretch receptors involved in volume control adapt as do those concerned with other regulatory mechanisms. However, this is not an unduly rapid process. Immersion and space flight experiments demonstrate that they remain active for at least 24–48 hours. This is sufficient time to correct for acute changes induced by other more common conditions such as excessive ingestion of salt and water, dehydration in a hot climate, or diarrhea. Kaczmarczyk et al. (143) have shown in the conscious dog on high sodium intake a postprandial rise in left atrial pressure which is accompanied by excretion of water and sodium. This observation may be interpreted as an example of the reaction of the volume regulatory response to an everyday stimulus. At the opposite extreme of pathophysiology there is evidence that the congestion of gradual heart failure may have roots in a blunting of the volume regulatory mechanism.

The contribution of Goetz et al. (144) became available just prior to the completion of this review. To them the connection between left atrial distension and water excretion which appears to us to be firmly established still meets with some reservations. However, the major difficulty is that they identify the concept of volume regulation with ADH and left atrial receptors alone. For volume regulation to be recognized as an independent parameter three criteria must be met. There must be a well defined compliance of the low pressure system with appropriate receptor networks in the chambers of the heart and these networks must control not only water but also sodium excretion.

The present paper adds new data supporting the critical concept of the constancy of the compliance of the low pressure system. It also presents the striking effects of immersion upon long-term sodium balance. Not enough information is available on the sodium role of the newly discovered ventricular receptor network and of the left versus the right atria. We may, however, attribute the natriuretic effectiveness of immersion to the fact that the stretch receptors in all the chambers of the heart together with the baroreceptors are stimulated by this condition.

ACKNOWLEDGMENTS

The competent help in completion of the manuscript by Mrs. I. Busch and Miss E. Gaebel and the editorial assistance of Miss L. Schepeler are gratefully acknowledged.

Part of the work, done in the laboratories of the authors, has been supported for several years by grants from the U.S. Air Force Office of Scientific Research. This review is based on a series of seminar lectures, given by one of us (O.H.G.) at the University of Kentucky College of Medicine, Department of Physiology and Biophysics, Lexington, Kentucky.

REFERENCES

1. Gauer, O.H., Henry, J.P., Sieker, H.O., and Wendt, W.E. (1951). Heart and lungs as a receptor region controlling blood volume. Amer. J. Physiol. 167:786.
2. Gauer, O.H., and Henry, J.P. (1956). Beitrag zur Homöostase des extraarteriellen Kreislaufs. Volumenregulation als unabhängiger physiologischer Parameter. Klin. Wschr. 34:356.
3. Gauer, O.H., and Henry, J.P. (1963). Circulatory basis of fluid volume control. Physiol Rev. 43:423.
4. Gauer, O.H., Henry, J.P., and Behn, C. (1970). The regulation of extracellular fluid volume. Annu. Rev. Physiol. 32:547.
5. Peters, J.P. (1935). Body Water. Charles C Thomas, Springfield Ill.
6. Weber, E.H. (1851). Über die Anwendung der Wellenlehre auf die Lehre vom Kreislaufe des Blutes und insbesondere auf die Pulslehre. Arch. Anat. Physiol. 497.

7. Borst, J.G.G., de Vries, L.A., van Leeuwen, A.M., den Ottolander, G.J.H., and Cejka, V. (1960). The maintenance of circulatory stability at the expense of volume and electrolyte stability. Clin. Chim. Acta 5:887.

8. Guyton, A.C., and Coleman, T.G. (1967). Longterm regulation of the circulation: Interrelationships with body fluid volumes. In E.B. Reeve and A.C. Guyton (eds.), Physical Bases of Circulatory Transport; Regulation and Exchange, pp. 179–201. W.B. Saunders, Philadelphia-London.

9. Henry, J.P., Gauer, O.H., and Sieker, H.O. (1956). The effect of moderate changes in blood volume on left and right atrial pressures. Circ. Res. 4:91.

10. Echt, M., Düweling, J., Gauer, O.H., and Lange, L. (1974). Effective compliance of the total vascular bed and the intrathoracic compartment derived from changes in central venous pressure induced by volume changes in man. Circ. Res. 34:61.

11. Gauer, O.H., Henry, J.P., and Sieker, H.O. (1956). Changes in central venous pressure after moderate hemorrhage and transfusion in man. Circ. Res. 4:79.

12. Harlan, J.C., Smith, E.E., and Richardson, T.Q. (1967). Pressure-volume curves of systemic and pulmonary circuit. Amer. J. Physiol. 213:1499.

13. Ungewiss, U. (1969). Kreislaufuntersuchungen bei essentieller Hypertension unter besonderer Berücksichtigung des Niederdrucksystems. Inaugural-Dissertation.

14. Shoukas, A.A., and Sagawa, K. (1971). Total systemic vascular compliance measured as incremental volume-pressure ratio. Circ. Res. 28:277.

15. Strey, W., Schwartzkopff, W., Wurm, W., and Frisius, H. (1973). Bestimmung der Weitbarkeit des System-, des Intra- und Extrathorakalkreislaufes beim Menschen nach kontrolliertem Blutverlust. Verh. Dtsch. Ges. Freisl.-forsch. 39:248.

16. Koubenec, H.-J., Risch, W.-D., and Gauer, O.H. (1975). Effective compliance of the total vascular system of man sitting in air and immersed in a bath. Pflügers Arch. 355 (suppl.):R24.

17. Gauer, O.H., and Thron, H.L. (1965). Postural changes in the circulation. In W.F. Hamilton and P. Dow (eds.), Sec. 2: Circulation, Vol. 3, pp. 2409–39. American Physiological Society, Washington, D.C.

18. Shepherd, I.T. (1966). Role of the veins in the circulation. Circulation 33:484.

19. Paessler, H., Schlepper, M., Westermann, K.W., and Witzleb, E. (1968). Venentonus-reaktionen in kapazitiven Hautgefässen bei passiver und aktiver Orthostase. Pflügers Arch. 302:315.

20. Boréus, L.O., and Hollenberg, N.K. (1972). Venous constriction in response to head-up tilt in man. Can. J. Physiol. Pharmacol. 50:317.

21. Price, H.L., Deutsch, St., Marshall, B.E., Stephen, G.W., Behar, M.G., and Neufeld, G.R. (1966). Hemodynamic and metabolic effects of hemorrhage in man, with particular reference to the splanchnic circulation. Circ. Res. 18:469.

22. Abrahám, A. (1960). Microscopic Innervation of the Heart and Blood Vessels in Vertebrates Including Man. Pergamon Press, Oxford-London-Edinburgh-New York-Toronto-Sydney-Paris-Braunschweig.

23. Nonidez, J.F. (1937). Identification of the receptor areas in the venae cavae and pulmonary veins which initiate reflex cardiac acceleration (Bainbridge's Reflex). Amer. J. Anat. 61:203.

24. Nonidez, J.F. (1939). Studies on the innervation of the heart. Amer. J. Anat. 65:361.

25. Johnston, B.D. (1968). Nerve endings in the human endocardium. Amer. J. Anat. 122:621.

26. Paintal, A.S. (1971). Cardiovascular receptors. In E. Neil (ed.), Handbook of Sensory Physiology, Vol. III/1, pp. 1–45. Springer, Berlin-Heidelberg-New York.

27. Paintal, A.S. (1973). Vagal sensory receptors and their reflex effects. Physiol. Rev. 53:159.

28. Gauer, O.H., Henry, J.P., Sieker, H.O., and Wendt, W.E. (1954). The effect of negative pressure breathing on urine flow. J. Clin. Invest. 33:287.

29. Sieker, H.O., Gauer, O.H., and Henry, J.P. (1954). The effect of continuous negative pressure breathing on water and electrolyte excretion by the human kidney. J. Clin. Invest. 33:572.

30. Murdaugh, H.V., Jr., Sieker, H.O., and Manfredi, F. (1959). Effect of altered intra-thoracic pressure on renal hemodynamics, electrolyte excretion and water clearance. J. Clin. Invest. 38:834.

31. Henry, J.P., Gauer, O.H., and Reeves, J.L. (1956). Evidence of the atrial location of receptors influencing urine flow. Circ. Res. 4:85.

32. Linden, R.J. (1972). Function of nerves of the heart. Cardiovasc. Res. 6:605.

33. Linden, R.J. (1973). Function of cardiac receptors. Circulation 48:463.

34. Smith, H.W. (1957). Salt and water volume receptors. Amer. J. Med. 23:623.

35. Arndt, J.O., Brambring, P., Hindorf, K., and Röhnelt, M. (1974). The afferent discharge pattern of atrial mechanoreceptors in the cat during sinusoidal stretch of atrial strips in situ. J. Physiol. 240:33.

36. Coleridge, H.M., Coleridge, J.C.G., and Kidd, C. (1964). Cardiac receptors in the dog, with particular reference to two types of afferent ending in the ventricular wall. J. Physiol. 174:323.

37. Öberg, B., and Thorén, P. (1972). Studies on left ventricular receptors, signalling in non-medullated vagal afferents. Acta Physiol. Scand. 85:145.

38. Öberg, B., and Thorén, P. (1973). Circulatory responses to stimulation of left ventricular receptors in the cat. Acta Physiol. Scand. 88:8.

39. Öberg, B. (1975). Personal communications.

40. Henry, J.P., and Pearce, J.W. (1956). The possible role of cardiac atrial stretch receptors in the induction of changes in urine flow. J. Physiol. 131:572.

41. Szczepańska-Sadowska, E. (1972). The activity of the hypothalamo-hypophysial antidiuretic system in conscious dogs. II. Role of the left vagosympathetic trunk. Pflügers Arch. 335:147.

42. Mancia, G., Romero, J.C., and Shepherd, J.T. (1975). Continuous inhibition of renin release in dogs by vagally innervated receptors in the cardiopulmonary region. Circ. Res. 36:529.

43. Malliani, A., Recordati, G., and Schwartz, P.J. (1973). Nervous activity of afferent cardiac sympathetic fibers with atrial and ventricular endings. J. Physiol. 229:457.

44. Baumgarten, R. von, Koepchen, H.P., and Aranda, I. (1959). Untersuchungen zur Lokalisation der bulbären Kreislaufzentren. I. Mitteilung: Funktionelle Organisation der Vagus-Wurzeln. Verh. Dtsch. Ges. Kreisl.-forsch. 25:170.

45. Öberg, B., and Thorén, P. (1973). Circulatory responses to stimulation of medullated and non-medullated afferents in the cardiac nerve in the cat. Acta Physiol. Scand. 87:121.

46. Saito, T., Yoshida, S., and Nakao, K. (1969). Release of antidiuretic hormone from neurohypophysis in response to hemorrhage and infusion of hypertonic saline in dogs. Endocrinology 85:72.

47. Menninger, R.P., and Frazier, D.T. (1972). Effects of blood volume and atrial stretch on hypothalamic single-unit activity. Amer. J. Physiol. 223:288.

48. Dennhardt, R., Ohm, W.W., and Haberich, F.J. (1971). Die Ausschaltung der Leberäste des N.vagus an der wachen Ratte und ihr Einfluss auf die hepatogene Diurese - indirekter Beweis für die afferente Leitung der Leber-Osmoreceptoren über den N.vagus. Pflügers Arch. 328:51.

49. Bazett, H.C., Thurlow, S., Crowell, C., and Stewart, W. (1924). Studies on the effects of baths on man. II. The diuresis caused by warm baths, together with some observations on urinary tides. Amer. J. Physiol. 70:430.

50. Graveline, D.E., and Balke, B. (1960). The physiologic effects of hypodynamics induced by water immersion. USAF School of Aviation Med. Rept. 60.

51. Graveline, D.E., and Jackson, M.M. (1962). Diuresis associated with prolonged water immersion. J. Appl. Physiol. 17:519.

52. Echt, M., Lange, L., and Gauer, O.H., with the technical assistance of H. Dannenberg (1974). Changes of peripheral venous tone and central transmural venous pressure during immersion in a thermoneutral bath. Pflügers Arch. 352:211.

53. Arborelius, M., Jr., Balldin, U.I., Lilja, B., and Lundgren, C.E.G. (1972). Hemodynamic changes in man during immersion with the head above water. Aerospace Med. 43:592.

54. Lange, L., Lange, S., Echt, M., and Gauer, O.H. with the technical assistance of H. Dannenberg (1974). Heart volume in relation to body posture and immersion in a thermoneutral bath. A röntgenometric study. Pflügers Arch. 352:219.

55. Epstein, M., Pins, D.S., Arrington, R., Denunzio, A.G., and Engstrom, R. (1975). Comparison of water immersion and saline infusion as a means of inducing volume expansion in man. J. Appl. Physiol. 39:66.

56. Behn, C., Gauer, O.H., Kirsch, K., and Eckert, P. (1969). Effects of sustained intrathoracic vascular distension on body fluid distribution and renal excretion in man. Pflügers Arch. 313:123.
57. Öberg, B. (1964). Effects of cardiovascular reflexes on net capillary fluid transfer. Acta Physiol. Scand. 62 (Suppl.):229.
58. Fitzsimons, J.T. (1972). Thirst. Physiol Rev. 52:468.
59. Stricker, E.M. (1973). Thirst, sodium appetite, and complementary physiological contributions to the regulation of intravascular fluid volume. In A.N. Epstein, H.R. Kissileff, and E. Stellar (eds.), The Neuropsychology of Thirst: New Findings and Advances in Concepts, pp. 73–98. V.H. Winston and Sons, Washington, D.C.
60. Greenberg, T.T., Richmond, W.H., Stocking, R.A., Gupta, P.D., Meehan, J.P., and Henry, J.P. (1973). Impaired atrial receptor responses in dogs with heart failure due to tricuspid insufficiency and pulmonary artery stenosis. Circ. Res. 32:424.
61. Kaiser, D., Linkenbach, H.J., and Gauer, O.H. (1969). Änderungen des Plasma-volumens des Menschen bei Immersion in ein thermoindifferentes Wasserbad. Pflügers Arch. 308:166.
62. Boylan, J.W., and Antkowiak, D.E. (1959). Mechanism of diuresis during negative pressure breathing. J. Appl. Physiol. 14:116.
63. Eckert, P. (1964). Die Beeinflussung der Badediurese durch Adiuretin. Pflügers Arch. 281:31.
64. Gauer, O.H., Eckert, P., Kaiser, D., and Linkenbach, H.J. (1967). Fluid metabolism and circulation during and after simulated weightlessness. Proceedings of the Second International Symposium on Man in Space, Paris 1965, pp. 212. Springer, Vienna-New York.
65. Epstein, M., Pins, D.S., and Miller, M. (1975). Suppression of ADH during water immersion in normal man. J. Appl. Physiol. 38:1038.
66. Segar, W.E., and Moore, W.W. (1968). The regulation of antidiuretic hormone release in man. I. Effects of change in position and ambient temperature on blood ADH levels. J. Clin. Invest. 47:2143.
67. Moore, W.W. (1971). Antidiuretic hormone levels in normal subjects. Fed. Proc. 30:1387.
68. Henry, J.P., Gupta, P.D., Meehan, J.P., Sinclair, R., and Share, L. (1968). The role of afferents from the low-pressure system in the release of antidiuretic hormone during non-hypotensive hemorrhage. Can. J. Physiol. Pharmacol. 46:287.
69. Share, L. (1974). Blood pressure, blood volume, and the release of vasopressin. In R.O. Greep, E. Knobil, and S.R. Geiger (eds.), Sec.7: Endocrinology, Vol.4, pp. 243–55. American Physiological Society, Washington, D.C.
70. Szczepańska-Sadowska, E. (1973). Hemodynamic effects of a moderate increase of the plasma vasopressin level in conscious dogs. Pflügers Arch. 338:313.
71. Share, L., and Levy, M.N. (1962). Cardiovascular receptors and blood titer of antidi-uretic hormone. Amer. J. Physiol. 203:425.
72. Share, L. (1967). Vasopressin, its bioassay and the physiological control of its release. Amer. J. Med. 42:701.
73. Share, L., and Claybaugh, J.R. (1972). Regulation of body fluids. Annu. Rev. Physiol. 34:235.
74. Share, L., and Levy, M.N. (1966). Carotid sinus pulse pressure, a determinant of plasma antidiuretic hormone concentration. Amer. J. Physiol. 211:721.
75. Kumada, M., Schmidt, R.M., Sagawa, K., and Tan, K.S. (1970). Carotid sinus reflex in response to hemorrhage. Amer. J. Physiol. 219:1373.
76. Share, L. (1965). Effects of carotid occlusion and left atrial distention on plasma vasopressin titer. Amer. J. Physiol. 208:219.
77. Baisset, A. and Montastruc, P. (1957). Polyurine par distension auriculaire chez le chien; rôle de l'hormone antidiurétique. J. Physiol. (Paris) 49:33.
78. Brennan, L.A., Jr., Malvin, R.L., Jochim, K.E., and Roberts, D.E. (1971). Influence of right and left atrial receptors on plasma concentrations of ADH and renin. Amer. J. Physiol. 221:273.
79. Shu'ayb, W.A., Moran, W.H., Jr., and Zimmermann, B. (1965). Studies of the mechanism of antidiuretic hormone secretion and the postcommissurotomy dilutional syndrome. Ann. Surg. 162:690.

188   Gauer and Henry

4

80. Johnson, J.A., Moore, W.W., and Segar, W.E. (1969). Small changes in left atrial pressure and plasma antidiuretic hormone titers in dogs. Amer. J. Physiol. 217:210.
81. Kappagoda, C.T., Linden, R.J., Snow, H.M., and Whitaker, E.M. (1974). Left atrial receptors and the antidiuretic hormone. J. Physiol. 237:663.
82. Share, L. (1975). The role of cardiovascular receptors in the control of ADH release. Cardiology. In press.
83. Ledsome, J.R., Linden, R.J., and O'Connor, W.J. (1961). The mechanisms by which distension of the left atrium produces diuresis in anaesthetized dogs. J. Physiol. 159:87.
84. Lydtin, H., and Hamilton, W.F. (1964). Effect of acute changes in left atrial pressure on urine flow in unanesthetized dogs. Amer. J. Physiol. 207:530.
85. Ledsome, J.R., and Mason, J.M. (1972). The effects of vasopressin on the diuretic response to left atrial distension. J. Physiol. 221:427.
86. Arndt, J.O., Reineck, H., and Gauer, O.H. (1963). Ausscheidungsfunktion und Hämodynamik der Nieren bei Dehnung des linken Vorhofes am narkotisierten Hund. Pflügers Arch. 277:1.
87. Kinney, M.J., and DiScala, V.A. (1972). Renal clearance studies of effect of left atrial distension in the dog. Amer. J. Physiol. 222:1000.
88. Gupta, P.D., Henry, J.P., Sinclair, R., and von Baumgarten, R. (1966). Responses of atrial and aortic baroreceptors to nonhypotensive hemorrhage and to transfusion. Amer. J. Physiol. 211:1429.
89. Claybaugh, J.R., and Share, L. (1973). Vasopressin, renin, and cardiovascular responses to continuous slow hemorrhage. Amer. J. Physiol. 224:519.
90. Bonjour, J.P., and Malvin, R.L. (1970). Stimulation of ADH release by the renin-angiotensin system. Amer. J. Physiol. 218:1555.
91. Mouw, D., Bonjour, J.-Ph., Malvin, R.L., and Vander, A. (1971). Central action of angiotensin in stimulating ADH release. Amer. J. Physiol. 220:239.
92. Muers, M.F., and Sleight, P. (1972). Action potentials from ventricular mechanoreceptors stimulated by occlusion of the coronary sinus in the dog. J. Physiol. 221:283.
93. Karim, F., Kidd, C., Malpus, C.M., and Penna, P.E. (1972). The effects of stimulation of the left atrial receptors on sympathetic efferent nerve activity. J. Physiol. 227:243.
94. Clement, D.L., Pelletier, C.L., and Shepherd, J.T. (1972). Role of vagal afferents in the control of renal sympathetic nerve activity in the rabbit. Circ. Res. 31:824.
95. Bunag, R.D., Page, I.H., and McCubbin, J.W. (1966). Neural stimulation of release of renin. Circ. Res. 19:851.
96. Gordon, R.D., Küchel, O., Liddle, G.W., and Island, D.P. (1967). Role of the sympathetic nervous system in regulating renin and aldosterone production in man. J. Clin. Invest. 46:599.
97. Wågemark, J., Ungerstedt, U., and Ljungquist, A. (1968). Sympathetic innervation of the juxtaglomerular cells of the kidney. Circ. Res. 22:149.
98. Davis, J.O. (1973). The control of renin release. Amer. J. Med. 55:333.
99. De Quattro, V., and Miura, Y. (1973). Neurogenic factors in human hypertension: mechanism or myth? Amer. J. Med. 55:362.
100. Thron, H.L., Brechmann, W., Wagner, J., and Keller, K. (1967). Quantitative Untersuchungen über die Bedeutung der Gefässdehnungsreceptoren im Rahmen der Kreislaufhomoiostase beim wachen Menschen. I. Das Verhalten von arteriellem Blutdruck und Herzfrequenz bei abgestufter Veränderung des transmuralen Blutdrucks im Bereich des Carotissinus. Pflügers Arch. 293:68.
101. Eckert, P., Kirsch, K., Behn, C., and Gauer, O.H. (1967). Wasser- und Salzhaushalt bei langdauernder Immersion im Wasserbad. Pflügers Arch. 297:R70.
102. Gauer, O.H. (1968). Osmocontrol versus volume control. Fed. Proc. 27:1132.
103. Korz, R., Fischer, F., and Behn, C. (1969). Renin-Angiotensin-System bei simulierter Hypervolämie durch Immersion. Klin.Wschr. 47:1263.
104. Epstein, M., and Saruta, T. (1971). Effect of water immersion on renin-aldosterone and renal sodium handling in normal man. J. Appl. Physiol. 31:368.
105. Epstein, M., Katsikas, J.L., and Duncan, D.C. (1973). Role of mineralocorticoids in the natriuresis of water immersion in man. Circ. Res. 32:228.
106. Epstein, M., Pins, D.S., Sancho, J., and Haber, E. (1976). Suppression of plasma renin and plasma aldosterone during water immersion in normal man. J. Appl. Physiol. In press.

107. Lazzara, R., Carson, R.P., and Klain, G.J. (1970). Diuresis initiated by coronary stretch receptors: A neurogenic mechanism for intravascular volume homeostasis. Fed. Proc. 29:460.
108. Ledsome, J.R., and Linden, R.J. (1968). The role of left atrial receptors in the diuretic response to left atrial distension. J. Physiol. 198:487.
109. Kappagoda, C.T., Linden, R.J., and Snow, H.M. (1973). Effect of stimulating right atrial receptors on urine flow in the dog. J. Physiol. 235:493.
110. Kappagoda, C.T., Linden, R.J., and Snow, H.M. (1972). The effect of distending the atrial appendages on urine flow in the dog. J. Physiol. 227:233.
111. Anderson, C.H., McCally, M., and Farrell, G.L. (1959). The effects of atrial stretch on aldosterone secretion. Endocrinology 64:202.
112. Cryer, G.L., and Gann, D.S. (1974). Right atrial receptors mediate the adrenocortical response to small hemorrhage. Amer. J. Physiol. 227:325.
113. Goetz, K.L., Hermreck, A.S., Slick, G.L., and Starke, H.S. (1970). Atrial receptors and renal function in conscious dogs. Amer. J. Physiol. 219:1417.
114. Goetz, K.L., Bond, G.C., Hermreck, A.S., and Trank, J.W. (1970). Plasma ADH levels following a decrease in mean atrial transmural pressure in dogs. Amer. J. Physiol. 219:1424.
115. Zehr, J.E., Johnson, J.A., and Moore, W.W. (1969). Left atrial pressure, plasma osmolality, and ADH levels in the unanesthetized ewe. Amer. J. Physiol. 217:1672.
116. McCance, R.A. (1936). Experimental sodium chloride deficiency in man. Proc. R. Soc. Lond. (Biol.) 119:245.
117. Leaf, A., and Frazier, H.S. (1962). Some recent studies on the actions of neurohypophyseal hormones. In Ch. K. Friedberg (ed.), Heart, Kidney and Electrolytes, pp. 263–80. Grune and Stratton, New York-London.
118. Arndt, J.O. (1965). Diuresis induced by water infusion into the carotid loop and its inhibition by small hemorrhage. Pflügers Arch. 282:313.
119. Johnson, J.A., Zehr, J.E., and Moore, W.W. (1970). Effects of separate and concurrent osmotic and volume stimuli on plasma ADH in sheep. Amer. J. Physiol. 218:1273.
120. Dunn, F.L., Brennan, T.J., Nelson, A.E., and Robertson, G.L. (1973). The role of blood osmolality and volume in regulating vasopressin secretion in the rat. J. Clin. Invest. 52:3212.
121. Nielsen, M., and Smith, H. (1952). Studies on the regulation of respiration in acute hypoxia. With an appendix on respiratory control during prolonged hypoxia. Acta Physiol. Scand. 24:293.
122. Szczepańska-Sadowska, E. (1972). The activity of the hypothalamo-hypophysial antidiuretic system in conscious dogs. I. The influence of iso-osmotic blood volume changes. Pflügers Arch. 335:139.
123. Moses, A.M., and Miller, M. (1974). Osmotic influences on the release of vasopressin. In R.O. Greep, E. Knobil, and S.R. Geiger (eds), Sec. 7: Endocrinology, Vol, 4, pp. 225–42. American Physiological Society, Washington, D.C.
124. Zehr, J.E., Hawe, A., Tsakiris, A.G., Rastelli, G.C., McGoon, D.C., and Segar, W.E. (1971). ADH levels following nonhypotensive hemorrhage in dogs with chronic mitral stenosis. Amer. J. Physiol. 221:312.
125. Wood, P. (1963). Polyuria in paroxysmal tachycardia and paroxysmal atrial flutter and fibrillation. Brit. Heart J. 25:273.
126. Kilburn, K. (1964). Fluid volume control and induced arrhythmias. Clin. Res. 12:186.
127. Mersch, F.D., and Arndt, J.O. (1969). Der Dehnungszustand der Herzvorhöfe unter dem Einfluss künstlicher Herzfrequenzänderungen bei narkotisierten Katzen. Pflügers Arch. 311:55.
128. Zucker, I.H., and Gilmore, J.P. (1973). Left atrial receptor discharge during atrial arrhythmias in the dog. Circ. Res. 33:672.
129. Haynal, I., and Matsch, J. (1968). Paroxysmal Tachycardia, p. 111 (Spastic urine). Akadémiai Kiadó, Budapest.
130. Rosing, D.R., Grossman, R.A., Goldberg, M., and Szidon, J.P. (1973). Mode and mechanism of the diuretic response to paroxysmal tachycardia in conscious dogs. Circulation 48 (suppl.):31.
131. Goetz, K.L., and Bond, G.C. (1973). Reflex diuresis during tachycardia in the dog. Evaluation of the role of atrial and sinoaortic receptors. Circ. Res. 32:434.

190   Gauer and Henry

132. Goodyer, A.V.N., and Glenn, W.W.L. (1955). Observations on the hyponatremia following mitral valvulotomy. Circulation 11:584.
133. Bruce, R.A., Merendino, K.A., Dunning, M.F., Schibner, B.H., Donohue, E., Carlsen, E.R., and Cummins, J. (1955). Observations of hyponatremia following mitral valve surgery. Surg. Gynecol. Obstet. 100:293.
134. Gauer, O.H., Henry, J.P., and Sieker, H.O. (1961). Cardiac receptors and fluid volume control. Prog. Cardiovasc. Dis. 4:1.
135. Bartter, F.C., and Schwartz, W.B. (1967). The syndrome of inappropriate secretion of antidiuretic hormone. Amer. J. Med. 42:790.
136. George, J.M., Capen, C.C., and Phillips, A.S. (1972). Biosynthesis of vasopressin in vitro and ultrastructure of a bronchogenic carcinoma. J. Clin. Invest. 51:141.
137. Boasberg, P.D., Rosenbloom, A., and Henry, J.P. (1975). Intrathoracic volume receptors: Control of vasopressin, and sodium and water excretion. Clin. Res. 23:92A.
138. Thornton, W.E., Hoffler, G.W., and Rummel, J.A. (1974). Anthropometric changes and fluid shifts. In R.S. Johnston and L.F. Dietlein (eds.), Proceedings of the Skylab Life Sciences Symposium, Vol. 2, pp. 637–658, NASA TM X-58154. NASA, Houston, Texas.
139. Thornton, W.E., and Hoffler, G.W. (1974). Hemodynamic studies of the legs under weightlessness. In R.S. Johnston and L.F. Dietlein (eds.), Proceedings of the Skylab Life Sciences Symposium, Vol. 2, pp. 623–635, NASA TM X-58154. NASA, Houston, Texas.
140. Thornton, W.E., and Ord, C.J. (1974). Physiological mass measurements in Skylab. In R.S. Johnston and L.F. Dietlein (eds.), Proceedings of the Skylab Life Sciences Symposium, Vol. 1, pp. 373–386, NASA TM X-58154. NASA, Houston, Texas.
141. Johnson, P.C., Driscoll, T.B., and LeBlance, A.D. (1974). Blood volume changes. In R.S. Johnston and L.F. Dietlein (eds.), Proceedings of the Skylab Life Sciences Symposium, Vol. 2, pp. 495–505, NASA TM X-58154. NASA, Houston, Texas.
142. Leach, C.S. (1974). Biochemical responses of the Skylab crewmen. In R.S. Johnston and L.F. Dietlein (eds.), Proceedings of the Skylab Life Sciences Symposium, Vol. 2, pp. 427–454, NASA TM X-58154. NASA, Houston, Texas.
143. Kaczmarczyk, G., Reinhardt, H.W., Kuhl, U., Riedel, J., Eisele, R., and Gatzka, M. (1975). Postprandial changes of left atrial pressure in conscious dogs on a high sodium intake. Pflügers Arch. 355:R51.
144. Goetz, K.L., Bond, G.C., and Bloxham, D.D. (1975). Atrial receptors and renal function. Physiol. Rev. 55:157.

International Review of Physiology
Cardiovascular Physiology II, Volume 9
Edited by Arthur C. Guyton and Allen W. Cowley
Copyright 1976   University Park Press   Baltimore

5
The Myocardium

G. A. LANGER, J. S. FRANK, and A. J. BRADY
University of California at Los Angeles, Medical Center

Supported by Grant HL-11351-09 from the United States Public Health Service, by a grant from the Castera Foundation, and by Career Development Award HL-04362-10 from the National Institutes of Health.

INTRODUCTION

This review focuses on three major aspects of the myocardium: 1) its ultrastructure, 2) excitation-contraction coupling processes, and 3) subcellular mechanical events. Despite the fact that each of these areas has developed rapidly over the past decade many significant problems remain. It is our goal not only to review this progress but to present what we believe to be realistic models of myocardial structure and function. It should be emphasized that these models should be viewed as "working schematics" since much experimental validation remains to be done. They do, however, serve to summarize present knowledge and define the problems which remain.

We have made a particular effort to include what we feel to be pertinent information derived from tissues other than myocardial. We have stressed comparison with skeletal muscle since both similarities and differences can be highly instructive in an attempt to achieve a comprehensive concept of myocardial physiology.

ULTRASTRUCTURE

This section is intended to present a morphological basis for the discussion to follow on excitation-contraction coupling (see "Ionic Mechanisms for Control of Myocardial Contraction") and on cardiac mechanics (see "Mechanical Responses in Cardiac Muscle"). The focus then will be on 1) those structures of the myocardial cell directly involved in excitation-contraction coupling (namely, the cell membrane (sarcolemma) and its extension, the transverse tubular system (T system) and the internal membrane system, the sarcoplasmic reticulum) and 2) the structures involved in cardiac mechanics (namely, the contractile proteins).

Membrane Structure

Knowledge of the ultrastructure of the mammalian myocardium has been rapidly expanding in recent years. Numerous excellent reviews on the ultrastructure of the heart have been published (1, 2). Although considerable information has come from the standard techniques of electron microscopy, increasingly more information about cell structure has been derived from the newer three-dimensional techniques of freeze-cleaving and scanning electron microscopy. Because of these newer developments in microscopy, we are closer to the visualization of membrane architecture and to the determination of the differences in molecular structure of various regions of the cell membrane than ever before. However, it is probably no surprise to the reader to learn that most, if not all, of our newer information on cell membrane structure comes from data on cells other than the myocardial cell. Numerous studies have been done on red blood cell membranes (3, 4) or cells grown in tissue culture (5), as they afford a simpler system in which to work. The complexity of the cardiac muscle cell and the lack of a good isolated membrane preparation have hindered membrane structural studies on this tissue.

However, studies on red blood cell membranes and other tissues have produced a wealth of new information on cellular membrane structure that appears common to all mammalian cells and thus deserves discussion here, especially since the newer concepts of cell membrane structure have such important implications to excitation-contraction coupling.

The cell membrane of almost all mammalian cells appears to be more complex than previously realized and our morphological definition of the membrane as only the trilaminar "unit" membrane is too limited and should be expanded to include the surface coat layers (6). In fact this "greater membrane" concept was first suggested by Bennett (7) over 10 years ago. Unfortunately the terminology used to describe the various components of the sarcolemma is not consistent in the literature and as a result much confusion exists over these structures. The term sarcolemma will be used here to include both the unit membrane and the surface layers, but for descriptive purposes it is necessary to divide the sarcolemma into three components (8): 1) the 90-Å "unit" membrane; 2) the glycoprotein *cell coat* located on the immediate outer surface of the unit membrane; and 3) the *external lamina* or carbohydrate coating just superficial to the cell coat.

Although both types of surface layers have been included under the single term glycocalyx (7), they have been found in some tissues, at least, to differ in fine structure and most probably in chemical composition. They are treated as separate layers in the following discussion, keeping in mind that the glycoprotein cell coat is an integral part of the unit membrane and the external lamina is intimately related to the cell coat (8, 9). Most importantly these surface layers are probably responsible for some of the fundamental properties of the cell membrane.

Sarcolemma and Transverse Tubular System   The sarcolemma is considered to be the morphological outer limit of the cell (1). At many sites along the myocardial cell surface the sarcolemma is deeply invaginated in the form of long, slender tubules called the transverse tubular system (T system). The length of the T tubules may be as much as half the diameter of the fiber (1). In addition, it has been demonstrated that many adjacent T tubules are interconnected by axial tubules (10–12).

The important structural features of these tubules are: 1) they are in open continuity with the extracellular space (see Figure 1) and thus the intraluminal fluid is continuous with the fluid of the interstitial space. This continuity of the T tubules has been repeatedly demonstrated by many workers (13–15) and elegantly visualized in freeze-cleaved preparations by Rayns, Simpson, and Bertrand (16); 2) the T tubules are large in diameter (1,000–2,000 Å), especially when compared to similar structures in skeletal muscle (400 Å); 3) the peripheral sarcolemma and the T tubular membrane appear to be structurally identical in that both demonstrate the unit membrane and the surface carbohydrate layers (1, 2) (see Figure 1).

Unit Membrane Structure   In cells fixed in glutaraldehyde followed by osmic acid and stained with uranyl acetate and lead citrate the most obvious structural component of the membrane is the trilaminar appearance—two thin, dark lines separated by a central lighter zone. This triple-layered pattern averages about 75–90 Å in thickness (see Figure 2). This appearance of the membrane was first described by Robertson (17), and Farquhar and Palade (18) and forms the major evidence for the so-called "unit" membrane theory which is based on the now "classic" Danielli-Davson (19) model of membrane structure. The membrane is visualized as lipid bilayer covered on both sides by layers of protein. Interpreted in light of the unit membrane model the two outer dark lines represent the polar head groups of the lipids and the absorbed non-lipids (proteins) stained by the heavy metals. The central light region represents the unstained hydrocarbon chains of the lipid molecules in the interior of the membrane. Implied in this model is a conceptualization of cell membranes as static structures with the component lipid and non-lipid molecules assuming specific positions. Both chemical and structural evidence has been accumulating to challenge seriously the unit membrane model (20) and the significance of the trilaminar appearance of the membrane (21). For example, the uniform appearance of membranes prepared for routine electron microscopy is difficult to reconcile with the marked differences in chemical composition of different membranes. The lipid and protein content of red cell membranes as compared to myelin membranes is very different—60% protein and 40% lipid for red blood cells—while myelin membranes are 80% lipid and 20% protein (21). Yet the electron optical images of thin sections from these widely different membranes appear quite similar. It is quite obvious that routine electron microscopy is not

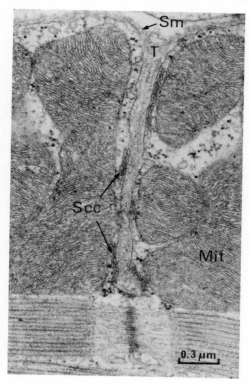

Figure 1. Longitudinal section of rabbit papillary muscle cell. The sarcolemmal membrane (*Sm*) invaginates into the cytoplasm at the level of the Z-band to form the transverse tubule (*T*). The surface carbohydrate layers extend into the lumen of the T tubule. Also present are subsarcolemmal cisternae (*Scc*) which abut on the T tubule membrane. Several mitochondria (*Mit*) are also visible. × 48,000.

particularly useful in understanding the membrane structure at the molecular level.

As new information on the chemistry of membrane proteins became available new models for membrane structure were needed. In 1973 Singer and Nicholson (22) described their fluid-mosaic model for the structure of biological membranes. Their model fits with thermodynamic principles and has considerable experimental evidence of both a chemical and structural nature for its support. They see the membrane proteins as globular and amphipathic with their hydrophilic ends protruding from the membrane and their hydrophobic ends embedded in the lipid bilayer. Some proteins, termed integral proteins, are simply embedded on one or the other side of the membrane; others pass entirely through the bilayer and are presumed to be involved in the formation of transmembrane pores. Such pores are envisioned as formed by groups of protein subunits in which the subunits are lined with negatively charged subgroups. The

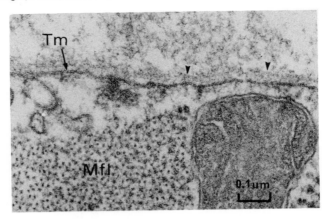

Figure 2. Cross-section of a portion of rabbit myocyte. The trilaminar "unit" membrane (*Tm*) is evident. The surface carbohydrate layers (*arrowheads*) are evident just superficial to the "unit" membrane. The cut ends of the myofilaments (*Mfl*) are also present. × 105,000.

theory has likened the proteins to icebergs in lipid sea, inferring then that the proteins are free to move in the lateral direction of the membrane (22). The structure of the membrane is conceived of as dynamic in which component proteins assume specific configurations that are in some way determined by the needs of the cell.

Edidin and Fambrough (23) attempted to measure the fluidity of the surface of cultured skeletal muscle fibers. They applied fluorescent-labeled antimembrane antibody to small parts of the surface area of a single muscle cell and then observed its spread from the point of application. The patches of antiplasma membrane antibody were observed to enlarge with time and this spread was inhibited by lower temperature. These data are consistent with the fluid nature of the lipid component of the membrane.

Strong structural support for the Singer-Nicholson model has come from the new ultrastructural technique of freeze-cleaving. With this technique fragments of tissue or isolated membranes are rapidly frozen to below −100°C and then cleaved at the low temperature. Cleavage planes follow natural paths of least resistance in frozen tissues and under most conditions this is through the interior of the membrane—between the two layers of lipid (24). As a result the internal molecular architecture of the cell membrane is exposed during the cleaving process. After the frozen tissue is cleaved the fractured face is replicated with carbon and platinum in a vacuum. The tissue is then thawed and washed away from the overlying replica which is then studied in the electron microscope. When material prepared this way is viewed in the electron microscope, clusters of particles are seen protruding above the lipid layer (25, 26). These particles are believed to represent the integral protein molecules of the membrane (22, 27). Membranes from a wide variety of cells show the presence of similar particles although the number and size of the particles vary with different membranes. It is important that red blood cell membranes and myelin membranes when prepared by the freeze-cleave technique have a very different appearance. The red blood cell membrane is covered with clusters of particles while myelin appears smooth and lacks these particles.

Many investigators are currently attempting to explain the different appearances of freeze-cleave membranes in molecular terms. Again most of the advances have been on the red blood cell membrane where several of the integral proteins have been isolated and characterized (28). However, some exciting freeze-cleave studies on striated muscle have shown the feasibility of identifying specific molecular components of this complex tissue. Rash and Ellisman (29) in a study on skeletal muscle were able to identify two classes of intramembrane particles—one which probably corresponds to the acetylcholine receptor-ionophore complexes of the neuromuscular junction and a second which may correspond to some of the molecules responsible for the electrogenic properties of the sarcolemma.

In cardiac muscle the studies on the molecular architecture of the unit membrane have been limited. The major work focused on the specialized portion of the membrane, the intercalated disc (30). The membrane at the periphery of the myocardial cell does, however, demonstrate many 60–120-Å particles that appear to be randomly distributed (see Figure 3). As yet freeze-cleave studies have not demonstrated any consistent differences in internal structure between the peripheral sarcolemma and the transverse tubular sarcolemma (2). However, this does not appear to be the case in skeletal muscle. Franzini-Armstrong (31) working with skeletal muscle from the tarantula spider found that the T tubular sarcolemma has noticeably fewer small and intermediate particles than the peripheral sarcolemma. The functional significance of this difference is not known.

Differences in the internal membrane architecture have been encountered in areas where the cell membrane from adjacent cardiac cells forms specialized junctions—the intercalated discs which are unique to cardiac cells. Each intercalated disc is formed by three types of junctional specialization designated fascia adherens, macula adherens (desmosomes), and the nexus (gap junction) (2). Freeze-cleave studies of these specialized regions of the sarcolemma are under active investigation. The internal membrane structure at the nexus is definitely different from the internal structure of the fascia adherens and the desmosome (30). At the nexus the interior of the membrane has packed arrays of 60-Å particles. At the desmosomes and fasciae adherens fine filaments appear to be irregularly distributed (2). The particles and filaments presumably represent the tails of the protein molecules seen anchored to the lipid bilayer, and some of the particles that penetrate the full diameter of the membrane may represent sites of pore formation (32).

The nexus functionally, as well as structurally, represents a specialized transport site which electrically couples myocardial cells by permitting the low

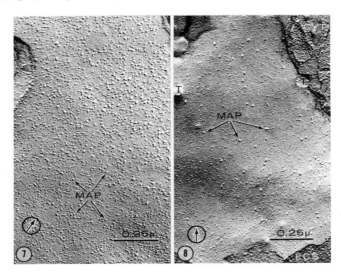

Figure 3. Freeze-cleave replica of sarcolemmal membrane of myocardial cell. Cleaving the membrane reveals a face A (*left*) and a face B (*right*) which are studded with many 60–100-Å particles. Reproduced with permission from N.S. McNutt and R.S. Weinstein (1970) (30).

resistance passage of ions from cell to cell. It is not entirely clear how this low resistance coupling is achieved but the fact that the membrane structure at the nexus is very different from that of the ordinary sarcolemma surely is important in its function.

While the newer information on the molecular structure of the cell membrane has come mainly from tissue other than the heart the applicability of this information to complex tissue such as the heart appears to be realistic. Freeze-cleave studies on cardiac cells have demonstrated a substructure of the myocardial cell membrane that is compatible with the newer models for the lipoprotein unit membrane.

Surface Coats There is now general agreement among electron microscopists that all cell membranes have surface layers superficial to the unit membrane. But until specialized methods were devised for the ultrastructural detection of carbohydrates the pressure of surface coats in multicellular organisms was difficult to detect. The staining techniques developed represent adaptations of well known histochemical procedures used by light microscopists to detect surface carbohydrates and polysaccharides. Most of the stains used, such as alcian blue, ruthenium red, lanthanum, and colloidal iron are polycationic colloids which bind to negatively charged surface coats and in so doing render the surface coat more electron dense and thus more visible in the electron microscope.

Cell coats have been demonstrated as carbohydrate-containing components of the cell membrane. The carbohydrates of the cell surface include glyco-proteins, glycolipids, and free polysaccharides (9). The glycoproteins of the surface coat contain many sialic acid-COOH groups which are negatively charged at physiological pH. The use of the polycationic stains has been useful in identifying the cell coat in the study of its distribution and its chemistry (8). Ruthenium red and alcian blue have been shown to react with polysaccharides (33). The colloidal iron staining technique indicates the presence of a large number of ionizable acidic groups, thought to be the COOH groups of sialic acid (34). Treatment with the enzyme neuraminidase, which hydrolyzes sialic acid moieties, results in the removal of most of the colloidal iron staining material.

To date there is no adequate way to visualize the molecular architecture of the surface layers and new methods are required. Unfortunately the freeze-cleaving technique has not been useful in this endeavor for, as pointed out, the membrane fractures along an internal hydrophobic plane so that transverse sections through the projecting carbohydrate structures are not frequently seen. As well, the high degree of hydration and filamentous structure of most surface coats makes it difficult to demonstrate its fine structure by freeze-cleaving (9).

The study of myocardial surface coatings is again largely incomplete. Fawcett and McNutt (1) in the comprehensive study of the myocardial ultra-structure describe the cell coat as a uniformly thick (500 Å) mat or feltwork of fine filaments along the peripheral and T tubular membrane. In some methods of preparation the surface coating appears layered with a less dense inner zone ($\sim$ 200 Å) adjacent to the unit membrane and a slightly denser outer zone (300 Å). Whether this layering is indeed representative of a structural and chemical difference is not known. Parsons (9) points out that the glycoprotein chains of the cell coat probably extend out from the lipoprotein membrane about 200–300 Å and that under conditions of tissue preservation the filaments may collapse and form the mat-like network seen in the electron microscope.

In skeletal muscle Zacks et al. (35) describe the different staining properties of the surface layers of the membrane. They found that an area (200 Å) immediately adjacent to the unit membrane (cell coat?) stained with colloidal iron while the rest of the external lamina did not except for a small band of stain at the most superficial region of the lamina. Ruthenium red on the other hand reacted with the whole surface layer in this tissue.

Colloidal iron staining and thus sialic acid localization have been done in the myocardial cells of the adult sheep and crayfish (36) and in neonatal rat heart cells grown in tissue culture (37). In the crayfish myocardium (36) the pattern of colloidal iron staining appears very similar to that described for skeletal muscle (35). In myocardial cells grown in tissue culture the particles of colloidal iron reaction are localized immediately adjacent to the unit membrane and extend a distance of approximately 200 Å (see Figure 4). Neuraminidase applied to this preparation almost completely eliminates the staining reaction.

Ruthenium red staining of the myocardial cell on the other hand covers the outer leaflet of the trilaminar plasma membrane, the cell coat and the external

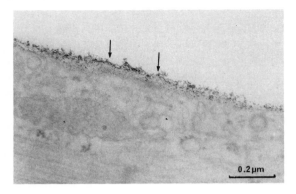

Figure 4. Electron micrograph of a portion of neonatal rat myoblast grown in tissue culture and exposed to colloidal iron. Section is unstained except for colloidal iron reaction. Note accumulation of stain on surface of the cell (*arrows*). × 95,000.

laminar with dense material. This pattern of ruthenium red staining has been demonstrated in the adult mammalian myocardium (33, 36). Lanthanum (La) has also been used to demonstrate the presence of surface layers in the myocardium and has a staining pattern similar to ruthenium red. Lesseps (38) first demonstrated this in embryonic chick hearts. Langer and Frank (39) studied the effect of ionic La on the function of neonatal rat hearts grown in tissue culture. When the living myocardial cells were exposed to ionic La solutions contractile function declined rapidly and La was deposited on sarcolemma surface in a very similar pattern as described for ruthenium red. A similar staining pattern for La was found in the adult myocardial cell of cat, dog, and rabbit by Martinez-Palomo (40). Unfortunately the precise role of La in staining of the cell surface is uncertain. Lesseps considered La staining material to contain lipid and probably protein polysaccharide. It certainly seems likely that La is reacting with the high density of negatively charged sites present in the surface layers of the myocardial cell.

As mentioned earlier the glycoprotein surface coat material of the peripheral sarcolemma continues down into the T tubules and thus is brought in close association with the contractile filaments (1, 2). The surface coat of the T system demonstrates the same staining reactions with ruthenium red, colloidal iron, and lanthanum as the peripheral sarcolemma, indicating that the cell coat and lamina lining the T tubules also contain a high density of negatively charged sites (2, 36, 41).

It is obvious that the myocardial membrane is a complex structure, and before we can begin to understand its function a greater understanding of the interactions between proteins, lipids, and carbohydrates in membranes is needed. As the further details of structure unfold, correlation with function becomes our most exciting challenge. Further attention should be paid to the possibility that the surface coat concentrates and holds ions such as calcium and thus plays a crucial role in E-C coupling (see "Ionic Mechanisms for Control of Myocardial Contraction").

Sarcoplasmic Reticulum The sarcoplasmic reticulum (SR) in mammalian cardiac muscle forms a lace-like network of tubules that range in diameter between 200–400 Å. The reticulum is closely applied to the myofilament mass and courses over many sarcomeres without interruption. Portions of the sarcoplasmic reticulum form flattened cisternae that are closely apposed to the sarcolemma either at the periphery of the cell or at the T tubules. The term subsarcolemmal cisternae has been applied to these cisternae by Fawcett (1). In contrast to skeletal muscle the cisternae of cardiac cells are considerably less capacious. For example, the fractional cell volume of the cisternal portion of the SR in the mammalian myocardium is 0.3% (42) while in frog fast twitch fibers it is 4.1% (43), in mammalian fast twitch fibers approximately 1.4% (44), and in mammalian slow twitch 0.92% (44). The terminal cisternae are thought to contain internal stores of Ca and the ultrastructural data given above suggest that skeletal muscle (especially frog fast twitch) has a much larger capacity for storage of Ca than the mammalian myocardium (see "Ionic Mechanisms for Control of Myocardial Contraction").

Most of our information on the ultrastructure of membranes from the sarcoplasmic reticulum has come from studies on isolated fractions (microsomes) and most of the work has been done on skeletal muscle. Thin sectioning shows that fragmented sarcoplasmic reticulum consists of vesicles bounded by a single trilaminar membrane. Freeze-cleaved experiments on skeletal muscle sarcoplasmic reticulum reveal particles of about 90–100 Å in diameter in the hydrophobic interior of the membrane (45). Heart microsomes appear to have different freeze-cleave images in that few vesicles have the concentration of particles that is seen in skeletal microsomes (46). The numerous particles on the skeletal sarcoplasmic reticulum leaflets are believed to be calcium-activated ATPase (47). The sarcoplasmic reticulum is a membrane specialized for transport and binding of calcium and because of its high specialization has relatively few proteins. The major one is ATPase (48). The contribution of this protein to the structure of the membrane is under active investigation.

The presence of a surface coat on membranes of the sarcoplasmic reticulum has not yet been demonstrated. However, the surfaces of membranes from isolated vesicles of sarcoplasmic reticulum from skeletal muscle are known to have an acidic protein fraction with a high Ca^{2+} affinity on the exterior of the membrane (48). It is very possible that these acidic residues contain sialic acid, since both cardiac and skeletal muscle microsomes stain on their external surfaces with colloidal iron (see Figures 5 and 6). Attention should be focused on the areas where the sarcoplasmic reticulum and the surface of the myofibers are closely associated, since presumably it is here that the exchange of material between the internal membrane system and the cell surface occurs. The subsarcolemmal cisternae, whether they abut on the peripheral sarcolemma or the T tubular sarcolemma, are separated from the surface membrane by about 150–

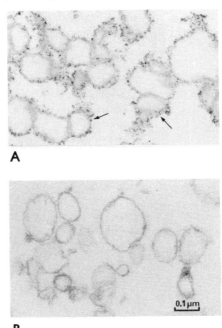

A

B

Figure 5. *A,* isolated sarcoplasmic reticulum from rabbit that was exposed to colloidal iron staining. Note stain reaction on surface of the vesicles. × 114,000. *B,* isolated sarcoplasmic reticulum from rabbit myocardium. Control vesicles lightly stained with uranyl acetate and lead citrate. × 114,000.

200 Å. The flattened lumen of the cisternae contains a granular dense material about which little is known. The cisternae appear to be connected to the sarcolemma by poorly visible globular densities. It is interesting that in skeletal muscle that junction between the terminal cisternae—T system membrane contains discrete junctional spots or "feet." It was postulated that these feet may represent functional pores or channels between the sarcoplasmic reticulum and the extracellular milieu, but more recent evidence obtained from freeze-cleave experiments fails to give evidence of these pores (31). The polycationic stains (La, ruthenium red, and colloidal iron) when used as routine extracellular traces, or stains for the cell surface, do not appear to enter intact cells and thus do not stain any portion of the sarcoplasmic reticulum. In a recent study in skeletal muscle, ruthenium red was reported in the lumen of the terminal cisternae, but this only occurred when the tissue was left overnight in ruthenium red and even then only in a few fibers (49). In another study La was reported in the sarcoplasmic reticulum of the mouse diaphragm, but again this occurred with only one of many specimens similarly stained and was not reproducible. La was presumed to enter the sarcoplasmic reticulum through an artificial communication to the extracellular space (50). At the present time there is little evidence to demonstrate continuity between the sarcoplasmic reticulum and the cell surface in the in vivo state.

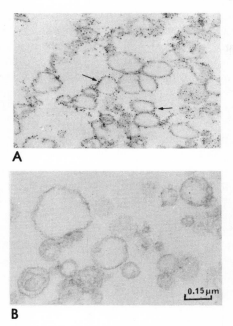

A

B

Figure 6. *A,* skeletal muscle sarcoplasmic reticulum stained with colloidal iron. Particles of colloidal iron are located exclusively in surface of the vesicles. × 78,000. *B,* isolated sarcoplasmic reticulum from skeletal muscle. Control vesicles lightly stained with uranyl acetate and lead citrate. × 78,000.

Contractile Proteins

Background   The most distinctive feature of the cardiac cell cytoplasm is the more or less continuous mass of myofilaments in which mitochondria and other cellular organelles are interspersed. When longitudinal sections of muscle are looked at under low magnification it is obvious that the myofilament mass is arranged in a regular alternating array of transverse bands. Light microscopists originally named these transverse bands by their birefringent properties. The strongly birefringent band (anisotropic) was termed the A-band while the weaker birefringent band (isotropic) was termed the I-band. Each I-band is bisected by a dense dark line called the Z-line. The segment between two successive Z-lines has been termed the sarcomere (1). In longitudinal sections it is obvious that the sarcomere is a repeating structural unit and indeed experimentation has proven it to be the functional unit of contraction. Within each sarcomere the contractile material, the myofilaments, are arranged in a definite pattern. Under the electron microscope it is obvious that the A- and I-bands are made up of thick and thin filaments (51). The I-band consists of thin filaments (50–70 Å) about 1.0 μm long. The thick filaments (100–150 Å) are about 1.5 μm long and are found in the A-band (51). Actually the length of the A-band is determined by the length of the thick filaments. The two sets of thin filaments are connected end to end at the Z-line while their opposite ends interdigitate with the thick filaments of the A-band (see Figure 7). The degree of interdigitation depends on the length of the muscle and in cardiac muscle fixed at a sarcomere length of 2.2 μm there is maximum overlap of thick and thin filaments with the A-band (52). It is this region of overlap of the thick and thin filaments that generates the force of muscle contraction by means of cross-bridges which project from the thick filaments and contact the thin filaments (see "Mechanical Responses in Cardiac Muscle"). In the center of the A-band there are only thick filaments and this region, called the H-zone, appears lighter in routinely stained sections. The H-zone will vary in its width with the sarcomere spacing. The more stretched the fiber the less overlap between thick and thin filaments and the more prominent the H-zone.

In the central region of the A-band the thick filaments are held in position by slender cross-links. The material forming the cross-links or bridges produces a thin band of increased electron density called the M-band.

The thick and thin filaments within the various bands of the sarcomere have a characteristic appearance when they are viewed in cross-sections (see Figure 8). Electron micrographs show four different patterns for the cut ends of the filaments: thick filaments are seen only in the H- and M-bands, both thick and thin filaments are present in the regions of overlap within the A-band, in the I-band only thin filament ends are seen, and in the Z-band a complex and poorly understood pattern of thin filaments is seen. In regions of overlap each thick filament is surrounded by six thin filaments (see Figure 8). Bacaner et al. (53) in a study on the microstructure of deep-frozen unfixed, unstained, hydrated muscle cells found morphological features that differ from those described above for conventionally prepared tissue. In frozen muscle viewed in the scanning electron microscope, myofibrils appeared to be composed of previously undescribed longitudinal structures between 400 and 1,000 Å wide that were termed "macromyofilaments." How these macromyofilaments relate to the now classic

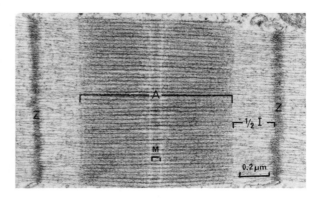

Figure 7. Electron micrograph of typical sarcomere unit from rabbit myocardium. This sarcomere is slightly contracted and as a result no H-zone is present. A-, I-, Z-, and M-bands are quite evident and the actin filaments can be seen interdigitating with the myosin filaments in the A band. × 72,000.

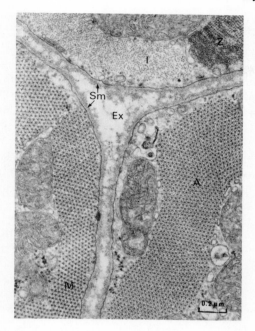

Figure 8. Cross-section of three rabbit papillary myocytes. The cut ends of the filaments produce distinct arrays in the A-, I-, Z-, and M-bands. Extracellular space (*Ex*); sarcolemmal membrane (*Sm*). × 75,000.

morphology of the myofilaments gleaned from fixed tissue or to the morphology of muscle in the living state is uncertain and awaits further investigation.

Contractile Proteins—Substructure   Most of the work on the substructure of the myofilaments has been done on proteins extracted from rabbit skeletal muscle since contractile proteins of the heart are more difficult to purify. Cardiac and white skeletal myosins have similar physiochemical and electron microscopic properties and the actin and tropomyosins from cardiac and skeletal muscle are also very similar. The differences that do exist between skeletal and cardiac muscle myosin involve: 1) different amino acid composition of myosin; and 2) a difference in ATPase activity of the two myosins. None of the above represent any major qualitative functional difference, so much of the data gained from skeletal muscle proteins have been applied to the heart (54).

It is now well established that myosin is the chief constituent of the thick filament. When myosin protein is selectively extracted from muscle the thick filament of the A-band disappears. When purified myosin is allowed to form aggregates the myosin monomers form elongated filaments that are almost indistinguishable from the thick filaments (55). In preparations where the myosin protein is extracted and filaments allowed to reform, the globular ends of the individual myosin molecules project from the filament and appear to be identical with the cross-bridge of the thick filaments of intact muscle (56). Both skeletal and cardiac myosins are extremely large molecules, indicating that these proteins are composed of several subunits. Tryptic digestion of skeletal myosin produces two products, readily separated in the analytical ultracentrifuge, called meromyosins. The more rapidly sedimenting one has been designated heavy meromyosin (HMM). It possesses both ATPase activity and the ability to bind to actin (see "Mechanical Responses in Cardiac Muscle"). The more slowly sedimenting fraction is called light meromyosin (LMM). HMM exists primarily in a globular form to which a highly helical rod is attached. Thus HMM represents the "head" of the myosin molecule and LMM represents the tail of the molecule. LMM is entirely α helices wound around each other (51, 55). But Squire (57) has proposed a three-stranded model for the thick filament, so the exact structure is not definitely determined. The structure of the myosin molecule should relate to the morphology of the thick filament seen in the electron microscope. Indeed all of the microscopic evidence indicates that the backbone of the thick filament is composed of the interlaced tails of LMM, while the globular head portion of HMM projects outward from the filament and constitutes the cross-bridges. Recent structural information yields an estimate of between 254 (56) and 264 (57) myosin molecules per thick filament and probably not more than 1 myosin molecule per cross-bridge (58).

The thin filaments of both cardiac and skeletal muscle consist mainly of three proteins: actin, tropomyosin, and troponin (59). The predominant protein of the thin filament is polymerized actin (F-actin). Two strands of F-actin wrapped in an α helix form the backbone of the thin filament. The double helical arrangement of actin has been seen in the electron microscopic examination of both cardiac and skeletal muscle (55). Tropomyosin is a highly helical protein and the elongated tropomyosin molecule appears to lie in the grooves between the two strands of actin. Troponin was originally thought to be a single globular protein but has recently been shown to be a complex of three different proteins: troponin-T[3] which interacts with tropomyosin, troponin-I which inhibits actomyosin ATPase, and troponin-C, the calcium-binding protein (55, 60). Both the troponin complex and tropomyosin are bound to actin filaments with troponin binding sites spaced 400 Å apart along the actin strands.

In cross-section, the M-band appears as bridges connecting each thick filament to six neighboring thick filaments (see Figure 9). Antibody staining studies by Pepe (61) have demonstrated that the M-band consists of a protein other than myosin, actin, or tropomyosin. The so-called M-band material has been resolved into two components and is known to be very loosely attached to the thick filaments, since it can be easily removed from the thick filaments by homogenization (62).

The distance between the thick filaments is known to increase as thin filaments slide into the A-band and the sarcomere shortens during contraction. This implies that the M-band cross-link must be flexible and have the ability to elongate and shorten as sarcomere length changes. Just how this dynamic structural change would occur is not clear at the present time. No change in

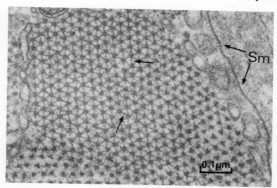

Figure 9. Cross-section through the M-band. The thick filaments are seen interconnected by radial cross-links (*arrows*) and are held in a hexagonal array. Sarcolemmal membrane (*Sm*).

M-band pattern at different sarcomere lengths has been observed in routine electron microscopy of either longitudinal or cross-sections (63). However, the extractability of M-band protein and the antibody staining characteristics of this protein complex are different at different sarcomere lengths (64). The relationship between these observations on changes in M-band properties at different sarcomere lengths and the morphology of M-band protein components needs further exploration.

The Z-bands at either end of the sarcomere are considered to form the fixed points to which the actin filaments attach and upon which they exert tension during contraction. The Z-bands also serve as links between successive sarcomeres within the myofibril. Any model of the structure of the Z-band must take into account its mechanical role of anchoring the thin filaments and in addition the fact that the Z-band is most probably involved in the production of new contractile units within the fiber.

At least three different interpretations of Z-line fine structure have been proposed. One test of these models is their ability to explain the images of the Z-band obtained by electron microscopy. However, one of the features of the Z-band is the diversity of its appearance under the electron microscope, both in longitudinal and transverse sections. In addition, the Z-band in cardiac muscle is more difficult to analyze since there is a greater density of "Z material" which tends to mask the Z-band lattice (1). No model yet proposed of Z-band structure is completely satisfactory, but two of the many models proposed have the widest following. The model of Knappes and Carlsen (65) suggests that each I-band filament is connected to four neighboring filaments by special Z-filaments. The other model put forth by Kelly (66) is based on the fact that the two strands of F-actin are wrapped in an α helix and suggests that as actin filaments enter the Z-band each F-actin strand uncoils, loops around another F-actin strand and recoils with an F-actin strand belonging to the adjacent I-band filament. This looping model argues that actin strands could weave a meshwork strong enough to transmit tension during contraction.

The question of Z-band composition is an important unsolved problem. It was suggested by Huxley (51) that tropomyosin is a major component of the Z-band. However, attempts to localize tropomyosin in the Z-line using labeled antibody techniques have so far failed. However, it has been pointed out by Pepe (61) that failure to demonstrate tropomyosin with this technique may be due to the blocking of reactive sites with other Z-band material. In any case the question of whether tropomyosin exists in the Z-line remains open. However, the protein α-actinin in the Z-band has been established (67). In addition, it has been suggested that the Z-band may have lipid and carbohydrate components (68). At this point there still is not enough information about Z-band components to explain why this band stains more intensely with heavy metal stains than the remainder of the I-band (see Figure 7).

A number of investigators have noticed the similarity of Z-band density in the electron microscope to the electron-dense material associated with the inner surface of the sarcolemma at portions of the intercalated discs, namely, the desmosomes and the fascia adherens (67, 68). Though the concept that Z-band material is involved myofibrillogenesis is still unresolved, many studies have pointed out that myofibrillar assembly is most frequently observed at the cell periphery (69). It has been suggested that Z-bands are derived from these specialized portions of the cell surface. The fact that Z-band material is abundant at the cell periphery in growing myoblasts and in adult hypertrophying cells (70) is support for an active role of the Z-band in the generation of new sarcomeres.

IONIC MECHANISMS FOR CONTROL OF MYOCARDIAL CONTRACTION

Currently, there remains a good deal of controversy with respect to the specific mechanisms by which the ions affect the cardiac contractile response. A large part of the controversy stems from the fact that diverse techniques are used to measure ionic movements. The techniques are more or less sensitive for different fractions of total flux and therefore the models derived from data obtained may not be comprehensive or may present apparent paradoxes. In addition there is a tendency to extrapolate data defined in skeletal muscle to the modeling of excitation-contraction processes of heart muscle and this is clearly unrealistic. Finally, data from nonphysiological preparations have been used as the basis for definition of fundamental physiological processes and this has presented problems. The goal of this section is to review the diverse experimental results and their interpretation and attempt to reconcile them into a more cohesive picture of ionic control mechanisms in the mammalian myocardium.

Background

It is now accepted that ionic control of contraction focuses upon control of calcium delivery to myofilamentous troponin sites (see "Mechanical Responses in Cardiac Muscle"). The quantity of Ca required for myofibrillar activation of dog heart has been defined in a recent excellent study by Solaro et al. (71).

Complete activation requires slightly more than 90 μmoles of Ca/kg wet wt. The relationship between activation of tension and total Ca required is, however, sigmoid and quite steep through a major portion of the activation range. For example an increment of only 13 μmoles of Ca/kg wet wt per beat is required to augment tension from 50 to 90% maximum and of only slightly more than 5 μmoles/kg to increase tension from 20 to 50%. By comparison skeletal muscle, as Solaro et al. (71) point out, requires approximately three times the amount of calcium, i.e., 300 μmoles/kg wet wt for maximal activation. This is due to the fact that the skeletal muscle cell contains three times the myofibrillar protein of the mammalian cardiac cell (72).

Therefore, in heart muscle, small variations in the delivery of Ca to myofilamentous sites will produce large changes in force development. Evidence of enhanced Ca influx has been demonstrated to occur in association with the positive inotropism associated with increased frequency of contraction (73–75), catecholamine administration (76, 77), digitalis (78, 79), and extracellular and intracellular sodium (Na) alteration (80–83). There is, then, general agreement that a large number of inotropic responses are dependent upon augmentation of Ca delivery to the myofilaments. The source of this Ca and the mechanisms by which the quantity delivered is controlled are not agreed upon.

Source and Movement of Contractile-dependent Calcium

If Ca is removed from the medium perfusing skeletal muscle with caution to preserve electrical activity of the cell membrane, force development is preserved for as long as 4½ hours (84). By contrast, Ca-free perfusion of mammalian heart muscle causes force to decline with a half-time of 45–70 s (85, 86). These relatively simple experiments point clearly to the fact that the source and/or route of Ca is very different in the two contractile tissues. Despite this, a mechanism for the release of Ca was recently proposed which was believed to be operative for both skeletal and cardiac muscle. The mechanism involved a "regenerative release" of Ca from the sarcoplasmic reticulum.

"Regenerative" or Ca-induced Ca Release   The possibility that concentrations of Ca that are subthreshold for direct contractile activation are capable of inducing or "triggering" the release of activating amounts of Ca from the sarcotubular system was put forth in 1970 from two laboratories (87, 88). The experimental preparations were skeletal muscle fibers from which the sarcolemma had been removed or "skinned" (Natori preparation) (89). In order that release from the sarcotubules be triggered, the fibers had to be heavily preloaded with Ca. Evidence for a similar process of triggered release was found in cardiac fibers with disrupted sarcolemmal membrane (90–93). These fibers demonstrate no action potential activity but are capable of rhythmical mechanical activity when Ca is added to EGTA-buffered solution in an amount sufficient to produce a Ca concentration of between 5×10^{-8} and 5×10^{-7} M (92, 93). The demonstration of triggered or regenerative release in these fibers which were skinned, or in which the sarcolemma was otherwise disrupted, formed the basis for proposals that the mechanism could be extrapolated to skeletal and cardiac

fibers with sarcolemma intact and was, therefore, a physiological component of excitation-contraction (E-C) coupling (94, 95). Subsequent investigation has proven this not to be the case.

As indicated above, Ca concentrations below 10^{-8} M are insufficient to produce regenerative release and intact muscles exposed to $[Ca]_0$ below this level would be expected to cease contraction. Armstrong, Bezanilla, and Horowicz (96) exposed single, intact fibers from frog semitendinosus muscle to EGTA-buffered solutions in which the Ca concentration was not more than 2×10^{-9} M. Force development continued for many minutes in this solution until the resting potential began to decline and excitability was lost. Andersson and Edman (84) used lanthanum to preserve action potential activity and noted that the twitch response of single semitendinosus fibers from the frog was preserved up to 4½ hours in Ca-free solution. Sandow, Pagala, and Sphicus (97) have also recently reported maintenance of essentially normal force development by frog sartorius fibers in 2×10^{-9} M Ca solution. In addition, Endo (98, 99) has recently reevaluated his original experiments and finds that Ca-induced Ca release is a property only of a sarcotubular system which is loaded with Ca to clearly nonphysiological levels. Triggered release could not be demonstrated to occur when the levels of cellular Ca are within the normal physiological range. More recent work by Fabiato and Fabiato (100) also indicates that triggered release is not a physiological mechanism in ventricular myocardium of the dog, cat, rabbit, or human. In these preparations nonphysiologically high Ca loading was necessary to demonstrate cyclic contractions in the skinned cells.

Therefore, Ca-induced release of Ca from the sarcotubular system is not physiologically operative in either skeletal or heart muscle.

Surface Ca Stores in Heart   Studies recently completed (101) show that the major component of force developed by skeletal muscles declines in zero $[Ca]_0$ at the same rate as isotopically (^{45}Ca) labeled Ca washes from cellular sites. These sites are almost certainly the Ca stores in the lateral cisternae of the sarcoplasmic reticulum (102) with release from these stores prompted in some way by depolarization and not by a Ca-induced Ca release.

The source of contractile-dependent Ca in the heart is clearly different. As noted previously in the cat (85) and rabbit (86) heart the rate at which force declines in the pressence of zero $[Ca]_0$ correlates most closely with a store of Ca which exchanges at a rate indistinguishable from that of interstitial exchange. The same relationship has also been defined for the guinea pig heart (101).

Despite the fact that decline of force and interstitial Ca exchange follow a similar time course it cannot be assumed that contractile-dependent Ca is homogeneously distributed in free solution in the interstitial space in heart muscle and simply flows across the sarcolemma to the myofilaments following each excitation. The action of E-C uncoupling agents indicates that such is almost certainly not the case.

Uncoupling Agents: (a) Cation Uncouplers   The ability of various cations to break specifically the link between excitation and contraction is well recognized. Their action, as might be expected, differs in heart and skeletal muscle. Those most extensively used in the heart are nickel (103, 104), zinc (105), manganese (104, 106), and lanthanum (107). The most potent uncoupler of this group of cations is La, a trivalent electron-dense ion. It has been used extensively over the past 5 years to provide further significant insight into the process of E-C coupling in the heart.

The addition of as little as 40 μM $[La]_o$ to the otherwise unaltered perfusion medium of mammalian heart muscle is capable of completely uncoupling excitation from contraction within 2–3 min (107). Coincident with its uncoupling action La produces a displacement of Ca from the tissue. The interpretation of its effects is largely dependent upon whether or not it crosses the sarcolemmal membrane and enters the cell. Results from a number of laboratories indicate that it does not (39, 40, 108–111). There is one study, however, which concludes that La penetrates intracellularly to the mitochondria and sarcoplasmic reticulum (112). In this study rat myometrium was exposed to radioactive ^{140}La for periods of 60 min and then homogenized. Cellular fractions including plasma membrane, endoplasmic reticulum, and mitochondria were prepared and counted for ^{140}La activity. It was noted that exposure, during homogenization, to 10 μM nonradioactive La did not affect the amount of ^{140}La in the intracellular fractions and from this it was concluded that the ^{140}La in the fractions was not released during homogenization and subsequently taken up. Since the original exposure to ^{140}La was in 5 mM La and this concentration was present in the extracellular space at the start of homogenization, it is not surprising that the addition of 10 μM La (altering the ^{140}La specific activity by only 0.2%) was not discernible. The evidence is strongly in favor of the conclusion that La does not gain access to intracellular structures.

If it is accepted that La does not cross the sarcolemma and that regenerative Ca release is not operative physiologically it follows that the major component of contractile-dependent Ca is located in La-accessible surface sites in the myocardium. As indicated under "Ultrastructure," La binds to the cellular coat. As it binds it has marked effects on cellular Ca exchange (39). It displaces a large component of rapidly exchangeable Ca and essentially eliminates Ca influx and efflux. The La-displaceable Ca is as much as 500 μmoles/kg wet tissue and this Ca exchanges at a rate indistinguishable from that in the interstitial space. It may, therefore, represent a "store" of superficial Ca—associated with the cell surface but in rapid equilibrium with Ca in the interstitial space per se.

The protein-polysaccharide surface coat has a high concentration of acidic residues which endows it with a high density of fixed negatively charged sites. It should, therefore, act as a cation exchanger with a graded affinity for various cations dependent on their size and charge. This was found to be the case for the Ca surface sites in myocardial tissue culture (113). The affinity for these sites was in the order La>Cd>Zn>Mn>Mg and this sequence was in the same order as the rate at which the ions diminished contractile force in ventricular tissue. Both the affinity and effect on tension correlated best with the nonhydrated radius of the ion. Those ions with radii closest to Ca (0.99 Å) were the most effective in displacement of Ca and in their ability to uncouple excitation from contraction.

The kinetic characteristics of the Ca displacement by these ions were most consistent with an action at the cellular surface.

The cation uncouplers, to varying degrees, act to displace a component of surface bound Ca and to inhibit trans-sarcolemmal Ca exchange. Their uncoupling action is not dependent upon excitation and contraction, i.e., exposure to the cations during a period of quiescence results in marked diminution of force evident in the first post-quiescent beat (114). This is in contrast to a group of pharmacological uncouplers.

(b) Pharmacological Uncouplers An organic compound, Verapamil, and its methoxy derivative D-600, have been found to be potent uncoupling agents (115, 116). Their mode of action is, however, quite distinct from the cations previously discussed in that the uncoupling effect does not develop unless repetitive contraction occurs. The effect on contractile force of these agents is proportional to the rate of beating (114, 117, 118). In addition, Verapamil does not directly displace a component of cellular Ca as do the cation uncouplers but acts to inhibit Ca influx. It is effective in both heart and skeletal muscle (114). These agents seem to inhibit movement of Ca to the membrane sites from which release to the myofilaments occurs. They have little effect on force development of the first beat after quiescence since the sites remain filled in the absence of contraction. Upon repetitive contraction the sites supply Ca to the myofilaments but their replenishment is blocked by these drugs.

(c) Uncoupling by Hydrogen Ion An acidosis produced by an increase in p_{CO_2} (respiratory acidosis) has a marked effect on E-C coupling and causes a large, rapid and readily reversible decline of myocardial contractility. A comparable metabolic acidosis, induced by a reduction in extracellular bicarbonate, has a much smaller effect (119–122). The uncoupling action secondary to a respiratory acidosis is, like the cations, not contraction dependent. In addition, the effect is strikingly different in heart and skeletal muscle (122, 123). Perfusion of fast twitch skeletal muscle with solution equilibrated to 20% CO_2 (pH = 6.8) has virtually no effect on contractile tension. By contrast, exposure of rabbit papillary muscle to the same p_{CO_2} rapidly reduces force development by 60–70% (122). The acidosis probably does not actively displace Ca from cardiac muscle, but inhibits both influx and efflux (122). Its action is, therefore, quite distinct when compared to either the cation or pharmacological uncouplers.

(d) Interpretation of Uncoupling Experiments The action of the various uncoupling agents reviewed above is most consistent with the placement of the major source of the heart's contractile-dependent Ca on cellular surface sites or, more specifically perhaps, in the surface coat which covers the sarcolemma and fills the transverse tubular system (see "Ultrastructure"). The cationic agents (e.g., La, Mn) most likely displace, in a competitive manner, Ca from the surface coat. In addition they block both inward and outward movements of Ca. These agents are relatively ineffective in skeletal muscle since it is unlikely that the deeper release sites (lateral cisternae) are readily accessible to them. Verapamil and D-600 can be visualized to block replenishment of the sites from which Ca is

released whether the sites be superficial, as in heart, or more deeply placed, as in skeletal muscle. A metabolic acidosis produces a weak uncoupling effect in heart presumably due to the fact that the H^+ ion is restricted in its diffusion through the membrane. Respiratory acidosis, on the other hand, produces a striking effect which is probably based on the high diffusivity of carbon dioxide through the membrane complex to a site at which formation of H^+ occurs and exerts its E-C uncoupling effect in the heart. The fact that equivalent degrees of respiratory acidosis are virtually without uncoupling effect in fast-twitch skeletal muscle is an important finding. Since carbon dioxide diffusion would not be expected to be significantly different for heart and skeletal muscle and since the intracellular buffering capacity of skeletal muscle is, if anything, less than in heart (124), the source and movement of coupling Ca are clearly different in the two tissues.

Although the evidence supports the conclusion that the major store of coupling Ca is located "superficial" to the lateral cisternae in mammalian heart, the exact site or sites and the detailed sequence of movement remain to be defined.

Voltage Clamping   There exists, at present, difficulty in reconciliation of the evidence for a surface source of Ca (reviewed above) with results obtained from voltage clamp experiments. The core of the difficulty centers about the values derived, in clamp experiments, from integration of the so-called slow inward current which is ascribed to Ca influx. If most of the coupling Ca were derived from sources outside the sarcolemmal membrane leaflet and its movement were electrogenic it should be measured as inward current. Integration of the area under the Ca current trace should represent total charge which then can be expressed in terms of Ca influx. From the study by Solaro et al. (71) 35 μmoles of Ca/kg are required to produce 90% of maximum force development. Voltage clamp studies indicate a maximum of 5–10 μmoles Ca influx/kg per beat (77, 125)—barely sufficient to raise intracellular free Ca concentration to threshold (71).

Beeler and Reuter (125) also noted that, following a quiescent period, the slow inward current was fully activated with the first depolarization and did not increase during subsequent depolarizations. This, despite the fact that contractile force increased steadily (staircase or Bowditch treppe) over the next five to eight depolarizations. Ochi and Trautwein (126) confirmed this finding. In addition, clamp studies have been done after exposure of the mammalian cardiac muscle to solutions low in $[Na]_0$ which produces marked increases in force development (77, 125). As with the staircase, there is no discernible increase in slow inward current associated with the increased force.

On the basis of these studies it is necessary to propose the existence of a store of Ca which supplies the myofilaments but the flux from which does not produce identifiable current, i.e., is not electrogenic. Therefore, it has been assumed that the major fraction of calcium must be released from stores inside the intracellular sarcoplasmic reticulum. It would follow that the model of E-C coupling for heart muscle differs little from that proposed for skeletal muscle (127). As indicated under "Interpretation of Uncoupling Experiments" such a model for heart is inconsistent with a number of experimental facts.

It is possible that, given the quantitative uncertainties of the voltage clamp technique (128) and the difficulties in accurate correction for simultaneous outward currents, the clamp approach might underestimate the slow inward current attributed to Ca. If, however, it is assumed that the values are essentially correct there remains a possible mechanism by which the results from the voltage clamp experiments can be reconciled with those derived from the other techniques. The mechanism proposed involves the operation of a Ca carrier.

Control of Calcium Movement

The heart contracts in an "all or none" manner and does not have the facility to recruit more or less fibers to produce variable amounts of force. Therefore, force has to be modulated and graded at the cellular level. One form of response is based on the ability of the cell to develop different levels of active force dependent upon the resting length of the cell. This is the classic "Frank-Starling" response. As far as is known at the present time this response is not dependent upon variation in the amount of Ca released to the myofilaments or upon alterations in ionic exchange. On the other hand there are a number of important inotropic responses which are based upon augmentation of Ca at the myofilaments. In the previous discussion the possibility of a Ca carrier system was raised and such a system would be expected to be operative in the determination of transmembrane Ca movement to the myofilaments. The question of control of Ca movement would then focus upon the factors important in control of carrier movement. If, of course, movement of contractile-dependent Ca were derived primarily from an intracellular store, a carrier-type system would be much less important. As reviewed, however, electrogenic movement of the major fraction of coupling Ca through a pore is unlikely and the uncoupling experiments are not consistent with the existence of a significant intracellular store for release. It is appropriate then to consider the control of Ca movement in terms of control of the movement of a carrier.

A Calcium Carrier Early studies done in frog heart (129) clearly demonstrated a competition between Na and Ca ions at the cellular surface and later studies in both frog (130) and mammalian heart (81, 83) were consistent with a carrier molecule involving Na and Ca. The experiments indicated binding on the carrier of 2Na or 1Ca, such that force was proportional to $[Ca]_0/[Na]_0^2$ (129). Studies on both frog and mammalian heart in which $[Na]_0$ was reduced gave clear evidence that Ca uptake was increased. Though a component of the net increment in Ca could be attributed to an effect of reduced $[Na]_0$ to diminish Ca efflux (131), a significant component was due to augmented Ca influx (81). The fact that increased Ca influx could be induced by decreased $[Na]_0$ was consistent with the existence of the proposed carrier which when operative would result in net outward movement of Na and net inward movement of Ca. Though the experiments which employed a reduction in $[Na]_0$ indicated the presence of a transmembrane carrier they did not mimic, in any realistic manner, a physiological stimulus to the Na-Ca carrier. Reductions of $[Na]_o$ of the extent required to stimulate Ca uptake do not occur under in vivo conditions. The findings did, however, suggest a mechanism which could very well be important in the physiological control of Ca movement.

It was suggested by Repke (132) and by Langer (133) that an increase of Na intracellularly, $[Na]_i$, might be a primary stimulus for augmentation of Ca influx. This suggestion was consistent with the operation of the carrier since it would be expected that it would be stimulated not only by decreased $[Na]_o$ but by increased $[Na]_i$—with net outward movement of Na (and net inward movement of Ca) produced in either case. Direct evidence for the existence of a Na-Ca carrier in an excitable biological system has been found in the squid axon membrane by Baker et al. (134). Using the internally perfused axon it was clearly demonstrated that, upon an increase in Na concentration at the inner side of the membrane, Ca influx was significantly augmented. Indirect evidence for the same sequence has now been found in heart muscle (83).

The operation of the carrier with its "uphill" transport of Na requires energy. The energetic requirements might be linked to the large outside-to-inside concentration gradient for Ca that exists across the cell membrane. This gradient is approximately 10^4 during diastole and 2×10^2 at the peak of contraction. The movement of Ca inward down its concentration gradient could then be tied, energetically, to the movement of Na outward, against its gradient.

The presence and operation of the carrier could reconcile some of the differences which arise from comparison of voltage clamp results, flux data, and function referred to previously. The carrier would operate such that an equivalent or nearly equivalent positive charge (Na^+) cycles outward as Ca^{2+} cycles inward. Thus a significant component of Ca influx in the myocardium would be undetected by the voltage clamp technique.

A Model for Calcium Movement

As reviewed above there is abundant evidence for both electrogenic and nonelectrogenic trans-sarcolemmal movement of Ca. This indicates two distinct mechanisms for Ca entry. On the basis of knowledge developing about membrane and cell surface structure (see "Ultrastructure") it is not premature to suggest a conceptual model which incorporates two distinct pathways for Ca movement into the cell—a "pore" pathway and a "carrier" pathway (see Figure 10). A description of the model serves to summarize much of what has been presented in this section.

Abundant amounts of Ca are bound to negatively charged sites located on the glycoproteins which constitute the external lamina. This Ca is in rapid equilibrium with that in the interstitial space. Ca from the external lamina supplies two systems which span the surface coat-unit membrane region. The system designated "pore" is modeled on the basis of the mosaic structure (reviewed under "Ultrastructure") in which integral proteins "float" in the lipid bilayer and abut to form hydrophilic channels or pores. These proteins possess

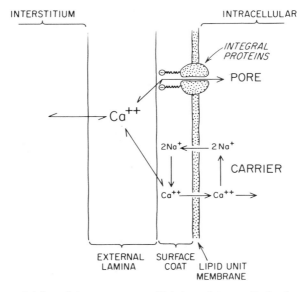

INTERSTITIUM   INTRACELLULAR

INTEGRAL PROTEINS

PORE

Ca^{++}

$2Na^+$ — $2Na^+$

CARRIER

Ca^{++} — Ca^{++} →

EXTERNAL   SURFACE
LAMINA   COAT   LIPID UNIT
                    MEMBRANE

Figure 10. A model for calcium movement. Ca is bound to negatively charged sites in the external lamina which sites are in rapid equilibrium with Ca in the interstitium. The external lamina is proposed to supply the Ca which moves across the sarcolemma via two routes: 1) through pores formed by the integral proteins embedded in the lipid membrane. Movement through this system would be electrogenic; 2) with a carrier (probably coupled to outward Na movement) such that movement via this system is electroneutral. Movement through both systems would occur upon depolarization. See text for further description.

negatively charged sialic acid end groups which extend into the surface coat. It is likely that these groups bind Ca and they may function to orient the proteins in the formation of the pores. The pores are visualized to function as selective channels for the electrogenic movement of ions, including Ca. It is the ionic movement through this system which is recorded with the voltage clamp technique.

The other system, designated "carrier," is based on the data derived from experiments in which Na-Ca interaction was evaluated. Movement of the carrier results in net transport of Na outward and Ca inward without net charge transfer—hence it is not electrogenic. It is easiest to think of carrier activation in terms of recruitment of more molecules with the characteristics of Na and Ca transport indicated. Since $[Ca]_o$ and $[Na]_o$ are quite invariant and since $[Ca]_i$ is under the control of the sarcotubular Ca pump, the level of $[Na]_i$ may be dominant in the determination of the number of active Na-Ca carriers under most physiological conditions. It is most convenient to visualize the carrier as moving Na outward and Ca inward during systole but it is clear from experimental results that the carriers are accessible to ions during diastole. The model is consistent with the experimental data reviewed: 1) elevation or lowering of $[Ca]_o$ will alter Ca in the external lamina with subsequent increase or decrease, respectively, at both the pore and carrier sites. Therefore, Ca current, as well as flux via the carrier, will change and be followed by a change in contractility. 2) Lowering of $[Na]_o$ will augment Ca influx via the carrier but have little or no effect on Ca at the pore. This will produce a positive inotropy without an increase in the slow inward (Ca) current. 3) The cation uncouplers (La, Mn, etc.) will displace Ca from the external lamina as well as compete for Ca at both the pore and carrier with inhibition of Ca movement at both sites. Ca current through the pore and influx via the carrier will decrease with a negative inotropic result. 4) The pharmacological uncouplers (Verapamil, D-600) act to inhibit Ca movement to the external lamina or from the external lamina to the pore and carrier sites. This action diminishes refilling of these sites in the beating muscle thereby causing a decrease of Ca current at the pore and Ca influx via the carrier. 5) The action of the hydrogen ion (induced by a respiratory acidosis) is best interpreted as an inhibition of the movement of the carrier which would reduce Ca influx via this route. The effect at the pore is, to our knowledge, not known since voltage clamp studies in the presence of respiratory acidosis have not been done.

The model, as proposed for heart muscle, places the control of Ca movements at sites which are influenced, quite directly, by changes in the interstitial space. Both pore and carrier are accessible, through external lamina and cell coat, at sites external to the lipid unit membrane. Skeletal muscle would have to be modeled quite differently since the source of contractile-dependent Ca is predominantly intracellular within the lateral cisternae of the sarcotubular system (102) and cycles internally. The very different action of the various uncoupling agents on skeletal muscle is consistent with this internal location (101, 114, 122). There are a few studies which indicate the possibility of a direct communication of the sarcotubular system with the extracellular space (50, 136) but the action of the uncoupling agents is not consistent with this.

Control of Contractility There exists considerable controversy with respect to the ionic mechanisms responsible for contractility control. It is generally agreed that augmented Ca influx is common to most ionically controlled inotropies but opinion differs on the mechanism by which increased Ca influx is stimulated. The literature pertaining to the increased contractility which follows upon increased frequency of contraction (the Bowditch "staircase") and upon administration of digitalis (137–140) serves to indicate some of the differences. One group (140) favors linking Ca influx to K efflux while others (138) relate the Ca movements to cellular levels of Na according to the model described (Figure 10).

Upon an abrupt increase in frequency of stimulation or upon administration of a positively inotropic dose of digitalis there occurs a net loss of K from the heart muscle (78, 133). It is likely that this is associated with a net increase of intracellular Na (78). Morad and Greenspan (140) have proposed a Ca-K exchange mechanism such that the outward movement of K is tied to an inward movement of Ca. They discount the role of Na on the basis that glycoside inotropy persists under voltage clamp conditions despite the blockade of the "fast" Na channel by a very specific poison, tetrodotoxin (TTX). They argue that since 95% of the "fast" Na channel is blocked there can be little, if any, net Na accumulation by the cell and, therefore, no stimulation of the proposed Na-Ca system. Not considered, however, is the existence of a "slow" Na channel (106, 141) through which considerable Na flows with each excitation (142). A study recently completed (143) quite clearly rules out K-Ca exchange as being of importance, at least with respect to the action of the digitalis glycosides. A decrease of pH in the medium perfusing heart muscle, induced by increase of p_{CO_2}, inhibits K efflux by quite specifically blocking a passive "leak" channel (144). The induction of the respiratory acidosis does not affect glycoside inhibition of active Na-K transport nor the production of increased contractility by the glycosides. It virtually eliminates, however, the increased K efflux observed in the absence of acidosis. Similar experiments have not, as yet, been done to determine the effect of respiratory acidosis on K efflux and contractile response following induction of a staircase.

There is, then, considerable reason to believe that the proposed Na-Ca carrier system may be of importance in the ionic control of contractility. Upon an abrupt increase in frequency Na influx rapidly increases. This causes $[Na]_i$ to rise until the Na-K pump and the Na-Ca carriers respond to affect an increased Na efflux with establishment of a new higher steady state active Na-K and Na-Ca exchange. A new higher steady state $[Na]_i$ would persist at the higher frequency of contraction (145), Ca influx via the Na-Ca carriers would be augmented and a positive inotropy would result. Application of a digitalis glycoside specifically inhibits active Na-K transport and $[Na]_i$ rises. The increased $[Na]_i$ feeds back to augment Na-K transport (to whatever degree possible in the presence of glycoside) and increase transport via the Na-Ca carriers (not influenced by glycoside). The augmented Ca influx and contractile response are then based on the same mechanism as indicated for the staircase (138).

McCans et al. (117) have employed Verapamil (see "Pharmacological Uncouplers") and concluded that the mechanism of the staircase can be dissociated from that of the digitalis compounds. This conclusion was based on the demonstration that Verapamil converted the response to increased heart rate from a positive to negative inotropism, whereas Verapamil did not reduce the positive inotropic response to ouabain (though it was considerably delayed). Based on the mode of action of Verapamil, this conclusion is not warranted. Verapamil acts to prevent replenishment of "release" sites (see Figure 10) and, for any given dose, its negative inotropic effect is enhanced by increased contractile frequency. This is because, in the presence of Verapamil, the sites cannot refill rapidly enough to maintain their Ca level (per beat) in the face of the increased rate of influx to the cell (per unit time) which normally follows upon increased rate of stimulation. Therefore the effect of Verapamil is to reduce Ca influx *per beat* and its negative staircase effect is predictable. On the other hand, ouabain was administered at fixed rate of stimulation and, given enough time, would be expected to augment Ca influx per beat enough to demonstrate a positive inotropism despite the fact that refilling of "release" sites is inhibited. The point is that Verapamil, a contraction-dependent uncoupling agent (see earlier), is not the appropriate drug to compare a frequency-dependent and a basically nonfrequency-dependent process. The appropriate experiment to use in comparison of the effect of Verapamil or D-600 on force-frequency to digitalis effect is paired pacing. Paired pacing will double the excitation rate but will not nearly double the rate of utilization of Ca. If the inotropy resulting from increased frequency and digitalis are based on the same mechanism then Verapamil or D-600 should not prevent the augmented contractility secondary to paired pacing. Willerson et al. (118) have shown that paired pacing produces significant increases in force development in the presence of D-600.

MECHANICAL RESPONSES IN CARDIAC MUSCLE

With the development of techniques with which an accurate relation between sarcomere length and force development can be established (146) the focus of muscle mechanics is now concentrated on the kinetics of cross-bridge interactions. Indeed, conceptual and experimental interests have shifted to 1) considerations of attachment and detachment rate constants for the cross-bridges; 2) a multicomponent chemical cycle involving the ATPase of the myofilaments and their concurrent conformational changes (related to isometric and isotonic contraction); and 3) more specific correlations between the time courses of heat evolution and high energy phosphate utilization.

Consistent with its historical development and anatomical simplicity, skeletal muscle physiology, in these three respects, leads cardiac muscle physiology by perhaps a decade. However, a few studies in cardiac muscle do exist in which the parallels and dissimilarities between skeletal and cardiac muscle can be discussed at the cross-bridge level.

This section of this review emphasizes the first of these three interests and presents the similarities and contrasts of cardiac and skeletal muscle responses to a variety of perturbations. The implications of these responses are discussed in regard to cardiac cross-bridge properties and the validity with which cardiac cross-bridge mechanics can be inferred from measurements on whole cardiac preparations. Since the cardiac literature on this subject is extremely limited a heavy reliance is placed on the few publications that do exist. In any case, the focus is on cardiac muscle rather than muscle in general.

Comparison of Living Skeletal
and Cardiac Muscle Responses to Mechanical Perturbations

General Responses Manifestations of cardiac force-length-velocity-time relations (determined mainly from excised mammalian papillary muscle) differ from those in skeletal muscle in a number of ways: 1) cardiac muscle has a higher resting tension in its functional range (section "Resting Muscle"); 2) at the same temperature activation of cardiac muscle is at least 1 order of magnitude slower in its onset than skeletal muscle (section "Maximum Activation"); 3) the compliance of excised active heart muscle appears to be 3- to 5-fold greater than active skeletal muscle (section "Compliance"); 4) this large compliance appears to be highly nonlinear, actually nearly exponential (section "Characterization of Cardiac Compliance—Quick Release"), and, according to some reports (147, 148), may possess a time dependency (section "Time-dependent Series Elasticity"); 5) the stiffness-force relation (measured with small perturbations) varies with initial muscle length in heart muscle differently than in skeletal muscle (149–151) (section "Small Perturbations—Stiffness-Force Relations").

At the present time we have no information to indicate that the cross-bridges of cardiac muscle are structurally or mechanically different from those of skeletal muscle. The rate constants for the development and relaxation of active force are obviously slower in heart muscle but it would appear that the primary differences between the perturbation responses of cardiac and skeletal muscle may lie in the manner in which cross-bridge activity in transmitted to the ends of the muscle preparation. Data available in the literature do not permit us to construct a unique mechanical analog with which to relate whole cardiac muscle mechanical responses to cross-bridge kinetics; but, one three-elastic-element system proves rather promising. It is discussed in some detail. However, an exhaustive discussion of models is beyond the scope of this review.

Resting Muscle   It is well known that cardiac muscle exhibits resting tension over most of its functional range of active force development while in most skeletal muscles, even at L_{max} (where maximal active force is evident), resting tension is negligible. We have generally assumed that resting muscle tension in heart muscle is borne by passive nonlinear viscoelastic elements in parallel with the active force generating elements, i.e., the cross-bridges of the muscle. Undoubtedly cross-bridge attachments can exist in resting cardiac muscle as indicated by the appearance of partial contractures under some conditions. Also the weak myosin layer-line pattern in x-ray diffraction studies suggests the presence of residual cross-bridge attachments (152). In the latter study, however, force development was not monitored during the study, nor was the physiological state of the muscle fully assessed before fixation. Hence, a normally relaxed preparation was not assured. Similarly, the failure of the reduction of external Ca^{2+} to reduce resting cardiac tension to negligible values sustains the general concept that some passive structures, evident in heart muscle but not apparent in skeletal muscle, bear resting tension at muscle lengths below L_{max}.

Furthermore, this resting elastic property in heart muscle is complex in that, in response to step length changes, the force changes subsequently with time, suggesting the presence of a viscous element. However, an even more complex structure is implicated by the fact that, in response to sinusoidal perturbations, the force-length plot of resting heart muscle is not stable at a fixed frequency of oscillation, but drifts in time towards a steady state relation in the course of several cycles (153, 154). This creep phenomenon is characteristic of long chain molecules with cross-linking such as is found in collagen, the major candidate for the parallel elastic force bearing element in resting cardiac muscle. The relative absence of collagen in the frequently cited preparations of skeletal muscle is consistent with this interpretation.

Maximum Activation A glance at the twitch force of cardiac and skeletal muscle, or more particularly the active stress, i.e., active force per unit cross-section at comparable sarcomere lengths, suggests that skeletal muscle is capable of greater force development than cardiac muscle. This observation would suggest that perhaps the cross-bridge of skeletal muscle can develop more force than cardiac muscle or that more cross-bridges can attach in the skeletal muscle myofilaments. However, when contractile force is referred to the myofilaments (accounting for the differences in extracellular space, mitochondria and SR) at maximal activation the differences largely disappear. For example, Page et al. (42) gives a myofilament volume for rat ventricular muscle of 48% of fiber volume and Mobley and Eisenberg (43) found a myofilament volume in skeletal muscle of 83%. Taking contractile force in cardiac muscle at maximum activation of 850 g/cm^2 fibers (stimulus 10/s in 10 mM caffeine and 7.5 mM Ca^{2+}, but uncorrected for series compliance) (155) and in frog sartorius at 1800 g/cm^2 (156) these values correct to about 1.8 kg/cm^2 myofilaments for heart muscle and 2.2 kg/cm^2 for sartorius. It is possible (although not experimentally verified) that correction of the heart muscle data for series compliance, i.e., the reduction of myofilament overlap during contraction against the series compliance, may make the values even closer. Thus, as far as myofilament force generation is concerned, the differences in the two types of muscle would appear to be more in their activation and inactivation rates and not in the number of cross-bridges that can attach to actin sites.

Compliance

Skeletal Muscle Huxley and Simmons (157) and Cleworth and Edman (158) have shown that the major series compliance of single skeletal muscle lies in the end sarcomeres of the muscle near the tendons. When this compliance is compensated by appropriate stretch the resulting force-length relation (in response to fast perturbations) is nearly linear. Furthermore, the stiffness or slope of this force-length relation is directly proportional to myofilament overlap. In other words, the stiffness as well as the steady state force is directly proportional to the number of attached cross-bridges and these cross-bridges appear as linear springs to fast length displacements. The compliance of the cross-bridge appears to be approximately 0.8%, i.e., a rapid release (< 1 ms) of 0.8% of sarcomere length will reduce cross-bridge force to zero.

Cardiac Muscle It is widely reported that a quick release of 6–10% of a muscle length is required to reduce papillary muscle force to zero. This is 10-fold greater than in skeletal muscle and in the absence of any indications that cardiac cross-bridges or myofilaments are structurally greatly different from skeletal muscle (see "Ultrastructure") we must conclude that the greater compliance of heart muscle must lie outside the cross-bridges. Indeed it is highly probable that no individual sarcomere has this degree of compliance without the detachment of cross-bridges.

The question then is to what structures this large compliance of heart muscle might be attributed. There are several highly likely candidates. First, several studies indicate that a large compliance exists at the ends of the muscle where the mechanical attachments are made to the experimental equipment (159, 160). However, if care is taken to minimize tie or clamp induced compliance, papillary muscle compliance can often be reduced to 3–4%.

A second possibility is nonhomogeneity in the distribution of stress as force rises or falls in the muscle (159) concomitant with stretch or release. If the combination of parallel stress bearing elements and active sarcomeres in one section of a muscle is momentarily unequal to that in another section, the displacement of the two sections may not be proportionally distributed over the two when muscle length is altered. Thus the weaker (more compliant) section will act as series compliance to the other, with some cross-bridge detachment possibly occurring in the compliant sections. Since we are looking for a 2–3% compliance (out of 3–4%, 1% being attributed to the cross-bridges) in the form of sectional inhomogeneity we may be looking for perhaps 4–6% extensibility in half the muscle. This would be something like 0.1 μm of compliance per 2 μm of sarcomere in the compliant region of the muscle when force is dropped to zero with a quick release. Obviously, if more of the muscle were inhomogeneous the length change per sarcomere would be less with release. In any case, this degree of nonuniformity would be difficult to detect in a papillary muscle by surface observations, particularly if the nonhomogeneous regions were not continuous, i.e., if the inhomogeneity consisted of many localized segments of sarcomeres. In the latter case even the more sophisticated sarcomere length-measuring techniques of laser diffraction (159, 161) might miss this type of inhomogeneity.

Characterization of Cardiac Compliance–Quick Release   Quick release of active cardiac muscle by increments of force or length (Figure 11*A*) until zero force is reached yields a relation between force and elastic extension which is highly nonlinear (Figure 11*B*). This force-extension has been used by muscle physiologists to characterize series elasticity. For example, in the case of successively larger load releases from, say, peak twitch tension (P_0), the rationale is that the shortening increments, accompanying the increments in load reduction, represent the extension of the series elastic element by the active contractile element (Figure 11*C*). In relation to the cross-bridges in the quick release maneuver the contractile element is presumed to represent the force generating capacity of cross-bridges and further that the quick release displacement occurs sufficiently rapidly that no attachments or detachments take place. Thus cross-bridge elastic extension, for those cross-bridges which are attached at the moment of release, would appear as series elasticity (also in series with stray compliance or end compliance of the muscle preparation). Extending the concept of a linear cross-bridge elasticity from skeletal muscle we could then have a somewhat stiff linear elastic element in series with a more compliant nonlinear elastic element (see Figure 13, *A* and *B*). This combination of elastic elements also in conjunction with the elastic elements bearing resting tension would contribute to the form of the force-extension relation, revealed for the muscle by the quick release technique. Presumably the same force-extension relation of

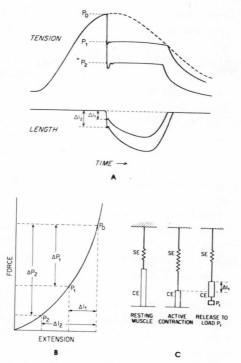

Figure 11. *A*, responses of cardiac muscle to quick releases at the peak of isometric contraction. *Upper traces*, tension responses for a release from P_0 to P_1 and P_2. *Lower traces*, concomitant length changes showing undamped length changes Δl_1, Δl_2, inertial oscillation, and subsequent shortening with each load. *B*, plot of elastic force-extension relation derived from data obtained in *A*. *C*, sketch showing the extension of series elasticity (*SE*) by the contractile element (*CE*) during isometric contraction and the length change in SE when the muscle is released from P_0 to load P_1.

the muscle would be derived by releasing the muscle by successive increments of length and measuring the simultaneously occurring undamped force changes.

Obviously, the same elastic elements could be evaluated by releasing the muscle at other times than at the peak of the twitch (however, cross-bridge stiffness would be less because fewer cross-bridges would be attached at the instant) and the initial muscle length at which the releases were given could be altered. Each of these initial states would evaluate the combination of elastic elements under individually different initial conditions so that it might be possible to sort out their separate characteristics. This possibility is discussed later.

Time-dependent Series Elasticity There have been several reports (147, 148) that the series elasticity of heart muscle may not be represented as a passive elastic element whose extension depends solely on the force extending it. Comparing the force-extension relations of papillary muscles obtained by quick release extant at different times during the rising phase of the twitch, indicated that the series elastic element becomes stiffer at given forces during the course of contraction.

Such a stiffening of elasticity might be expected due to at least three possible conditions: 1) the number of cross-bridges attached and, therefore, the collective stiffness of the cross-bridge, would be different at different times during contraction. This is probably easiest to appreciate if it is recognized that in an elastic element represented by the parallel combination of cross-bridges the length change necessary to reduce the force of one cross-bridge (extended by attachment but not further displaced by external length changes) to zero would be the same as for 1,000 attached cross-bridges. Therefore, the cross-bridge element length change accompanying a release to zero force at the peak of the isometric twitch would be the same as the length change, say, at half peak twitch tension. Thus the compliance of the cross-bridges ($\Delta l/\Delta P$) and so the compliance of the muscle at peak twitch tension would appear less than at half peak tension. This peculiarity arises from the fact that cross-bridge stiffness (collectively) increases by parallel addition of elements rather than the continuous elongation of a nonlinear elastic element which presumably represents stray compliance and tie or clamp-induced compliance at the ends of the muscle. 2) The force-extension relation of the muscle would appear to get stiffer with time during the contraction if inhomogeneity were the major source of series elasticity and this inhomogeneity were to decrease during contraction, i.e., if releases given at the peak of contraction characterized a more ordered distribution of sarcomeres than releases given at half peak tension. This possibility is definitely feasible but cannot be evaluated until an accurate assessment of sarcomere length distribution in the contracting muscle can be performed. 3) An artifactual problem occurs in quick release experiments which is not always recognized or properly evaluated. The quick release obviously takes a finite time. It is presumed that the release is sufficiently fast that no cross-bridge detachment or reattachment occurs which would increase the magnitude of the release necessary to reach zero tension. However, if the release is slow, particularly if appreciable resting tension is present (which tends to creep with slow release), an excessive release will be necessary both to overcome the passive element creep and the release of force in reattached cross-bridges.

Furthermore, whether the release is a force or length increment, there are always inertial oscillations accompanying the release and these are either damped by external means further slowing the release time, or the measurement is made after the oscillations subside (a further delay), or a linear extrapolation is made through the oscillations. All of these methods are subject to considerable error which leaves the evaluation of compliance by the quick release method, at best, obscure; therefore, other techniques are required.

Small Perturbations—Stiffness-Force Relations If the perturbations, either force or length, are kept small (say, $<0.5\% L_{max}$) then the ratio $\Delta P/\Delta l$ gives an approximate measure of muscle stiffness. Both step stretches and releases and low amplitude sinusoidal perturbations have been used in these studies (151,

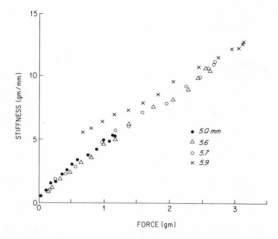

Figure 12. Plot of stiffness-force relation of a living papillary muscle contracting isometrically at four different initial lengths. Both resting stiffness (left-most point in each sequence) and stiffness at peak contractile force increase with initial length up to 5.7 mm length; thereafter, resting stiffness and force increase more rapidly than active stiffness and force such that the slope of the stiffness-force relation tends to decline. (Modified from Brady (151)).

162, 163). These studies show that the stiffness-force relation of the cardiac muscle (Figure 12) is linear or nearly so as would be expected if muscle stiffness and force were both directly proportional to the number of attached cross-bridges. However, note that the system is not simple. The slope of the stiffness-force relation declines at long muscle lengths even though total stiffness is greater at all forces. This reduction of slope is a factor which must be inherent in any useful mechanical model of heart muscle (see "A Three-Elastic-Element Model for Excised Cardiac Muscle Preparations").

The small perturbations analysis of glycerol-extracted heart muscle (164), where the level of activation of the muscle can be set by the Ca^{2+} in the activating solution, gives a compliance of about 2% for mammalian papillary muscle at 20°C. Compliance by similar techniques in skeletal muscle is about 0.8% (157). In the glycerol-extracted cardiac preparation muscle end compliance was minimized by glueing the fiber strands to the transducer and muscle puller. The differences in compliance between the two muscles has not been reconciled. Nonuniformity in the sarcomere of the cardiac preparation cannot be excluded since the sarcomere lengths in these studies were not clamped, as in the skeletal muscle studies. On the other hand, it cannot be ruled out at this point that the cross-bridges of heart muscle are more compliant than in skeletal muscle.

The perturbation studies on living papillary muscle are more difficult to interpret and, indeed, the results are somewhat different. The slope of the linear stiffness-force relation, obtained by the application of small perturbations at varied times during the rising phase of contraction, gives a measure of the length constant of the exponential force-extension relation. In other words, if the force-extension relation is expressed as:

$$P = C(e^{Kl} - 1) \tag{1}$$

where P is the muscle force; l is the extension of the elastic element from its zero force length; C is a force constant; and K the reciprocal length constant. The stiffness of such an element is given by:

$$\Delta P/\Delta l = K(P + C) \tag{2}$$

The product KC gives the stiffness of the element at zero force. Reciprocal length constants in the range of $50-100$ mm^{-1} have been reported (151, 162, 163). (These values are referred to a unit length of muscle). These values are comparable with those obtained by fitting an exponential force-extension relation to the quick release data described earlier. Thus the large and small perturbation data seem to be reasonably compatible. However, each method infers a muscle compliance of $5-10\%$ (i.e., the release necessary to reduce force to zero at the peak of a twitch). This is severalfold greater than the compliance values obtained for the same glycerol-extracted preparations.

The conclusion evident from these studies is that the compliance of the glycerol preparation, while greater than in length clamped skeletal muscle, is less than in living muscle with current mounting methods. Therefore, to attempt to deduce cross-bridge kinetics from living muscle preparations requires an accurate analog with which to relate the whole muscle to the cross-bridges. The perturbation studies of the less compliant glycerol-extracted cardiac muscle preparation and the analysis of a probable mechanical analog of heart muscle (with linear cross-bridge elasticity) give some insights in this direction.

Perturbation Studies of Glycerol-extracted Cardiac Muscle

Time Course of Responses to Step Changes in Length Following the step increase in force in a stretch perturbation (Figure 13A) tension declines in a time-dependent manner analogous to a viscoelastic response. Abbott (165), in insect fibrillar muscle, Rüegg et al. (166), in mammalian skeletal muscle, and Steiger (164), in cardiac muscle, have used this technique and propose that this apparent viscoelastic response is, in part, a cross-bridge detachment property such that the rate constant for the second phase of the tension fall (Figure 13A) gives a measure of the rate constant for detachment of the stressed cross-bridges. The first phase is presumably a viscoelastic tension fall which has only a small temperature dependence. The second phase is most predominant at lower temperatures showing the greatest temperature dependence of all the phases. The third phase of the perturbation response is the redevelopment of force with a relatively slow time dependence. This latter phase has been presumed to reflect the increased rate constant for cross-bridge attachment concurrent with, and following, the stress relaxation and cross-bridge detachment period.

The general character of the responses is similar in all types of glycerinated preparations so far studied as well as in their living counterparts. (Recently, Steiger has shown similar responses in living cardiac muscle under conditions of

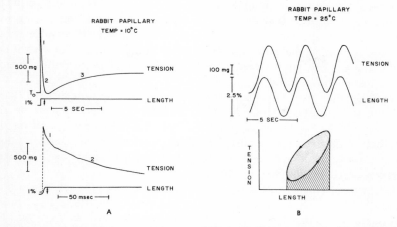

Figure 13. *A*, response of a papillary muscle with steady activation T_0 to a small stretch length change of $t = 0$. *Upper traces*, slow time base; *lower traces*, same responses displayed on a fast time base; T_1 is the force resulting from the length change. The numbers represent: *1*, the first phase of rapid decline in force indicative of stress relaxation; *2*, the second slower phase of tension fall associated with cross-bridge detachment; and *3*, stress-induced delayed tension. *B*, tension (*upper trace*) and length (*middle trace*) responses to sinusoidal length perturbations of small amplitude. Lower Lissajou figure is the force-length plot of the same responses showing counterclockwise rotation of the loop at this frequency of oscillation. Shaded areas show the difference between the work done in the stretch and the release portions of the cycle.

sustained contractile activity, i.e., high Ca, caffeine, low Na) (unpublished observations). The fact that these responses to perturbation are seen in both living and glycerol-extracted cardiac muscle, and furthermore, except for the slower rate constants of the cardiac responses (about 1 order of magnitude less than rabbit psoas muscle; Abbott and Steiger, in press), the similarity of cardiac muscle responses to those of skeletal and insect flight muscle, indicates that, in spite of the mechanical differences between these muscles, cardiac cross-bridge properties are amenable to the same type of experimental study.

Responses to Sinusoidal Perturbations The delayed tension manifest in muscle following stretch indicates that the muscle is capable of doing work on the system in response to mechanical perturbation. This can be more fully demonstrated by driving the activated preparation with a sinusoidal length change. Under these conditions and at an appropriate frequency both living and glycerol-extracted cardiac muscle (as also skeletal and insect muscle) will do work on the driving system. This can be seen from Figure 13*B*. The upper trace shows the sinusoidal tension changes that result from small amplitude sinusoidal length changes (second tracing) applied to a living papillary muscle in a steady state of activation.

When this tension record is plotted against the length record the Lissajou figure (lower tracing) results. The size of the loop depends on the phase shift between the tension and length responses. Obviously if tension and length are in phase, as they would be for an elastic spring, the loop would be closed. If the muscle were visoelastic the loop would be open but rotation would be in the clockwise direction such that the work done in stretching the muscle (the area under the upper curve) would be greater than the area under the curve described by the release half of the cycle. Under these conditions then, more work is put into the system with stretch than is done by the muscle with release, hence, energy is lost to the muscle.

However, when the muscle is cycled at the optimal frequency (~ 0.5 Hz, at $20°C$) the loop rotates counterclockwise so that more work is done on the release half of the cycle (upper portion of the Lissajou figure) than on the stretch half of the cycle (lower portion). Under these conditions the muscle does work on the external system and, as shown in insect fibrillar muscle, additional ATP is consumed in doing so (167, 168).

The interpretation of this result is that stretch increases the attachment rate constant (f) (169) such that more cross-bridges subsequently attach. The stressed cross-bridges previously attached now have an increased rate constant (gx) for detachment so that these cross-bridges detach at a more rapid rate as indicated by the fall in tension following the step stretch or the decline in tension in the early stretch phase of the sinusoidal perturbation. If the frequency of stretch and release of the sinusoidal waveform coincides with the rate constants, work can be performed against the driving system.

These observations form the possible basis for some of the following classically observed muscle properties: 1) the Fenn effect, i.e., the increase in muscle metabolism associated with an increase in muscle work as indicated by the counterclockwise rotation of the tension-length loop of Figure 13B; 2) inactivation with shortening (the duration of an isotonic contraction is shorter than an isometric contraction (170). This is also seen in the release responses (164) where tension falls after a delay, i.e., the counterpart of delayed tension rise after a stretch; 3) stress relaxation. The first phase of tension fall after a stretch (Figure 13A) is not strongly temperature dependent, as is the second phase, so that stress relaxation effects appear separable from stressed related cross-bridge detachment; 4) the temperature dependence of force development and shortening. The rate constants for the various phases following step length changes provide new insight into the processes that underlie the temperature dependence of cardiac mechanics; 5) the inhibition of contractility by ADP. Abbott and Mannherz (171) have shown that elevation of ADP concentration reduces the rate constant of phase 2 in glycerinated myofibrils, indicating that conditions which result in high myoplasmic ADP levels will reduce the detachment rate of cross-bridges and thus slow relaxation or produce partial contracture; 6) the force-velocity relation. New insight has been given into the dependence of force and shortening velocity on myoplasmic Ca^{2+}. In a most interesting study of glycerol-extracted mammalian papillary muscle, Herzig (172) showed that above an activation level of about 20% P_0([Ca] $= 5 \times 10^{-7}$) Ca^{2+} still influenced isometric tension strongly but had little effect on shortening velocity.

It is evident that these studies on glycerinated and living cardiac muscle preparations provide a new approach to the study of fundamental mechanisms of cardiac contraction. However, it remains to show how these responses in heart muscle are modified by resting and nonlinear series elastic elements in the whole muscle preparation.

A Three-Elastic-Element Model for Excised Cardiac Muscle Preparations

By the introduction of more elements into the model, a more realistic representation can be achieved. For example, in Figure 14A, Se_1 represents the stray compliance of the equipment and tie or clamp-induced end compliance. SE_2 represents compliance within the sarcomere (but not cross-bridge compliance) or the compliance of one section of the muscle with respect to a more strongly contracting section, i.e., tissue inhomogeneity. The elastic properties and force generating elements of the cross-bridges are indicated by the box. Within the box the elements are represented as increments of force generation (and cross-bridge element stretching occurs sequentially in time (arrow)). Parallel elasticity (PE) is shown shunting both SE_2 and the cross-bridge system. An even more realistic model might include separate parallel elastic elements across the cross-bridge element and SE_2. However, to assign or measure meaningful values for this large number of elements in a model becomes hopeless. No unique combination of values could be found to describe the muscle responses.

On the other hand a simplification can be made which, although still not uniquely definable, is useful and gives some insight into the relation between muscle responses and cross-bridge characteristics. The available data and some feasible presumptions suggest the arrangement shown in Figure 14B. As in Figure 13A the cross-bridge force generators are represented by parallel units with separate elastic elements. Let SE represent the combination of stray compliance (equipment, ties end sarcomeres) and inhomogeneity. Parallel elasticity (PE) now represents the element bearing resting tension and shunting the

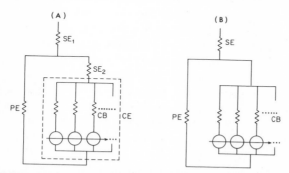

Figure 14. Models of mechanical properties of papillary muscle. A, general model representing stray and preparation end compliances as SE_1, inhomogeneity of active mechanical forces between segments of the muscle as SE_2, parallel elasticity (*PE*), and contractile element (*CE*) (in the box). The elastic properties of the cross-bridges are represented in series with each force generating unit. The *arrow* indicates the time dependence of activation of the cross-bridges. It is assumed that with activation of a cross-bridge attachment each cross-bridge elastic element is extended by a fixed amount ($\sim$ 1% of a half sarcomere length). CE gets stiffer (more attached cross-bridge units) as activation proceeds. B, simplification of model in A to effect a three-elastic-element model where the series compliance outside the cross-bridges has been lumped together in SE. CE, i.e., the cross-bridge system (*CB*), is the same as in A.

force-generators and their elastic units. Furthermore, let PE be compressible, i.e., a negative force can be supported by PE when the active elements shorten sufficiently (this characteristic represents the tendency for heart muscle preparations to prefer an unloaded length longer than that to which they can actively shorten). The cross-bridge elastic elements are assumed to be linear but, with activation, they form in parallel so that the collective cross-bridge element stiffness increases in direct proportion to active force development. Thus, SE bears resting tension and the linear (but active force dependent) cross-bridge elastic elements are in parallel with PE.

This model is consistent with a decline in slope of the stiffness-force relation with increased muscle length (149, 151, 162), i.e., as resting stiffness increases (Figure 12). This is evident if total cross-bridge stiffness is less than resting PE stiffness. Thus, as CE shortens, stretching SE, PE is unloaded (or compressed) and a less stiff element, i.e., collective cross-bridge elasticity, tends to replace the original resting PE, making the new series combination of SE and cross-bridge elasticity increase at a lesser rate (with active force development) than if PE were not unloaded, as in the PRT (Maxwell) model (151). In the latter model, the stiffness-force relation is shifted vertically with increased resting length, but the stiffness-force curves remain parallel. The three-elastic-element combination of parameters, then, has the right qualitative characteristics to explain the stiffness-force data. If, however, PE is more compliant than the cross-bridge elastic, then the slope of the active muscle stiffness-force relation increases with an increase in initial muscle length. This is a possible, but as yet unreported, observation.

From these considerations it is apparent that, if the major compliance is at the ends of the muscle, then cross-bridge activity in these preparations must be transmitted to the ends of the muscle through the relatively compliant and nonlinear series elasticity. In this case it would be inaccurate to simply subtract resting tension from total force in order to deduce cross-bridge force. Indeed, it is quite possible that, in quick release responses of the three-elastic-element model, total muscle force might be transiently zero with the cross-bridge stretched and developing an active force against the compressed PE.

Obviously, small perturbations are going to be more revealing of cross-bridge properties than larger perturbations to zero force, where the origin of the force-extension relation is obscure. If the perturbations are sufficiently small, SE compliance may be approximated as linear changes, PE force or length changes should be small, and the contributions of the cross-bridges should be more apparent. A more detailed analysis of this model will appear elsewhere with particular emphasis on the relation between the whole muscle and the cross-bridges.

SUMMARY

A comparison of some of the mechanical properties of cardiac with other types of muscle has been made, showing that, except for the speed of some responses, cardiac muscle is similar to other types of muscle. Furthermore, the techniques used in both living and glycerol-extracted insect fibrillar and verte-brate skeletal muscle are applicable to heart muscle, where the focus of the technique is now on cross-bridge mechanics and energetics.

It is particularly encouraging to see many well known phenomena such as inactivation with shortening, stress related increases in active force, and the Fenn effect begin to find some more specific relation to cross-bridge mechanical and chemical activity. The high compliance of cardiac preparations still clouds the interpretation of data obtained from whole muscle preparations; however, the reduced compliance of glycerol-extracted cardiac muscle offers some hope of obviating some series compliance. Indeed, the correspondence in mechanical responses of living and glycerol-extracted preparations shows that glycerol prepa-rations are of great utility since the time dependence of activation also can be removed in these studies.

A more complete analysis of muscle models, in which the cross-bridge contribution to muscle elasticity is more realistically evaluated, should help in relating muscle measurements to cross-bridge activity. Furthermore, studies on both living and glycerol-extracted cardiac muscle, particularly if sarcomere length can be controlled, offer new hope of closing the perpetual gap in our understanding of cardiac muscle physiology relative to skeletal muscle.

REFERENCES

1. Fawcett, D.W., and McNutt, N.S. (1969). The ultrastructure of the cat myocardium. J. Cell Biol. 42:1.
2. McNutt, N.S., and Fawcett, D.W. (1974). Myocardial Ultrastructure. In G.A. Langer and A.J. Brady (eds.), The Mammalian Myocardium, pp. 1–49. John Wiley and Sons, New York.
3. Tillack, T.W., and Marchesi, V.T. (1970). Demonstration of the outer surface of freeze-etched red blood cell membranes. J. Cell Biol. 45:649.
4. Steck, T.L. (1974). The organization of proteins in the human red blood cell membrane. A Review. J. Cell Biol. 62:1.
5. Winzler, R.J. (1970). Carbohydrates in cell surfaces. Int. Rev. Cytol. 29:77.
6. Revel, J.P., and Ito, S. (1967). The surface components of cells. In B.D. Davis and L. Warren (eds.), The Specificity of Cell Surfaces, pp. 221–234. Englewood Cliffs, N.J., Prentice-Hall.
7. Bennett, H.S. (1963). Morphological aspects of extracellular polysaccharides. J. Histochem. Cytochem. 11:14.
8. Martinez-Palomo, A. (1970). The surface coats of animal cells. Int. Rev. Cytol. 29:29.
9. Parsons, D.F., and Subjeck, J.R. (1972). The morphology of the polysaccharide coat of mammalian cells. Biochim. Biophys. Acta 265:85.
10. Forssmann, W.G., and Girardier, L. (1970). The study of the T system in rat heart. J. Cell. Biol. 44:1.
11. Simpson, F.O., and Rayns, D.G. (1968). The relationship between the transverse tubular system and other tubules at the Z disc levels of myocardial cells in the ferret. Amer. J. Anat. 122:193.
12. Sperelakis, N., and Rubio, R. (1971). An orderly lattice of axial tubules which interconnect adjacent transverse tubules in guinea-pig ventricular myocardium. J. Mol. Cell. Cardiol. 2:211.
13. Simpson, F.O., and Oertelis, S.J. (1962). The fine structure of sheep myocardial cells; sarcolemmal invaginations and the transverse tubular system. J. Cell Biol. 12:91.
14. Nelson, D.A., and Benson, E.S. (1963). On the structural continuities of the trans-verse tubular system of rabbit and human myocardial cells. J. Cell Biol. 16:297.

232   Langer, Frank, and Brady

15. McCallister, L.P., and Hadek, R. (1970). Transmission electron microscopy and stereo ultrastructure of the T-system in frog skeletal muscle. J. Ultrastructure Res. 33:360.
16. Rayns, D.G., Simpson, F.O., and Bertrand, W.S. (1967). Transverse tubular apertures in mammalian myocardial cells: Surface Array. Science 156:656.
17. Robertson, J.D. (1960). The molecular structure and contact relationship of cell membrane. Prog. Biophys. Biophys. Chem. 10:303.
18. Farquhar, M.G., and Palade, G.E. (1965). Cell junctions in amphibian skin. J. Cell Biol. 26:263.
19. Danielli, J.F., and Davson, H. (1935). Contribution to theory of permeability of thin films. J. Cell Comp. Physiol. 5:495.
20. Singer, S.J. (1971). The molecular organization of biological membranes. In L.I. Rothfield (ed.), Structure and Function of Biological Membranes, p. 145. Academic Press, New York.
21. Weinstein, R.S. (1969). The structure of cell membranes. New Eng. J. Med. 281:86.
22. Singer, S.J., and Nicholson, G.L. (1972). The fluid mosaic model of the structure of cell membrane. Science 175:720.
23. Edidin, M., and Fambrough, D. (1973). Fluidity of the surface of cultured muscle fibers. J. Cell Biol. 57:27.
24. Branton, D. (1966). Fracture faces of frozen membranes. Proc. Natl. Acad. Sci. USA 55:1048.
25. Branton, D. (1969). Membrane structure. Annu. Rev. Plant Physiol. 20:209.
26. Weinstein, R.S. (1969). Electron microscopy of surface faces of red cell membranes. In G.A. Jamieson and T.J. Greenwalt (eds.), Red Cell Membrane. Structure and Function, p. 36. J.B. Lippincott Co., Philadelphia.
27. Marchesi, V.T., Tillack, T.W., Jackson, R.L., Segrest, J.P., and Scott, R.E. (1972). Chemical characterization and surface orientation of major glycoprotein of human erythrocyte membrane. Proc. Natl. Acad. Sci. USA 69:1445.
28. Steck, T.L. (1974). The organization of proteins in the human red blood cell membrane. J. Cell Biol. 62:1.
29. Rash, J.E., and Ellisman, M.H. (1974). Studies of excitable membranes. I. Macromolecular specializations of the neuromuscular junction and the nonjunctional sarcolemma. J. Cell Biol. 63:567.
30. McNutt, N.S., and Weinstein, R.S. (1970). The ultrastructure of the nexus. A correlated thin-section and freeze-cleave study. J. Cell Biol. 47:666.
31. Franzini-Armstrong, C. (1974). Freeze-fracture of skeletal muscle from the Tarantula spider. Structural differentiations of sarcoplasmic reticulum and transverse tubular membranes. J. Cell Biol. 61:501.
32. McNutt, N.S., and Weinstein, R.S. (1973). Membrane ultrastructure at mammalian intercellular junctions. Prog. Biophys. Mol. Biol. 26:45.
33. Luft, J.H. (1971). Ruthenium red and violet. II. Fine structural localization in animal tissues. Anat. Rec. 171:369.
34. Wetzel, M.G., Wetzel, B.K., and Spicer, S.S. (1966). Ultrastructural localization of acid mucosubstances in mouse colon with ion-containing stains. J. Cell Biol. 30:299.
35. Zacks, S.I., Sheff, M.F., and Saito, A. (1973). Structure and staining characteristics of myofiber external lamina. J. Histochem. Cytochem. 21:703.
36. Howse, H.D., Ferrans, V.J., and Hibbs, R.G. (1970). A comparative histochemical and electron microscope study of the surface coating of cardiac muscle cells. J. Mol. Cell. Cardiol. 1:157.
37. Langer, G.A., and Frank, J.S. (1975). Calcium exchange in cultured cardiac cells. In M. Lieberman and T. Sano (eds.), Developmental and Physiologic Correlates of Cardiac Muscle. Raven Press, New York.
38. Lesseps, R.J. (1967). The removal by phospholipase C of a layer of lanthanum staining material external to the cell membrane in embryonic chick cells. J. Cell Biol. 34:173.
39. Langer, G.A., and Frank, J.S. (1972). Lanthanum in heart cell culture. Effect on calcium exchange correlated with its localization. J. Cell Biol. 54:441.
40. Martinez-Palomo, A., Benitz, D., and Alanis, J. (1973). Selective deposition of lanthanum in mammalian cardiac cell membranes. Ultrastructure and electrophysiological evidence. J. Cell Biol. 58:1.
41. Frank, J.S., and Langer, G.A. (1974). The myocardial interstitium: Its structure and its role in ionic exchange. J. Cell Biol. 60:586.

42. Page, E., McCallister, L.P., and Power, B. (1971). Stereological measurement of cardiac ultrastructure implicated in excitation-contraction coupling. Proc. Natl. Acad. Sci. USA 68:1465.

43. Mobley, B., and Eisenberg, B. Sizes of components in frog skeletal muscle measured by methods of stereology. J. Gen Physiol. In press.

44. Eisenberg, B., and Kuda, A. (1975). Stereological analysis of mammalian skeletal muscle. II. White vastus of adult guinea pig. J. Ultrastructure Res. 51.

45. Deamer, D.W., and Baskin, R.J. (1969). Ultrastructure of sarcoplasmic reticulum preparations. J. Cell Biol. 42:296.

46. Baskin, R.J. and Deamer, D.W. (1969). Comparative ultrastructure and calcium transport in heart and skeletal muscle microsomes. J. Cell Biol. 43:610.

47. Stewart, P.S., and MacLennan, D.H. (1974). Surface particles of sarcoplasmic reticulum membranes. J. Biol. Chem. 299:985.

48. MacLennan, D.H., and Yip, C.C. (1972). Isolation of sarcoplasmic reticulum proteins. Symp. Quant. Biol. 38:469.

49. Howell, J.N. (1974). Intracellular binding of ruthenium red in frog skeletal muscle. J. Cell Biol. 62:242.

50. Waugh, R.A., Spray, T.L., and Sommer, J.R. (1973). Fenestrations of sarcoplasmic reticulum. Delineation by lanthanum ion, as a fortuitous tracer and in situ negative stain. J. Cell Biol. 59:254.

51. Huxley, H.E. (1963). Electron microscopic studies on the structure of natural and synthetic proteins filaments from striated muscle. J. Mol. Biol. 7:281.

52. Huxley, A.F., and Huxley, H.E. (1964). A discussion in the physical and chemical basis of muscular contraction. Proc. R. Soc. (Lond.) B160:433.

53. Bacaner, M., Broadhurst, J., Hutchinson, T., and Lilley, J. (1973). Scanning transmission electron microscope studies of deep-frozen unfixed muscle correlated with spatial localization of intracellular elements by fluorescent X-ray analysis. Proc. Natl. Acad. Sci. USA 70:3423.

54. Katz, A.M. (1970). Contractile proteins of the heart. Physiol. Rev. 50:63.

55. Katz, A.M. (1974). Contractile proteins. In G.A. Langer and A.J. Brady (eds.), The Mammalian Myocardium, pp. 51–79. John Wiley and Sons, New York.

56. Pepe, F.A. (1967). The myosin filament. I. Structural organization from antibody staining observed in electron microscopy. J. Mol. Biol. 27:203.

57. Squire, J.M. (1973). General mode of myosin filament structure. III. Molecular packing arrangements in myosin. J. Mol. Biol. 77:291.

58. Potter, J.D. (1974). The content of troponin, actin and myosin in rabbit skeletal muscle myofibrils. Arch. Biochem. Biophys. 162:436.

59. Ebashi, S., Endo, M., and Ohtsuki, I. (1969). Control of muscle contraction. Quart. Rev. Biophys. 2:351.

60. Greaser, M.L., and Gergely, J. (1971). Reconstitution of troponin activity from three protein components. J. Biol. Chem. 246:4226.

61. Pepe, F.A. (1966). Some aspects of the structural organization of the myofibril as revealed by antibody-staining methods. J. Cell Biol. 28:505.

62. Eaton, B.L., and Pepe, F.A. (1972). M band protein, two components isolated from chicken heart muscle. J. Cell Biol. 55:681.

63. Pepe, F.A. (1971). Structure of the myosin filament of striated muscle. In J.A.V. Bulter and D. Noble (eds.), Progress in Biophysics and Molecular Biology, Vol. 22, p. 77. Pergamon Press, New York.

64. Kundrat, E., and Pepe, F.A. (1971). The M band studies with fluorescent antibody staining. J. Cell Biol. 48:340.

65. Knappes, G.G., and Carlsen, F. (1962). The ultrastructure of Z disc in skeletal muscle. J. Cell Biol. 13:323.

66. Kelly, D.E. (1967). Models of muscle Z-band fine structure based on a looping filament configuration. J. Cell Biol. 34:827.

67. Etlinger, J.D., and Fischman, D.A. (1972). M and Z components and the assembly of myofibrils. Symp. Quant. Biol. 38:511.

68. Rash, J.E., Shay, J.W., and Biesele, J.J. (1968). Urea extraction of Z-bands, intercalated discs and desmosomes. J. Ultrastructure Res. 24:181.

69. Fischman, D.A. (1970). The synthesis and assembly of myofibrils in embryonic muscle. In A.A. Moscona and A. Monroy (eds.), Current Topics in Developmental Biology, p. 235. Academic Press, New York.

234   Langer, Frank, and Brady

70. Legato, M. (1970). Sarcomerogenesis in human myocardium. J. Mol. Cell. Cardiol. 1:425.
71. Solaro, R.J., Wise, R.M., Shiner, J.S., and Briggs, F.N. (1974). Calcium requirements for cardiac myofibrillar activation. Circ. Res. 34:525.
72. Hasselbach, W., and Schneider, G. (1951). Der l-myosin und Aktingehalt des Kaninchenmuskels. Biochem. Z. 321:462.
73. Langer, G.A., and Brady, A.J. (1963). Calcium flux in the mammalian ventricular myocardium. J. Gen. Physiol. 46:703.
74. Neidergerke, R. (1963). Movements of calcium in beating ventricles of the frog heart. J. Physiol. (Lond.) 167:551.
75. Sands, S.D., and Winegrad, S. (1970). Treppe and total calcium content of frog ventricle. Amer. J. Physiol. 218:908.
76. Reuter, H. (1965) Über die Wirkung von Adrenalin auf den cellulären Ca-Umsatz des Meerschweinvorhofs. Neun. Schmied. Arch. Exp. Pathol. Pharmakal. 251:401.
77. New, W., and Trautwein, W. (1972). Inward currents in mammalian myocardium. Pflügers Arch. 334:1.
78. Langer, G.A., and Serena, S.D. (1970). Effects of strophanthidin upon contraction and ionic exchange in rabbit ventricular myocardium: relation to control of active state. J. Mol. Cell. Cardiol. 1:65.
79. Bailey, L.E., and Sures, H.A. (1971). The effect of ouabain on the washout and uptake of calcium in the isolated cat heart. J. Pharmacol. Exp. Ther. 178:259.
80. Niedergerke, R. (1957). The rate of action of calcium ions on the contraction of the heart. J. Physiol. (Lond.) 138:506.
81. Langer, G.A. (1964). Kinetic studies of calcium distribution in ventricular muscle of the dog. Circ. Res. 15:393.
82. Glitsch, H.G., Reuter, H., and Scholz, H. (1970). The effect of the internal sodium concentration on calcium fluxes in isolated guinea pig auricles. J. Physiol. (Lond.) 209:25.
83. Tillisch, J.H., and Langer, G.A. (1974). Myocardial mechanical responses and ionic exchange in elevated sodium perfusate. Circ. Res. 34:40.
84. Andersson, K.E., and Edman, K.A.P. (1974). Effects of lanthanum on the coupling between membrane excitation and contraction of isolated frog muscle fibers. Acta Physiol. Scand. 90:113.
85. Bailey, L.E., and Dresel, P.E. (1968). Correlation of contractile force with a calcium pool in the isolated cat heart. J. Gen. Physiol. 52:969.
86. Shine, K.I., Serena, S.D., and Langer, G.A. (1971). Kinetic localization of contractile calcium in rabbit myocardium. Amer. J. Physiol. 221:1408.
87. Endo, M., Tanaka, M., and Ogawa, Y. (1970). Calcium induced release of calcium from the sarcoplasmic reticulum of skeletal muscle fibers. Nature 228:34.
88. Ford, L.E., and Podolsky, R.J. (1970). Regenerative calcium release within muscle cells. Science 167:58.
89. Natori, R. (1954). Property and contraction process of isolated myo-fibrils. Jikeikai Med. J. 1:119.
90. Bloom, S. (1970). Spontaneous rhythmic contraction of separated heart muscle cells. Science 167:1727.
91. Bloom, S. (1971). Requirements for spontaneous contractility in isolated mammalian heart cells. Exp. Cell Res. 69:17.
92. Fabiato, A., and Fabiato, F. (1972). Excitation-contraction coupling of isolated cardiac fibers with disrupted or closed sarcolemmas. Circ. Res. 31:293.
93. Bloom, S., Brady, A.J., and Langer, G.A. (1974). Calcium metabolism and active tension in mechanically disaggregated heart muscle. J. Mol. Cell. Cardiol. 6:137.
94. Ford, L.E., and Podolsky, R.J. (1972). Calcium uptake and force development by skinned muscle fibers in EGTA buffered solutions. J. Physiol. 223:1.
95. Kerrick, W.G.L., and Best, P.M. (1974). Calcium ion release in mechanically disrupted heart cells. Science 183:435.
96. Armstrong, C.M., Bezanilla, F.M., and Horowicz, P. (1972). Twitches in the presence of ethylene glycol bis(β-aminoethylether) N, N'-tetraacetic acid. Biochim. Biophys. Acta 267:605.
97. Sandow, A., Pagala, M.K.D., and Sphicus, E.C. (1974). "Zero" Ca effects on excitation contraction coupling (EEC). Fed. Proc. 33:1259.

98. Endo, M. (1975). Conditions required for calcium-induced release of calcium from the sarcoplasmic reticulum. Proc. Jap. Acad. 51:467.
99. Thorens, S., and Endo, M. (1975). Calcium-induced calcium release and "depolarization" induced calcium release: their physiological significance. Proc. Jap. Acad. 51:473.
100. Fabiato, A., and Fabiato, F. (1973). Activation of skinned cardiac cells. Subcellular effects of cardioactive drugs. Eur. J. Cardiol. 1/2:143.
101. Rich, T.L., and Langer, G.A. (1975). A comparison of excitation contraction coupling in heart and skeletal muscle: An examination of "calcium-induced calcium release." J. Mol. Cell. Cardiol. 7:747.
102. Winegrad, S. (1970). The intracellular site of calcium activation of contraction in frog skeletal muscle. J. Gen. Physiol. 55:77.
103. Ong, S.S., and Bailey, L.E. (1973). Uncoupling of excitation from contraction by nickel in cardiac muscle. Amer. J. Physiol. 209:17.
104. Kohlhardt, M., Bemer. B., Krause, H., and Fleckenstein, A. (1973). Selective inhibition of the transmembrane Ca conductivity in mammalian myocardial fibers by Ni, Co and Mn ions. Pflügers Arch. 38:115.
105. Nayler, W.G., and Anderson, J.E. (1965). Effects of zinc on cardiac muscle contraction. Amer. J. Physiol. 209:17.
106. Ochi, R. (1970). The slow inward current and the action of manganese ion in guinea pig's myocardium. Pflügers Arch. 316:81.
107. Sanborn, W.G., and Langer, G.A. (1970). Specific uncoupling of excitation and contraction in mammalian cardiac tissue by lanthanum. J. Gen. Physiol. 56:191.
108. Doggenweiler, C.F., and Frenk, S. (1965). Staining properties of lanthanum on cell membranes. Proc. Natl. Acad. Sci. USA 53:425.
109. Revel, J.P., and Karnovsky, M.J. (1967). Hexagonal array of subunits in intercellular junctions of mouse heart and liver. J. Cell Biol. 33:C7.
110. Shea, S.M. (1971). Lanthanum staining of the surface coat of cells. Its enhancement by the use of fixatives containing Alcian blue or cetylpyridinium chloride. J. Cell Biol. 51:611.
111. Forbes, M.S., Rubio, R., and Sperelakis, N. (1972). Tubular systems of limulus myocardial cells investigated by use of electon-opaque tracers and hypertonicity. J. Ultrastructure Res. 39:580.
112. Hodgson, B.J., Kidwai, A.M., and Daniel, E.E. (1972). Uptake of lanthanum by smooth muscle. Can. J. Physiol. Pharmacol. 50:730.
113. Langer, G.A., Serena, S.D., and Nudd, L.M. (1974). Cation exchange in heart cell culture; correlation with effects on contractile force. J. Mol. Cell. Cardiol. 6:149.
114. Langer, G.A., Serena, S.D., and Nudd, L.M. (1975). Localization of contractile dependent calcium; comparison of manganese and verapamil in cardiac and skeletal muscle. Amer. J. Physiol. 229:1003.
115. Fleckenstein, A. (1971). Specific inhibition and promoters of calcium action in the excitation-contraction coupling of heart muscle and their role in the prevention or production of myocardial lesions. In P. Harris and L.H. Opie (eds.), Calcium and the Heart, pp. 135–188. Academic Press, New York.
116. Kohlhardt, M., Bauer, B., Krause, H., and Fleckenstein, A. (1972). New selective inhibitors of the transmembrane Ca conductivity in mammalian myocardial fibers. Studies with the voltage clamp technique. Experientia 28:288.
117. McCans, J.L., Lindenmayer, G.E., Munson, R.G., Evans, R.W., and Schwartz, A. (1975). A dissociation of positive staircase (Bowditch) from ouabain-induced positive inotropism. Circ. Res. 35:439.
118. Willerson, J.T., Crie, J.S., Adcock, R.C., Templeton, G.H., and Wildenthal, K. (1974). Influence of calcium on the inotropic actions of hyperosmotic agents, norepinephrine, paired electrical stimulation and treppe. J. Clin. Invest. 54:957.
119. Smith, H.W. (1926). The action of acids on turtle heart muscle with reference to the penetration of anions. Amer. J. Physiol. 76:411.
120. Pannier, J.L., and Leusen, I. (1968). Contraction characteristics of papillary muscle during changes in acid-base composition of the bathing fluid. Arch. Int. Physiol. Biochim. 76:624.
121. Cingolani, H.E., Mattiazzi, A.R., Blesa, E.S., and Gonzales, N.C. (1970). Contractility in isolated mammalian heart muscle after acid base changes. Circ. Res. 26:269.

236   Langer, Frank, and Brady

122. Poole-Wilson, P.A., and Langer, G.A. The effect of CO_2 on myocardial function and Ca^{2+} exchange. In preparation.
123. Pannier, J.L., Weyne, J., and Leusen, I. (1970). Effects of PCO_2, bicarbonate and lactate on the isometric contractions of isolated soleus muscle of the rat. Pflügers Arch. 320:120.
124. Poole-Wilson, P.A., and Cameron, I.R. (1973). A comparison of the control of intracellular pH in cardiac and skeletal muscle. Clin. Sci. 44:15P.
125. Beeler, G.W., Jr., and Reuter, H. (1970). The relation between membrane potential, membrane currents and activation of contraction in ventricular myocardial fibers. J. Physiol. (Lond.) 207:211.
126. Ochi, R., and Trautwein, W. (1971). The dependence of cardiac contraction on depolarization and slow inward current. Pflügers Arch. 323:187.
127. Kaufmann, R., Bayer, R., Fürniss, T., Krause, H., and Tritthart, H. (1974). Calcium-movement controlling cardiac contractility. II. Analog computation of cardiac excitation-contraction coupling on the basis of calcium kinetics in a multi-compartment model. J. Mol. Cell. Cardiol. 6:543.
128. Johnson, E.A., and Lieberman, M. (1971). Heart: excitation and contraction. Annu. Rev. Physiol. 33:479.
129. Lüttgau, H.C., and Niedergerke, R. (1958). The antagonism between Ca and Na ions on the frog's heart. J. Physiol. (Lond.) 143:486.
130. Niedergerke, R. (1963). Movements of Ca in frog heart ventricles at rest and during contractions. J. Physiol. (Lond.) 167:515.
131. Reuter, H., and Seitz, N. (1968). The dependence of calcium efflux from cardiac muscle on temperature and external ion composition. J. Physiol. (Lond.) 195:451.
132. Repke, K. (1964). Uber den biochemischen Wirkungsmodus von Digitalis. Klin. Wochenschr. 41:157.
133. Langer, G.A. (1965). Calcium exchange in dog ventricular muscle: relation to frequency of contraction and maintenance of contractility. Circ. Res. 17:78.
134. Baker, P.F., Blaustein, M.P., Hodgkin, A.L., and Steinhardt, R.A. (1969). The influence of calcium on sodium efflux in squid axons. J. Physiol. (Lond.) 200:431.
135. Glitsch, H.G., Reuter, H., and Schultz, H. (1970). The effect of the internal sodium concentration on calcium fluxes in isolated guinea pig auricles. J. Physiol. (Lond.) 209:25.
136. Birks, R.I., and Davey, D.F. (1969). Osmotic responses demonstrating the extracellular character of the sarcoplasmic reticulum. J. Physiol. (Lond.) 202:171.
137. Gilmore, J.P., Nizolek, J.A., Jr., and Jacob, R.J. (1971). Further characterization of myocardial K^+ loss induced by changing contraction frequency. Amer. J. Physiol. 221:465.
138. Langer, G.A. (1971). Physiology in medicine: the intrinsic control of myocardial contraction-ionic factors. New Eng. J. Med. 285:1065.
139. Lee, K.S., and Klaus, W. (1971). The subcellular basis for the mechanism of inotropic action of cardiac glycosides. Pharmacol. Rev. 23:193.
140. Morad, M., and Greenspan, A.M. (1973). Excitation-contraction coupling as a possible site for the action of digitalis in heart muscle. In L. Dreifus and W. Likoff (eds.), pp. 479–489, Cardiac Arrhythmias, Green and Stratton, New York.
141. Rougier, O.G., Vassort, G., Garnier, D., Gargouil, Y.M., and Coraboeuf, E. (1969). Existence and role of slow inward current during the frog atrial action potential. Pflügers Arch. 308:91.
142. Langer, G.A. (1967). Sodium exchange in dog ventricular muscle: relation to frequency of contraction and its possible role in the control of myocardial contractility. J. Gen. Physiol. 50:1221.
143. Poole-Wilson, P.A., and Langer, G.A. (1975). Glycoside inotropy in the absence of an increased K^+ efflux. Circ. Res. 37:390.
144. Poole-Wilson, P.A., and Langer, G.A. (1975). Effect of pH on ionic exchange and function in rat and rabbit myocardium. Amer. J. Physiol. 299:570.
145. Langer, G.A. (1972). Myocardial K^+ loss and contraction frequency. J. Mol. Cell. Cardiol. 4:85.
146. Gordon, A.M., Huxley, A.F., and Julian, F.J. (1966). Tension development in highly stretched vertebrate muscle fibers. J. Physiol. 184:143.

147. Pollack, G.H., Huntsman, L.L., and Verdugo, P. (1972). Cardiac muscle models: an overextension of series elasticity? Circ. Res. 31:569.

148. Nobel, M.I., and Else, W. (1972). Reexamination of the applicability of the Hill model of muscle to cat myocardium. Circ. Res. 31:580.

149. Sagawa, K., and Loeffler, L. (1975). A one-dimensional viscoelastic model of cat heart muscle studied by small length perturbations during isometric contraction. Circ. Res. 36:498.

150. Pinto, J.G., and Fung, Y.C. (1973). Mechanical properties of the stimulated papillary muscle in quick-release experiments. J. Biomechanics 6:617.

151. Brady, A.J. (1973). An analysis of mechanical analogs of heart muscle. Eur. J. Cardiol. 1/2:93.

152. Matsubara, I., and Millman, B.M. (1974). X-ray diffraction patterns from mammalian heart muscle. J. Mol. Biol. 82:527.

153. Fung, J.C. (1970). Mathematical representation of the mechanical properties of the heart muscle. J. Biomechanics 3:381.

154. Pinto, J.G., and Fung, Y.C. (1973). Mechanical properties of the heart muscle in the passive state. J. Biomechanics 6:597.

155. Bodem, R., and Sonnenblick, E.H. (1974). Deactivation of contraction by quick releases in the isolated papillary muscle of the cat. Circ. Res. 34:214.

156. Ritchie, J.M., and Wilkie, D.R. (1958). The dynamics of muscular contraction. J. Physiol. 143:104.

157. Huxley, A.F., and Simmons, R.M. (1972). Mechanical transients and the origin of muscular force. Cold Spring Harbor Symp. Quant. Biol. 37:669.

158. Cleworth, D.R., and Edman, K.A.P. (1972). Changes in sarcomere length during isometric tension development in frog skeletal muscle. J. Physiol. 227:1.

159. Krueger, J.W., and Pollack, G.H. (1975). Myocardial sarcomere dynamics during isometric contraction. J. Physiol. (Lond.) 251:627.

160. Winegrad, S. (1974). Resting sarcomere length-tension relation in living frog heart. J. Gen. Physiol. 64:343.

161. Nassar, R., Manring, A., and Johnson, E.A. (1974). Light diffraction of cardiac muscle: sarcomere motion during contraction. Ciba Foundation Symposium 24. ASP, Amsterdam.

162. Nilsson, E. (1972). Influence of muscle length on the mechanical parameters of myocardial contraction. Acta Physiol. Scand. 85:1.

163. Templeton, G.H., Donald, T.C., III, Mitchell, J.H., and Hefner, L. (1973). Dynamic stiffness of papillary muscle during contraction and relaxation. Amer. J. Physiol. 224:692.

164. Steiger, G.J. (1971). Stretch activation and myogenic oscillation of isolated contractile structures on heart muscle. Pflügers Arch. 330:347.

165. Abbott, R.H. (1973). The effects of fiber length and calcium ion concentration on the dynamic response of glycerol extracted insect fibrillar muscle. J. Physiol. (Lond.) 231:195.

166. Rüegg, J.C., Steiger, G.J., and Schadler, M. (1970). Mechanical activation of the contractile system in skeletal muscle. Pflügers Arch. 319:139.

167. Rüegg, J.C., and Tregear, R.T. (1966). Mechanical factors affecting the ATPase activity of glycerol-extracted insect fibrillar flight muscle. Proc. R. Soc. B 165:497.

168. Steiger, G.J., and Rüegg, J.C. (1969). Energetics and "efficiency" in the isolated contractile machinery of an insect fibrillar muscle at various frequencies of oscillation. Pflügers Arch. 307:1.

169. Huxley, A.F. (1957). Muscle structure and theories of contraction. Prog. Biophys. Biophys. Chem. 7:255.

170. Brady, A.J. (1965). Time and displacement dependence of cardiac contractility: problems in defining the active state and force-velocity relations. Fed. Proc. 24:1410.

171. Abbott, R.H., and Mannherz, H.G. (1970). Activation by ADP and the correlation between tension and ATPase activity in insect fibrillar muscle. Pflügers Arch. 321:223.

172. Herzig, J.W. (1972). Myogene autoregulation den kontraklilität in den zellen der papillarmuskatur des säugerherzens. Diplomarbeit. Universität. Bochum.

International Review of Physiology
Cardiovascular Physiology II, Volume 9
Edited by Arthur C. Guyton and Allen W. Cowley
Copyright 1976   University Park Press   Baltimore

6
Factors Influencing Cardiac Performance

V. S. BISHOP, D. F. PETERSON, and L. D. HORWITZ
The University of Texas Health Science Center at San Antonio

INTRODUCTION

Cardiac output can be varied over an enormous range in a normal animal or man. In trained athletes it has been observed to increase from resting values of approximately 5 liters/min to as high as 42 liters/min during maximal exercise. A multitude of factors are responsible for the remarkable ability of the circulatory system to regulate tissue perfusion in accordance with the requirements of the body. It is apparent that, regardless of the capacity of the heart to pump blood, a cardiac reserve of 6-fold or greater could not be achieved in the absence of changes in the peripheral circulation. These include various alterations in response to tissue metabolic needs which result in decreased peripheral resistance and increased venous return. Whether, in normal circumstances, the maximum cardiac output is determined by peripheral circulatory mechanisms or by the pumping capacity of the heart itself is controversial. Nevertheless, the heart is the prime energy source of the cardiovascular system and as such must be capable of accomplishing rapid adjustments in output to meet changes in tissue demands. If a pathological process impairs cardiac performance, the limitation in maximum cardiac output clearly lies within the heart itself.

In this review, we analyze the principal factors which regulate the performance of the heart and control of cardiac output in the intact circulation. Several factors are considered in this discussion: 1) the regulation of the stroke volume; 2) the control of the heart rate; 3) the interrelationship of factors which determine ventricular performance; and 4) evaluation of ventricular performance during exercise.

STROKE VOLUME

Cardiac performance is often regarded as synonymous with cardiac output. However, the two components of the cardiac output—heart rate and stroke volume—are influenced by separate groups of factors, which only partially overlap. Therefore, a realistic appraisal of cardiac pump performance requires consideration of the mechanisms which control each of these two variables individually, as well as an understanding of the nature of their interaction. In the case of stroke volume, the determinants are best described in terms of cardiac muscle mechanics.

Preload

In the late 19th century, several investigators concluded that, in skeletal muscle, the initial length and tension determined the magnitude of the subsequent contractile response. In 1895, Otto Frank (1) demonstrated that the same principle was applicable to the myocardium. Using isometrically contracting isolated frog hearts, he observed that the greater the length and tension of the cardiac muscle prior to contraction, the greater the developed tension during contraction. In 1914, Patterson and Starling (2) extended Frank's observations to an isolated canine heart-lung preparation, and noted that, provided the functional condition of the heart is unchanged, stroke volume varies directly with diastolic filling. More recently, the biophysical basis for this finding has been clarified by relating myocardial ultrastructure to length-tension characteristics of isolated cardiac muscle and pressure-volume characteristics of the intact heart (3–6).

The length-tension relationship for the cardiac sarcomere provides insight into the regulation of myocardial function. When muscle fibers contract isometrically, at a variety of initial fiber lengths, an active tension curve can be plotted. As the initial fiber length increases, active tension increases until a maximum is reached at 2.2 μm. As shown in Figure 1A, with further increase in length, active tension decreases. Resting tension, which represents the passive resistance of the muscle to stretch, increases progressively with increases in initial fiber length.

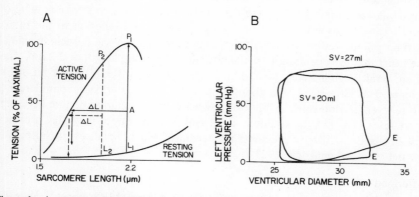

Figure 1. A, sarcomere length-tension curve. Isometric tensions P_1 and P_2 correspond to sarcomere lengths L_1 and L_2, respectively. Hypothetical shortening (ΔL), occurring at similar afterloads, is shown for the different sarcomere lengths. Redrawn from Sonnenblick et al. (5) by courtesy of *Circulation. B*, the pressure-diameter relationship of the left ventricle in a conscious animal. Note the dependence of diameter changes and, in this case, stroke volume on the initial end diastolic fiber length (*E*).

242   Bishop, Peterson, and Horwitz

In Figure 1*A*, at resting muscle lengths L_1 and L_2, the corresponding active tensions generated during isometric contraction are P_1 and P_2. When afterload is reduced, allowing external shortening to occur, less energy is expended in the generation of tension and more is available for shortening. The muscle shortens until it reaches a point on the active tension curve. At a constant contractile state, the extent of shortening for any given afterload is dependent upon the fiber length, or preload. Because the normal heart, under basal conditions, can be expected to have an initial or resting end diastolic sarcomere length below 2.2 μm (4, 5), it is operating on the ascending portion of the active tension curve, and there is considerable reserve for utilization of the Frank-Starling mechanism.

The effect of the Frank-Starling mechanism in the intact heart can be appreciated through comparison of plots of ventricular pressure versus ventricular volume, or dimensions, as shown in Figure 1*B*. These left ventricular pressure-diameter loops were obtained in a conscious dog, at rest and after elevation of filling pressure by rapid intravenous infusion of isotonic fluid (7, 8). They are similar to the length-tension curve in Figure 1*A* and may be interpreted in the same manner. At the end of diastole, the ventricle begins to develop tension. When the left ventricular pressure exceeds the aortic pressure, shortening of the fibers begins. Shortening continues until the active tension curve is reached. In this example, when preload is increased, the extent of shortening of the left ventricular diameter increases from 7 to 9 mm, resulting in an increase in stroke volume from 20 to 27 ml.

Since the Frank-Starling mechanism is firmly established as a basic biophysical property of cardiac muscle, it would be expected to be a factor in physiological settings. This is, in fact, the case, although in the intact animal its presence may be masked or overshadowed by other compensatory mechanisms. Nevertheless, its importance in the beat-to-beat regulation of stroke volume is easily documented. During sinus arrhythmias or postural changes, alterations in left ventricular end diastolic size result in predictable changes in stroke volume (8, 9). There is a physiological setting which provides an exception to the Frank-Starling generalization. At very low resting heart rates in the conscious animal, the heart may be operating near the optimum sarcomere length of 2.2 μm, since often, additional increments in filling pressure are without affect on stroke volume. This circumstance is discussed in more detail later. During moderate or severe exercise, increases in preload augment stroke volume, despite very high heart rates (10, 11). In addition, the Frank-Starling effect may play a critical role in sustaining life by providing a mechanism which equalizes the outputs of the right and left ventricles (12).

Afterload

As illustrated by the length-tension relationship for cardiac muscle, when the afterload is increased, a larger amount of energy is expended in the development of contractile tension. Consequently, the extent of shortening decreases. In the isolated heart, abrupt increases in aortic resistance during diastole reduce stroke volume in the next beat (13). The reduction in stroke volume varies inversely with the increase in load. Similar responses have been observed in the intact animal (14). Other studies (15, 16) provide a more quantitative analysis of the effects of arterial pressure on cardiac output. In these studies, the cardiac output was evaluated as a function of the right or left atrial pressure and aortic pressure under static conditions. The results were displayed by three-dimensional plots showing the effects of changing arterial pressure load on left ventricular output. Since reflexes to the heart were eliminated and heart rate was constant, changes in cardiac output were due entirely to changes in stroke volume. Within much of the physiological pressure range, the steady state cardiac output was highly dependent upon ventricular inflow and almost independent of arterial pressure. With right atrial pressure controlled, cardiac output was maintained until aortic pressure exceeded 150 mm Hg. When mean left atrial pressure was controlled, cardiac output declined when aortic pressure exceeded 120 mm Hg. Although the level of cardiac output was dependent upon the initial mean atrial pressures, the maximum pressure against which the heart could pump approached 250 mm Hg under all conditions. With right atrial pressure controlled, it appeared that the Frank-Starling mechanism was an important factor in the left heart, when the residual volume in the left atrium and ventricle varied. This was minimized when left atrial pressure was controlled at a constant level, as evidenced by a more rapid decline in cardiac output as the arterial pressure was increased. Thus, to maintain cardiac outflow against a higher pressure load, the heart must be filled with a greater volume of blood. It is noteworthy that it is the response of the left heart which is limited in the face of an elevated pressure load, whereas the right heart appears to limit the output at subnormal venous pressures. In the latter case, cardiac output was independent of right atrial pressures at levels above 8 mm Hg.

In the conscious dog, increases in arterial pressure due to intravenous infusion of angiotensin do not consistently reduce stroke volume until left ventricular end diastolic pressure exceeds 15 mm Hg; this corresponds approximately to a 30% increase in left ventricular peak systolic pressure (17). Until these levels of filling pressure and afterload are reached, the left ventricular end diastolic and end systolic internal diameter changes are of similar magnitude. Above these levels, the increase in left ventricular end systolic diameter exceeds the change in end diastolic diameter. These changes in left ventricular internal diameter with increasing preload are in marked contrast to those obtained during acute volume loading (7). Increasing left ventricular end diastolic pressure to similar levels by rapid intravenous infusions (which do not markedly increase arterial pressure) gives a 2-fold greater increase in left ventricular end diastolic diameter than end systolic diameter. Thus, in the intact conscious animal, an increase in arterial pressure above 120–150 mm Hg severely limits the extent of shortening as a result of the elevated end systolic ventricular size. It is noteworthy that stroke volume is almost independent of arterial pressure over the normal physiological range (80–120 mm Hg). Yet, even in this range, if it were not for the Frank-Starling mechanism, stroke volume would be expected to decrease rapidly with increments in arterial pressure.

Myocardial Contractility

For any given preload or afterload, the extent of shortening is a function of the contractile state of the muscle. In normal conditions, the contractile state, and consequently the extent of shortening, are determined by the quantity of ionic calcium available to the contractile proteins (18). In addition, stimulation of the β-adrenergic receptors by endogenous release of norepinephrine or administration of catecholamines results in increased myocardial contractility (19–21). Thus, the extent of shortening, or the volume ejected, from any end diastolic size, can be modulated through neural mechanisms.

As shown in Figure 2A, at equivalent end diastolic size and afterload, an inotropic stimulus causes a greater shortening, as reflected by a smaller end systolic size. The inotropic effect of neural reflexes or other interventions can be analyzed by estimation of changes in the maximum rate of pressure or force development during the isovolumic phase of systole (21, 22), the maximum velocity of contractile element shortening calculated from the isovolumic pressure changes (20), peak pressure in a ventricle which is contracting isovolumically (23), the rate of circumferential, or segmental, fiber shortening during ejection (7, 20), or stroke volume or stroke work as a function of the filling pressure (24). Levy and his co-workers (23, 25, 26) have reported that direct stimulation of the vagus nerve reduced peak developed pressure in the isovolumically contracting ventricle, the effect being more pronounced when sympathetic activity was high. These results support the studies of Sarnoff et al. (27) in which the carotid sinus reflex influenced the stroke work output of the heart. The specificity of the autonomic control of cardiac contractility was established by Randall et al. (28). Stimulation of the sympathetic nerves increases the force generated from a constant length of myocardium, whereas vagal stimulation reduces it. Furthermore, selective stimulation of specific cardiac nerves reveals specific local control in various regions of the myocardium.

In the anesthetized dog, the baroreceptor reflex may exert pronounced influences on cardiac performance through reflex alterations in heart rate and contractility or, indirectly, through reflex changes in preload or afterload (27). In the conscious animal, baroreceptor-mediated alterations in either myocardial contractility or stroke volume are small or absent (29). This may reflect differences in vagal and sympathetic control in the conscious, versus the anesthetized, state. In the conscious dog at rest, vagal restraint is high, and sympathetic tone is low, while the opposite is true during anesthesia. In addition, surgical interventions and anesthetic agents may alter the myocardial contractile state and neural regulation. Furthermore, in conscious dogs, other reflexes may prevent or mask the complete manifestation of the baroreceptor reflex. Nevertheless, because of the potential of the baroreceptor in the regulation of the circulation, it would seem probable that these receptors play a role in the beat-to-beat regulation of myocardial performance. Moreover, receptors located in the heart or in vascular sites other than the carotid sinus may exert more powerful reflex controls upon the myocardium. Finally, it should be recognized that lack of knowledge concerning the importance of small changes in myo-

A

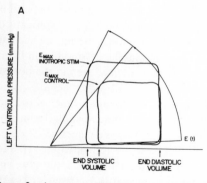

B

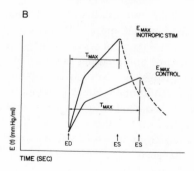

Figure 2. A, pressure-volume curves obtained in an isolated left ventricle with and without an inotropic stimulus. The pressure-volume ratio, the slope of the line $E(t)$, increases throughout systole reaching a maximum value of E_{max} at the end of systole. Although beginning at the same end diastolic volume, an inotropic stimulus increases the pressure-volume ratio as indicated by the slope of the line E_{max}. The arrow on the arch connecting $E(t)$ with E_{max} illustrates the progressive increase in the pressure-volume ratio moving in a counterclockwise direction. B, the pressure-volume ratio $E(t)$ as a function of time during a single cardiac cycle. End diastole (ED) and end systole (ES) are shown for a control beat and with an inotropic stimulus. The time (T_{max}) from end diastole to E_{max} is also illustrated. Redrawn from Suga et al. (35) by courtesy of $Circulation Research$.

cardial performance may limit our ability to appreciate the potential of neural reflexes which elicit only subtle changes.

The extent to which alteration in myocardial contractility can influence the stroke volume in the normal heart is not clear. In open-chest anesthetized dogs, an increase in myocardial contractility often augments stroke volume, particularly if the heart is depressed (27). In humans with diminished myocardial function due to heart disease, an increase in myocardial contractility also increases stroke volume (30). However, using conscious dogs, Noble et al. (31) found that inotropic stimuli confined to the myocardium increased dP/dt without significantly altering left ventricular end diastolic pressure or stroke volume. The inotropic stimuli increased the acceleration of blood flow, the rate of rise of pressure and the pulse pressure in the aorta. Presumably, increased stretch in the aorta during early systole may provide an opposing force which restricts the ability of the myocardium to augment stroke volume. A major assumption in the above study was that left ventricular end diastolic pressure is an accurate index of the preload (end diastolic volume) at rest and during the responses observed after injection of inotropic stimuli. It was also assumed that intracoronary injections of catecholamines or calcium had no other effects than the observed positive inotropic effect. During mild levels of exercise, in the absence of apparent preload changes or a fall in ejection pressure, there is a small increase in stroke volume which appears to be mediated by increased myocardial contractility (11). Also, β-adrenergic blockade appears to decrease stroke volume, through diminished ability of myocardial fibers to shorten, in conscious dogs during acute volume loading and exercise (32, 33). The concept that contractility changes cannot alter stroke volume unless accompanied by changes in preload and afterload is somewhat contradictory to the response observed in isolated cardiac muscle. Before accepting this observation without qualification, supportive data concerning changes in contractility on pump function are needed. In view of the responses observed in isolated cardiac muscle, it would seem likely that changes in contractility should influence stroke volume, even in the normal state. Additional studies in intact, conscious animals, with other hemodynamic factors controlled and direct measurements of ventricular dimensions, are needed to settle this issue, and to establish the role of preload and afterload on the regulation of stroke volume in the presence of inotropic changes.

Recently, Suga and his co-workers (34, 35) used the fundamental variables of the functioning ventricle, pressure and volume, to relate contractile state and pump performance of the left ventricle. The dynamics of the left ventricle were described in terms of the time varying ratio of instantaneous pressure $P(t)$ to instanteous volume $V(t)$. The time course of the instanteous $P{:}V$ ratio over the entire cardiac cycle was defined by

$$E(t) = P(t)/[V(t) - V_d]$$

where $P(t)$ = instantaneous intraventricular pressure; $V(t)$ = instantaneous intraventricular volume; and V_d = the fixed correction volume. Thus, at any time (t), $E(t)$ is the time varying slope of the pressure-volume regression line.

$E(t)$ has physiological importance because the elastance of the myocardium varies with time throughout systole. Starting with the initiation of systole, the elasticity or the pressure-volume ratio increases until end systole is reached. Elasticity then decreases during diastole. As shown in Figure 2A, during early systole, when volume is essentially unchanged, $E(t)$ increases with left ventricular pressure. During ejection, when the variation in left ventricular pressure is minimized, the reduction in ventricular volume further increases $E(t)$ which reaches a maximum value (E_{max}) at the end of systole. Consequently, the time variance of elasticity can be a useful indicator of the flow generating capacity of the ventricle. The maximum value (E_{max}) is related to the flow generating capacity of the ventricle under a given loading condition, while the time course of $E(t)$ from onset of systole affects the rate of ejection. E_{max} appears to be significantly affected by inotropic stimuli and independent of the loading conditions and heart rate (Figure 2, A and B). On the other hand, T_{max}, which is the time interval from end diastole to E_{max}, can be slightly increased as the ventricular contraction is changed from isovolumic to auxobaric. In addition, heart rate and isoproterenol shortened T_{max} (Figure 2B). The authors have postulated that $E(t)$ is a fundamental property of ventricular contraction. This was supported mathematically by the fact that a given $E(t)$ could be normalized with respect to E_{max} and T_{max} to yield a single curve with a unique shape regardless of the contractile state, heart rate, or loading conditions. Thus, E_{max} and T_{max} are the characteristic parameters of the instantaneous pressure-volume ratio. Referring to the length-tension or pressure-volume relationship shown in Figure 1, A and B, it becomes obvious that changes in the inotropic state would alter E_{max} through changes in the active tension curve. Since these studies were performed in isolated hearts, the relevance of this approach to the intact animal has not been determined. However, the model provides a refreshing new insight into the factors which determine stroke volume.

Effect of Alteration in Heart Rate on Stroke Volume and Cardiac Output

The extent to which cardiac output is dependent upon changes in heart rate varies with the physiological circumstances. Electrical pacing of the atrium or vagal block in an anesthetized animal or a conscious animal results in a linear decrease in stroke volume as heart rate increases (36–39). In recent studies in conscious dogs, end diastolic left ventricular dimensions decreased for a given increment in heart rate (40, 41). Therefore, the decline in stroke volume was due to reduction in end diastolic cardiac muscle fiber length.

Such a reduction in end diastolic size is difficult to explain solely on the basis of mechanical limitation in filling due to shortening of diastole during tachycardia. Since the greater portion of ventricular filling occurs early in diastole, increases of up to 100 beats/min in heart rate would be expected to reduce the period of diastasis primarily, and therefore have little direct influence on filling. It is possible, however, that shortening of ventricular diastole could influence atrial systole. The total contribution of the atrial contraction is small, probably no more than 20% of filling. However, this could account for the initial decline in stroke volume. Perhaps alterations in rates of myocardial relaxation influence early diastolic filling, even at heart rates near 100 beats/min. Regardless of the cause, a decline in end diastolic fiber length occurs, and this results in a reduction in stroke volume.

A number of studies have shown that cardiac output is a curvilinear function of increasing heart rate (37–39, 42). As heart rate increases, cardiac output increases to a maximum level, which has been termed the optimum heart rate. With additional increases in heart rate, the cardiac output declines (36, 42, 43). Since the decline in stroke volume is linearly related to increasing heart rate, the curvilinear relationship between cardiac output and heart rate can be predicted mathematically. This has been partially confirmed experimentally by Noble et al. (37), but, since the atrial pacing frequency was limited to physiological ranges (approximately 150 beats/min), they were unable to describe the descending limb of the curve relating cardiac output to heart rate.

During vagal blockade in conscious animals, cardiac output increased as heart rates approached 200 beats/min, although stroke volume declined as a linear function of the increase in rate (38, 39). Using the average data from these studies, the maximum cardiac output of 182 ml min^{-1} kg^{-1} occurred at 165 beats/min. Thus, doubling the resting heart rate (80–160 beats/min) increased cardiac output approximately 50 ml min^{-1} kg^{-1} or 27%. However, 90% of the maximum increase in cardiac output persisted at heart rates of 120–220 beats/min, indicating that the apex of the cardiac output response was very broad. Although only a limited amount of data is available from human studies, it appears that increments in heart rate have a more pronounced negative effect on stroke volume, when compared to the dog. Using the above approach for estimating the cardiac output/heart rate relationship, the maximum change in cardiac output was similar to the dog but occurred at a much lower heart rate (100 beats/min). The heart rate range corresponding to 90% of the maximum cardiac response was narrower and shifted to lower heart rates (70–120 beats/min). Thus, it is apparent that the stroke volume/heart rate relationship is an important determinant of the cardiac output response to tachycardia.

The results of these studies in conscious dogs vary somewhat with those obtained in anesthetized animals, or in conscious animals during ventricular pacing; in the latter two situations a reduction in cardiac output was found above 120 beats/min. This could be explained if anesthesia alters the stroke volume/heart rate (SV/HR) relationship and ventricular pacing reduces the effectiveness of the atrial contribution to ventricular filling. In anesthetized dogs, increases in atrial pressure shifted the apex of the cardiac output/heart rate curve to higher cardiac outputs and heart rates, indicating that ventricular filling was an important determinant of this relationship (42). However, in these studies the maximum stroke volume attainable by increasing filling pressure was reduced by tachycardia.

In conscious animals, the tachycardia from vagal blockade gave simultaneous decreases in stroke volume and left ventricular end diastolic and end systolic diameters (38, 39). Although heart rate was near 200 beats/min, the maximal stroke volume response to acute volume loading was not limited. Furthermore, the maximum cardiac output obtained during the intravenous infusion of saline was directly proportional to heart rate. The increase in left ventricular end diastolic and end systolic diameter was less during vagal blockade. Thus, the peak stroke volume was obtained through a greater extent of shortening from a smaller initial fiber length. When sympathetic innervation was abolished with β-adrenergic blockade, the maximum stroke volume obtained by volume loading was limited in both the normal and vagal blocked conditions. β-Adrenergic blockade decreased the maximum stroke volume by limiting the extent of myocardial fiber shortening, as reflected by an elevated end systolic diameter (41).

Based upon the infusion studies, it is apparent that the SV/HR relationship can be altered by changes in filling pressure. However, whether changes in initial filling pressure alter both the slope and intercept of the SV/HR relationship has not been determined experimentally. When filling pressure is constant, stroke volume is maintained despite changes in heart rate, which suggests that the initial filling pressure may determine the slope, as well as the intercept, of the SV/HR relationship. Because the initial sarcomere length is an important determinant of the Frank-Starling reserve, a corollary to the effects of heart rate on the stroke volume and end diastolic size is that the Frank-Starling reserve varies inversely with the heart rate. Increasing heart rate increases the reserve. In contrast, decreasing the heart rate decreases the reserve, as the end diastolic size and, consequently, stroke volume approach a maximum. Data are lacking concerning the effects of arterial pressure or inotropic intervention on this relationship. However, as discussed earlier, suppression of the sympathetic control to the heart limits the stroke volume response to volume loading. Similarly, β-adrenergic blockade reduces the decline in end systolic diameter during increases in heart rate, altering the SV/HR relationship.

Previous studies by Warner and Toronto (44) have demonstrated that cardiac output can be regulated to alter flow through changes in stroke volume alone if heart rate is fixed. Thus, in the conscious animal, this mechanism is operative where such regulation is attained through changes in ventricular filling pressure. On the other hand, if stroke volume is held constant by regulating the filling pressure, changes in cardiac output are proportional to the heart rate. Although cardiac output can be regulated through changes in both heart rate and stroke volume, the mechanisms which determine their relative contributions in normal individuals at rest and during stressful exertion, as compared to physically conditioned subjects or patients with limited cardiovascular reserve, are not known. Perhaps, in each of these three conditions, a different optimum end diastolic size (sarcomere length) is obtained as a result of changes in venous return and in heart rate.

Since the myocardial oxygen consumption is greater when cardiac output is increased through changes in heart rate rather than stroke volume, one may wonder why cardiac output is so frequently adjusted through changes in heart rate, an apparently less efficient mechanism. Perhaps it is because adequate changes in cardiac output, when going from a resting to a stressful state, require increases in both heart rate and stroke volume. Based upon the SV/HR relationship at rest, it becomes obvious that large changes in cardiac output cannot be attained through heart rate or stroke volume alone. Stroke volume must be maintained above the level which would normally result from a tachycardia induced in the resting state, and further increases in output would depend on increased heart rate.

REGULATION OF HEART RATE

Much of the reserve output of the heart lies in its ability to increase the rate at which blood is ejected. This increased heart rate provides an important contribution to cardiac output under numerous physiological circumstances. In most situations in which stress is applied to the cardiovascular system, heart rate changes play a key role in the physiological adaptation by increasing cardiac output. At constant filling pressure, without alteration in myocardial contractility, changes in cardiac output depend upon changes in heart rate. When stroke volume is initially relatively high and heart rate relatively low, increased cardiac output is predominantly a function of heart rate. Additionally, subtle heart rate changes may be an important factor in beat-to-beat cardiovascular adjustments; such alterations in the heart rate may provide a fine tuning mechanism for maintenance of cardiac output.

Neural influences produce the most dramatic and important alterations in heart rate, although numerous other factors, including chemical, humoral, or mechanical alterations, changes in body temperature or osmotic pressure, and administration of pharmacological agents may also modify cardiac pacemaker activity. The actual heart rate is due to the net effect of all these influences superimposed upon the spontaneous pacemaker rate.

Intrinsic Regulation

Spontaneous depolarization is characteristic of cardiac pacemaker cells. As in all spontaneously excitable tissue, the frequency of generation of an action potential depends primarily upon the relative movement of Na^+ and K^+ ions across the cell membrane during diastole. Since the details of this mechanism are discussed in several recent monographs (45—48), it will suffice to highlight only those basic principles which are most important to understand the regulation of heart rate. During diastole, the maximum membrane potential is largely determined by the rate of K^+ efflux. Throughout diastole, K^+ conductance decreases while a small inward conductance of Na^+ is maintained or increased. In late diastole, this causes a gradual depolarization of the membrane termed "diastolic depolarization." When diastolic depolarization reaches the threshold for impulse generation, Na^+ rapidly enters the cell, triggering the action potential. Change in the rate of diastolic depolarization is the primary means by which the time interval between action potentials is altered. Thus, factors which determine the conductance of K^+ in large measure determine heart rate. Changes in mechanical stretch, temperature, and the chemical environment can alter the intrinsic heart rate (49). Intrinsic heart rate is defined as the observed rate after total cardiac denervation and is generally agreed to be between 130 and 150 beats/min for both man and dog. For the purposes of our subsequent discussion, we will assume that the sinoatrial nodal cells, which in normal hearts exhibit the fastest rates of diastolic depolarization, comprise the cardiac pacemaker and establish the intrinsic rate.

Mechanical Factors   Mechanical deformation of pacemaker cells results in an increased heart rate. When the isolated cat sinoatrial node is stretched, the discharge rate is increased; the greater the magnitude, or rate, of stretch the greater the response (50). In open-chest anesthetized dogs, longitudinal and lateral stretch of the in situ sinoatrial node increases heart rate, whether or not autonomic blocking agents have been used (51). After blockade, increases in maximum heart rate were 5—15% above control. Within limits, as was true in the isolated sinoatrial node tissue, the greater the magnitude of the stretch, and the greater the rate at which the stretch was applied, the greater the resultant cardioacceleration.

Of great interest is the extent to which the increases in heart rate caused by mechanical deformation of the sinoatrial node can be produced in response to increased venous return. Conscious dogs which have undergone pharmacological cardiac denervation with atropine and propranolol respond to elevation of right and left heart filling pressures by acute volume loading with small, but consistent, increases in heart rate (52). The mean increase was 12 beats/min. This is less than the increases in rate induced by mechanical distention in the studies of Brooks et al. (51). It appears that, although mechanical distention of the sinoatrial node can induce an increase in intrinsic rate, the magnitude of the change in healthy, conscious animals is quite limited. If so, this mechanism probably plays only a minor role in physiological adaptation to stress. A potentially more important role of this mechanism may be the maintenance of cardiovascular homeostasis in response to transient changes in atrial filling pressure.

Temperature Effects Heart rate varies directly with body temperature. In 1912 Knowlton and Starling (53) described a linear relationship between heart rate and temperature with an increase of 9 beats/min/°C in the isolated dog heart-lung preparation. In studies of intact dogs, whose hearts have been denervated surgically or pharmacologically, increases of 6–7 beats/min/°C have been reported (54). The embryonic human heart in a tissue bath increases in rate by about 7 beats/min/°C between 20 and 40 °C (55). Normal men, who have received autonomic blocking agents, show similar changes (56). It is clear that temperature effects on the S.A. node are predictable. Since the physiological range for temperature variations is limited to 4–5°C, it is likely that maximal heart rate responses are limited to approximately 35 beats/min. During heat stress, increases in cardiac output are primarily due to increases in heart rate (57), and are important in increasing cutaneous blood flow. Intrinsic regulation may contribute significantly to this total response. It is also possible that the direct effect of temperature rise on the S.A. node during exercise contributes significantly to increased cardiac output secondary to tachycardia. Still, these direct effects appear to account for only a small portion of the total rate response to temperature change. Heart rate changes of 50 beats/min have been demonstrated to accompany a 1°C change in right atrial temperature (57). This is primarily a consequence of neural reflexes, however, as will be discussed later.

Chemical and Humoral Factors Catecholamines increase the frequency of sinoatrial node discharge, whereas acetylcholine slows the rate of discharge (58). These changes appear to be mediated through alterations in sodium and potassium fluxes across the cell membranes. Alterations in local concentrations of oxygen (54), calcium (59), and probably other electrolytes as well, will also influence the sinoatrial discharge rate. Thyroid hormone can directly increase the heart rate in isolated mouse embryos (60). It is apparent that many alterations in the chemical milieu of the sinoatrial node can affect heart rate. Thus, although not a topic in this discussion, one must recognize that the inter- and intracellular balance of ions and hormones may also substantially alter the influences of the neurally released transmitters acetylcholine and norepinephrine. Such alterations, however, are probably more indicative of pathological than physiological conditions.

Neural Factors

Although numerous intrinsic and extrinsic factors may influence heart rate, its regulation is primarily determined by neural influences which have efferent pathways in the vagus and cardiac sympathetic nerves. When a stress is imposed, changes in sympathetic and parasympathetic neural activity to the heart have been shown to produce rapid, dramatic changes in cardiac output due predominantly to changes in heart rate, especially when the filling pressure is constant or already somewhat elevated (61). In addition, the nervous system has been shown to elicit very subtle beat-to-beat modifications in heart rate (62). These modifications may be significant in preventing gradual shifts in cardiac output, arterial pressure, and filling pressure. In the following section we discuss potential mechanisms whereby neural influences produce adjustments in heart rate to either maintain a constant homeostatic state or to alter cardiac output in response to stress.

Sympathetic Influences   In most species, the sympathetic nerves to the heart arise from the first five thoracic roots of the spinal cord. The majority of sympathetic fibers pass through the stellate ganglia and traverse the ansa subclavia to the heart. In the dog all cardiac efferent sympathetic fibers appear to follow this route (63). Since the dominant sympathetic innervation of the right atrium (and S.A. node) is via the right cardiac sympathetic nerves (64), cardioacceleration is initiated primarily through activation of this pathway. Innervation from the left side exerts a dominant influence on inotropic function, and has less chronotropic influence.

Ordinarily, sympathetic nerves to the heart conduct tonic discharges which, presumably, influence resting heart rate (45). Direct electrical stimulation of the right stellate ganglion or ansa subclavia can elicit large increases in heart rate. As will be discussed later, however, these sympathetic increases may not be independent of resting vagal tone in the intact animal. Also, recent evidence has indicated that identifiable branches of the cardiac sympathetic nerves produce measurably different effects on heart rate and inotropic performance of the heart (65).

Parasympathetic Influences   The parasympathetic nerves to the heart all travel via the vagi. Efferent cardiac fibers in the right vagus terminate predominately in the sinoatrial node and are of major importance in regulation of heart rate. Efferent fibers from the left vagus also innervate the sinoatrial node, in part, but their cardiac effect on the atrioventricular node is dominant, and this nerve strongly influences conduction from the sinoatrial node to the ventricle.

Increased vagal activity produces bradycardia. Maximal electrical stimulation of the right vagus usually produces transient asystole. Obviously, the vagus can exert powerful influences on heart rate and cardiac output. Tonic activity in the vagus varies greatly among different species. Evidence that the relationship between resting heart rate and tonic vagal activity is species related can be obtained by blocking vagal innervation of the heart pharmacologically or surgically. For example, after atropine (0.1 mg kg^{-1}) heart rates of resting, conscious dogs increase from 90 to 220 beats/min (52) whereas heart rate in anesthetized rabbits is unchanged by bilateral vagotomy, averaging 292 beats/min (66). Thus, dogs are considered more "vagal" than rabbits and in this respect more closely resemble man (67). As we shall discuss later, one cannot unequivocally attribute tachycardia after vagal blockade to diminished vagal inhibition alone. Alterations in sympathetic input may occur simultaneously as vagal influence changes.

Paradoxically, recent studies have indicated that electrical stimulation of the vagus may not always produce bradycardia (68, 69). This has been explained by either phase dependency of the S.A. node responsiveness to vagal stimulation (68), or by activation of anomalous sympathetic fibers traveling in the vagus (69).

Efferent Autonomic Balance The means whereby sympathetic and parasympathetic influences interact on the heart to regulate heart rate has received considerable attention recently, but is by no means completely understood. A major problem is to characterize the interaction between the vagal and sympathetic efferent components. It has been a long-standing assumption that heart rate alterations are accomplished by either changes in vagal or sympathetic activity alone, or by some synergistic effect of these two neural inputs to the sinoatrial node. According to this formulation, reciprocal changes in activity magnify heart rate changes while parallel changes in neural activity minimize them. Accordingly, it has been assumed that heart rate is linearly related to the magnitude of the difference between cardiac sympathetic and vagal activity. However, it appears that the summation of the effects of vagal and sympathetic stimulation is not linear (70, 71). When the vagus and sympathetic fibers to the heart are electrically stimulated simultaneously, the vagal influence is dominant, since the net effect is a slowing of the heart rate (70, 71). Warner and Russell (70) have developed and tested a mathematical model which suggests that changes in heart rate, in either direction, can potentially be explained by changes in vagal tone associated with a constant, resting sympathetic tone. Since vagal influences dominate, if vagal activity were high, heart rate would remain low even in the face of high resting sympathetic tone. In contrast, vagal withdrawal could cause a substantial increase in rate (71). In a recent review, Levy (72) has emphasized that when the basal heart rate is increased by tonic sympathetic stimulation, reduction in heart rate due to vagal stimulation is enhanced. Therefore, heart rate changes are not always a simple summation of vagal and sympathetic activity.

Although the model of Warner and Russell provides insight into functional mechanisms in a system dominated by the vagus nerve, the fact remains that sympathetic efferent activity to the heart is not constant. Tachycardia may be a net result of increased sympathetic activity and simultaneous reduction in vagal activity. The converse can produce bradycardia. For example, it has been demonstrated recently that bradycardia elicited by afferent aortic nerve stimulation in the rabbit is affected by both increased vagal activation and withdrawal of sympathetic activity (66). These authors found that after section of the vagi the decrement in heart rate in response to aortic nerve stimulation was approximately 50% of the control response observed when all neural input was intact. After section of the cardiac sympathetic nerves, with the vagi intact, the response was again approximately 50% of the control response when all nerves were intact. It would appear that when both systems were intact they contributed equally to the observed bradycardia. However, this conclusion may not be entirely justified. In a study by Levy and Zieske (71), the left vagus and right cardiac sympathetic nerves of dogs were stimulated simultaneously. Heart rate response curves were obtained for changes in frequency of vagal stimulation, while sympathetic stimulation was held constant. A series of curves was obtained at different constant sympathetic stimulation frequencies. A second series of curves was then obtained by maintaining vagal stimulation constant, while altering frequency of sympathetic stimulation. The results are shown in Figure 3. It appears that vagal and sympathetic influences on the heart cannot be directly summed. Also, nerve sections do not necessarily reveal the relative influence of each nerve network on any given observed response. If, from Figure 3, one assumed that a given stimulus produced change in vagal activity from 2 to 6 cycles/s, while simultaneously sympathetic activity changed from 4 to 2 cycles/s, the net effect would be a fall in heart rate of about 55 beats/min. A simple summation of responses obtained by measuring changes due to alteration of sympathetic activity when the vagus is silent, and altered vagal activity when the sympathetic activity is silent, would result in the sympathetics being responsible for a 25-beat change and the vagus for a 30-beat change; this would indicate that each system exerts approximately equal influence. When one traces the curves described while both nerves are active, the sympathetics are now responsible for only 5—15 beats/min of the change, while the vagus apparently influences 40—50 beats/min of the change, since, from Figure 3, this portion of the surface is little influenced by changes in sympathetic activity, but is greatly influenced by changes in vagal activity. From this example, it is clear that quantification of the influence of different reflex pathways which control heart rate becomes quite complex. Although recent studies have provided significant advances in the qualitative identification of the efferent limbs of many important reflex adjustments in heart rate, interpretation of their relative contribution on the basis of results of neural ablation or pharmacological denervation techniques may be prone to quantitative error.

Substantial evidence is accumulating to support the foregoing example which suggests that variations in heart rate are determined primarily by changes in vagal activity. The extent of the variation for any given reduction in vagal activity seems to be determined by the existing sympathetic activity. Of course, when vagal activity is low, or silent, as is probably the case in exercise, heart rate would always be high although changes would be directly proportional to, and more substantially influenced by, changes in sympathetic activity. Conversely, in the resting dog, where sympathetic control is apparently minimum, heart rate would change in proportion to the vagal activity.

Neurotransmitter Influences The effects of vagal and sympathetic stimuli upon the heart are mediated through release of the neurotransmitters, acetylcholine and norepinephrine. Since both sympathetic and parasympathetic nerves are tonically active, both norepinephrine and acetylcholine are continuously released at the sinoatrial node. Knowledge of the relative influences of these two transmitters should provide insight into the observed vagal dominance over cardiac sympathetic influences. In fact, when applied directly to the heart, in equimolar concentrations, acetylcholine dominates the response and effectively

CHANGE IN HEART RATE

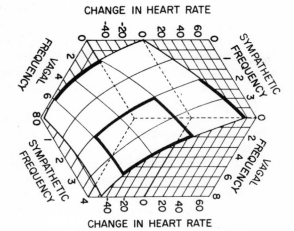

CHANGE IN HEART RATE

Figure 3. Response surface, representing the change in heart rate (in beats/min) as a function of the frequency of vagal and sympathetic stimulation (in pulses/s). Response surface represents the mean response of 10 animals. Redrawn from Levy and Zieske (71) by courtesy of the *Journal of Applied Physiology.*

inhibits the action of norepinephrine (58, 73). Unfortunately, the effects of exogenous acetylcholine and norepinephrine cannot be directly compared with vagal and sympathetic nerve stimulation. Levy and Zieske (74) showed that a given coronary arterial blood concentration of ACH is equivalent to a certain frequency of vagal stimulation at one site in the heart, whereas it corresponds to an entirely different frequency of vagal stimulation at other loci. However, the relative influences of transmitter substances on heart rate tend to support findings from nerve stimulation that change in vagal activity is a more important determinant of heart rate regulation than changes in sympathetic activity.

CNS Interactions In depth studies of baroreceptor reflexes in a variety of physiological states have provided insight into central nervous system (NS) autonomic interactions which regulate heart rate. Central interactions may change depending upon the physiological state of the animal. For example, bradycardia due to high parasympathetic tone may be attenuated at the CNS level during elevated sympathetic tone induced by exercise (75) or congestive heart failure (76). Additionally, the state of arousal or general anesthesia may influence autonomic interactions (77) at the CNS level. Such alterations in CNS bias would, of course, provide a means for eliminating baroreflex interference with cardiac output requirements in special circumstances, while assuring maintenance of homeostasis in normal, less stressful situations. The involvement of the central nervous system in baroreflex regulation is complex. The interested reader may wish to refer to the recent review by Sagawa, Kumada, and Schramm (78).

Beat-to-Beat Implications Reflex adjustments in heart rate may produce subtle beat-to-beat alterations in cardiac output. Levy et al. (79) demonstrated that a single electrical stimulation of the vagus during a cardiac cycle gave a slowing of the rate for several succeeding cycles. When repetitive discrete bursts of pulses were applied to the right vagus, the degree of slowing was related to the synchronization of the stimulus with the phase of the cardiac cycle (80). More recently, it has been demonstrated that alterations in heart rate can be produced by beat-to-beat changes in neural activity of afferent fibers from arterial baroreceptors. When a single burst of electrical pulses applied to the aortic nerve of the rabbit was confined to one cardiac cycle, slowing of the heart began within the second succeeding interval (62). Since the resting heart rate of the rabbit is normally above 250 beats/min, onset of reflexly initiated bradycardia occurred in less than 0.5 s. This illustrates the capacity for rapid cardiovascular adjustments to beat-to-beat changes in response to sensory signals reaching the central nervous system. Such a rapidly responding, subtle reflex mechanism may be an important contributor to the stabilization of cardiac output in the unstressed subject. Chronic dogs have been observed to undergo pronounced 24-hour fluctuations in blood pressure and heart rate after denervation of arterial baroreceptors (81). Exaggerated changes in cardiac output secondary to alteration in peripheral resistance were observed in these baroreceptor denervated dogs. Rapid reflex adjustments in the periphery and the heart were lacking, and resulted in an inability to compensate as rapidly or effectively as normal animals.

Sensory Origin of Reflexes   Sensory receptors are located at both discrete sites and, in some cases, diffusely throughout the body, sampling the physiological milieu. These receptors are critical for maintenance of cardiovascular homeostasis and for dealing adequately with environmental stresses, which would otherwise threaten the physiological state. Stretch sensitive cardiovascular receptors which respond to changes in pressure are considered most frequently as regulators of the heart rate. In addition, arterial chemoreceptors, thermoreceptors, and nociceptors can exert significant reflex influences on heart rate.

Stretch of the arterial baroreceptors, located in the aortic arch and carotid sinus, increases afferent neural activity and elicits reflex bradycardia. Characterization of these receptors has been among the most thoroughly documented of physiological phenomena (45, 82), yet many details of their functional significance are still unknown or debatable. There is disagreement regarding the relative functional significance of aortic versus carotid sinus baroreceptors. As pointed out in the foregoing discussion, little is definitively known about the implications of subtle changes in arterial baroreceptor activity.

The low pressure receptors located in the cardiopulmonary region are stretched at physiological pressures encountered in the venous circulation (pulmonary and systemic). The reflex heart rate response elicited by direct mechanical stretch of these receptors is characteristically tachycardia (83), and is most likely important in the regulation of venous pressure in the right and left atrium and, indirectly, the end diastolic size of the heart. Consequently, an increase in venous return could increase cardiac output as a result of the tachycardia reflexly mediated by stretch of these receptors. Increases in atrial pressure by rapid intravenous infusion of blood, or blood substitutes, in conscious dogs give a significant tachycardia and elevation in cardiac output (52). The heart rate response to direct stretch of the atrial-venous junction, or to acute intravenous infusions, probably involves both the sympathetic and parasympathetic inputs to the sinoatrial node (52, 84), although some investigators suggest that direct mechanical stretch of the atrial-venous junction involves only the sympathetic efferent nerves (85). Tachycardia associated with coronary occlusion in conscious dogs is related to a rise in left atrial pressure and is mediated primarily through the right cardiac sympathetic nerves (86). It is tempting to speculate that this response is a compensatory reflex to increase output and reduce filling pressure. Unfortunately, current evidence does not merit such a conclusion. Still the importance of "low pressure" cardiopulmonary receptors is currently undergoing extensive investigation. These receptors play a definite role in cardiovascular adjustments to stress. Their function is discussed in detail elsewhere in this volume (see Chapter 3).

Stimulation of arterial chemoreceptors by either hypercapnia or hypoxia has a complex net influence on heart rate. The primary, direct reflex produces bradycardia, but this is observed only when respiratory tidal volume is controlled (87). Tachycardia is seen in the spontaneously breathing dog, and is due to a secondary response elicited by reflex hyperpnea (87).

Neural influences product dramatic changes in cardiac output secondary to heart rate as a result of changes in body temperature. Receptors are located in the skin and hypothalamus and possibly in some abdominal visceral structures as well (57). Both external and internal receptors are probably important in eliciting heart rate responses. Internal receptors appear to be more critical, since subjects exposed to sudden changes in skin temperature exhibit heart rate changes which correlate more closely with blood temperature (88). Although most of the response is abolished by total autonomic blockage, no information is available to indicate relative significance of sympathetic or parasympathetic involvement.

Afferent impulses from pain receptors have been reported to produce reflex tachycardia or bradycardia (45). The nature of the response seems to depend on the location of the receptor and whether it is of the "pressor" or "depressor" type. Pressor type receptors stimulate the cardiac sympathetic nerves and inhibit the vagus to produce tachycardia (89). Depressor type receptors have the opposite effect. The physiological significance of these differences is not clear.

VENTRICULAR PERFORMANCE

Assessment of Ventricular Performance

Assessment of ventricular performance is difficult in the resting, conscious animal, or when induction of anesthesia has been the sole intervention. Conditions are rarely entirely basal, and measurements reflect the particular levels of a number of variables, including autonomic nervous system tone. To some extent, contributions of neural reflexes can be overcome through employment of pharmacological blocking agents, or surgical cardiac denervation, to neutralize effects of the autonomic nervous system. However, many other variables remain uncontrolled. These include heart rate, peripheral resistance, and factors influencing venous return (90). In addition, it is doubtful that estimates of ventricular function under near-basal conditions reflect the capacity of the heart to respond appropriately to stressful situations. It would seem that measurements actually obtained during stress would be the truest measure of the performance of the ventricles and the cardiovascular system as a whole.

Functional assessment under resting, or near-resting, conditions usually involves evaluation of myocardial contractility through measurement of the maximum rate of development of left ventricular pressure, maximum rate of shortening of left ventricular diameter or circumference, or ejection fraction. Efforts have also been made to apply the force-velocity relationships, which are demonstrable in isolated cardiac muscle, to the intact heart (91). However, inconstant time courses for the active state, the relatively high resting tension of cardiac muscle, the presence of significant viscous and inertial elements in the intact heart, and difficulties in extrapolating accurately to a theoretical zero load velocity of contractile element shortening make it doubtful that the maximum velocity of contractile element shortening (V_{max}) can be interpreted as a fundamental property of cardiac muscle (92). Therefore, the maximum rate of pressure development during isovolumic systole (dP/dt_{max}), and the maximum rate of diameter shortening (dD/dt_{max}), both of which are closely related to contractile element velocity and rate of muscle fiber shortening, have been utilized to assess myocardial function. These parameters can be measured directly, rather than estimated indirectly as is necessary for V_{max}. Recent studies have confirmed that dP/dt_{max} and dD/dt_{max} are reliable indices of the inotropic state of the myocardium (40). Changes in preload, afterload, and heart rate have little or no influence on these derivatives in the intact conscious animal. Furthermore, inotropic interventions elicit proportionate changes in both dP/dt_{max} and dD/dt_{max}. Ejection fraction or end systolic ventricular dimensions appear to reflect extent of myocardial fiber shortening and, therefore, to be indices of myocardial contractile function. However, the value of these measurements has not been systematically studied.

Myocardial contractility is only one factor in determining ventricular function and provides little information concerning pump performance (30). Venous return, pulmonary and systemic vascular resistances, blood volume, and other factors are as relevant to assessment of the ability of the heart to pump blood. As a resting measurement, cardiac output has little value in evaluation of ventricular performance, since the normal range is wide and it is dependent on so many variables that changes are difficult to interpret. Compensatory mechanisms, such as increases in preload and heart rate, may maintain stroke volume or cardiac output near normal levels, even in the presence of severe myocardial disease. However, during stress, cardiac reserve may be severely limited.

Stroke Work

For the reasons discussed earlier, the assessment of ventricular performance has been based principally on observations made during elevation of either preload or afterload. Evaluations have been based on the Frank-Starling mechanism. Sarnoff and his co-workers (93) measured stroke work of the left ventricle (the product of stroke volume and the difference between the mean aortic pressure and left ventricular pressure at end diastole) as a function of left ventricular end diastolic pressure. Mean left atrial pressure, mean end-diastolic left ventricular sarcomere length, or left ventricular end diastolic volume can also be designated at the independent variable.

Stroke work has been investigated extensively in open chest, cardiac bypass preparations (93). In such experiments, when heart rate was maintained constant, steady increments in ventricular filling pressure caused stroke work to increase rapidly to a maximum level. When the contractile state was enhanced by infusion of a catecholamine, the relationship between stroke work and filling pressure was shifted to the left on the pressure axis, and maximum stroke work increased. Conversely, when the heart was failing, at any given filling pressure, stroke volume and arterial pressure. Thus, it is apparent that extracardiac factors actually decreased. Such a decrease is termed a descending limb of the stroke work curve.

However, since stroke work is dependent upon the stroke volume and the relative change in arterial pressure with respect to the filling pressure, efforts to extrapolate these findings to interpretation of cardiac performance in intact, conscious animals or man are on insecure grounds. In the conscious animal, when filling pressure is rapidly increased by acute volume loading, stroke work normally increases to a peak level and then declines (94). With increasing filling pressure, stroke volume rises rapidly to a constant maximum value. Arterial pressure also increases to a maximum value, but the increment is small, and usually less than the change in filling pressure. Consequently, when both stroke volume and arterial pressure are constant, further increases in filling pressure will decrease the stroke work. This descending limb of the stroke work curve is not evidence of impaired ventricular function, but rather a consequence of the method of calculation. It is predictable as a result of the concomitant changes in stroke volume and arterial pressure. Thus, it is apparent that extracardiac factors can influence the stroke work response.

Because elevation of afterload is less hazardous than administration of large fluid loads, pressor agents have been used to obtain ventricular function curves in patients with cardiac disease (95). However, such curves have been found to be poorly reproducible in animal studies (17). This is not surprising in view of the number of neural and mechanical adjustments which occur with changes in arterial pressure. Furthermore, this stress reduces the rate and extent of myocardial fiber shortening, and limits stroke volume, despite increases in left ventricular end diastolic pressure or volume. Therefore, stroke work is dependent not only upon changes in arterial pressure relative to the left ventricular end diastolic pressure, but also upon the effects of arterial pressure on the extent of shortening.

In addition to the above limitations, a principal problem with stroke work curves, obtained by any method, is that they describe the functional state of the heart in a qualitative manner, but do not permit quantification of cardiac output reserve capacity. To surmount this difficulty, cardiac pump performance has been quantified by measurement of ventricular output and stroke volume curves during acute volume loading.

Ventricular Output Curves

Rapid intravenous infusions of blood or other isotonic solutions increase cardiac output to a maximum steady value due to the steady increase in ventricular filling pressure and ventricular size (7, 94). The relationship between cardiac output and filling pressure is shown in Figure 4. The maximum, or plateau, cardiac output is an index of cardiac performance. A reduced maximum cardiac output is indicative of impaired ventricular performance, while an elevated maximum level suggests an improved performance. This technique permits quantification of cardiac performance within a reproducible, stressful setting in terms of the ascending slope of the cardiac output curve and the cardiac output reserve (96). The difference between the maximum cardiac output and the resting or preinfusion cardiac output is the cardiac reserve. Although the maximum cardiac output obtained with acute volume loading can be exceeded during exercise, and perhaps with other stresses, it has been found to be highly reproducible from day to day in conscious dogs. While cardiac output reserve is dependent upon the preinfusion conditions, maximum cardiac output does not vary as in the conscious animal, the relationship of the cardiac output to the arterial pressure is controlled near normal levels during the infusion, which is

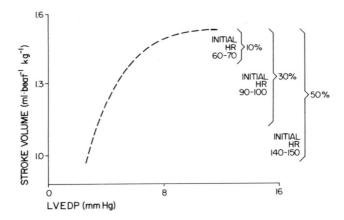

Figure 4. Stroke volume and left ventricular end diastolic pressure (*LVEDP*) relationship. Brackets to the right and percentages illustrate the probable stroke volume response to increases in filling pressure corresponding to the initial heart rates shown.

ordinarily the case in conscious animals, this approach is an effective tool for estimating the net effect of physiological and pathological variables on ventricular performance. Although similar in concept to the standard stroke work function curve, it has the advantage of being a quantitative method of evaluating pump performance. However, when heart rate and stroke volume are allowed to vary as in the conscious animal, the relationship of the cardiac output to the filling pressure, as well as the maximum value, must be interpreted in light of the respective changes in stroke volume and heart rate. The response of these two variables to acute volume loading will be discussed later.

The sensitivity of the ventricular output curves in evaluating changes in ventricular performance in conscious animals has been demonstrated during progressive heart failure produced by ionizing irradiation of the heart (32, 97, 98). As the heart deteriorates progressively during heart failure, the maximum ventricular output decreases and the curve relating cardiac output to filling pressure is shifted to the right along the pressure axis (Figure 5A). Equivalent responses were noted in the stroke volume curves.

The importance of sympathetic innervation to the heart can also be illustrated by the ventricular output curves (32, 98). The maximum cardiac output and initial ascending slope are reduced following either β-adrenergic receptor blockade or cardiac neural ablation. This is the result of approximately equal diminutions in the peak heart rate and stroke volume (32, 96).

Stroke Volume Curves

The stroke volume component of the ventricular output curve is a key to the understanding of the performance of the heart as a pump. Normally, the maximum stroke volume response to increases in ventricular filling range between 30 and 60% (32, 94, 98). However, this magnitude may be greatly altered by inotropic intervention, the initial end diastolic size (due to the initial heart ate), and the compliance of the ventricle. It would be useful to know more

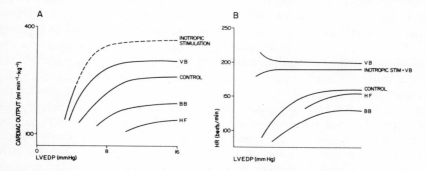

Figure 5. Cardiac output (A) and heart rate responses (B) as a function of left ventricular end diastolic pressure (LVEDP) under a variety of conditions. Vagal blockade (VB), β-adrenergic blockade (BB), and heart failure (HF). These curves were redrawn from previous studies by the authors (see refs. 32, 39, 52, 94, 96, 97) with permission from the American Journal of Physiology and Circulation Research.

about the interrelationship between indices of myocardial contractility and ventricular performance during acute volume loading. It appears that, since end diastolic dimensions are being increased to maximum effective levels, the ejection fraction, which expresses the ability of the myocardial cells to shorten from any given end diastolic size, must be a critical determinant of maximum stroke volume. During volume loading, both the end diastolic and end systolic sizes increase. The increase in end systolic size results from the additional energy required to develop tension in the enlarged ventricle. Accordingly, less shortening occurs for a given end diastolic size. Under normal conditions, the increment in end diastolic size during infusion greatly exceeds the change in end systolic size. Thus, stroke volume increases. Impairment in contractile state of the heart due to alterations in sympathetic activity or myocardial failure can increase the end systolic size relative to the end diastolic size, thereby reducing the maximum stroke volume attainable during volume loading (32).

The magnitude of the increase in stroke volume, during volume loading, is also limited by the pressure-volume characteristics of the left ventricle. If the initial end diastolic volume is large, as a result of a low heart rate, acute volume loading, although increasing the filling pressure substantially, may cause only small increases in the end diastolic volume (see Figures 1 and 4). At large initial end diastolic volumes, the heart may be operating near the apex of the length-tension curve (Figure 1). Under these circumstances, without a measurement of ventricular size, one might conclude the heart is depressed rather than realizing that the initial stroke volume is near maximum. However, it should be emphasized that similar maximum stroke volumes are obtained regardless of the initial heart rate. As illustrated in Figure 4, when the initial heart rate is low, only small or insignificant changes in stroke volume are achieved with increasing filling pressure; however, at higher initial heart rates the magnitude of the stroke volume change is increased, resulting in a steeper ascending slope. Because the maximum stroke volume attainable during infusion is unaltered by heart rate, the peak ventricular output obtained in normal animals is proportional to the heart rate at the peak of the curve. Consequently, heart rate becomes an important factor in determining the cardiac output reserve (Figure 4) (32, 94).

Most investigators, when evaluating cardiac performance, assume that the compliance of the heart is unchanged by the physiological or pathological state of the heart. Obviously, if the stiffness of the ventricular wall is altered, filling pressure is no longer an index to ventricular size. When the stiffness of the ventricular wall is increased, greater increments in ventricular filling pressures are required to stretch the myocardium. Consequently, although the myocardium may still be capable of developing force and shortening, the stroke volume response is less for any given increase in filling pressure. Recent studies have suggested that myocardial ischemia or infarction increases the stiffness of the myocardium (99, 100). If so, in this condition, ventricular output curves plotted as a function of the filling pressure may be unreliable. Of more importance, the increase in stiffness of the ventricular wall limits the Frank-Starling reserve, since extremely high filling pressures may be required to stretch the myocardium.

Thus, without measurements of ventricular dimension and pressures, the exact functional status of the ventricle may be difficult to determine.

Heart Rate

As discussed earlier, heart rate assumes a central role in the regulation of cardiac output, especially when venous return is maintained. Thus it is not surprising that the magnitude of the ventricular output response curve, as well as the cardiac output reserve, will be significantly influenced by the heart rate. In fact, both vary directly with the peak heart rate obtained during volume loading (Figure 4). In addition, it should be emphasized that when the initial heart rate is fixed at different levels, either by atrial pacing or vagal blockade, the slope of the ascending portion of the ventricular output and stroke volume curves increases with the heart rate (Figure 4). Consequently, the steepness of the performance curves is dependent upon heart rate as well as the inotropic state of the myocardium. The heart rate effect follows from the pressure-volume characteristics of the heart. At high heart rates, the end diastolic size is reduced, so that small changes in filling pressure can elicit larger changes in ventricular volume resulting in a greater increment in stroke volume for a given pressure alteration (32) (Figure 4).

The increase in heart rate induced by acute volume loading is probably due to a mechanism analogous to the Bainbridge reflex (52, 94), mediated by neural reflexes originating from receptors located in the cardiopulmonary region. The heart rate usually increases 30–50% over the control resting rate (Figure 5B). However, the magnitude of the change has been shown to be dependent upon the initial resting heart rate, decreasing as the resting heart rate is elevated (52). Although both vagal and sympathetic efferent innervation may influence the magnitude of the response, the tachycardia appears to be mediated principally through a reduction in vagal tone. Sympathetic activity usually remains fairly constant during the infusions as indicated by little or no change in $dP/dt_{(max)}$ or rate of myocardial shortening in the transverse plane. Thus, the magnitude of the heart rate response due to vagal withdrawal is modulated by the existing sympathetic activity (52).

Arterial Pressure

It is apparent from the previous discussion concerning the effects of afterload on the output of the heart that the ventricular output and stroke volume curves are significantly influenced by the arterial pressure. Indeed, in anesthetized animal preparations, arterial pressure frequently increases during acute volume loading. However, in conscious animals, the arterial pressure is well regulated as the cardiac output increases during the infusion; the increase averages 10–15 mm Hg. Consequently, the total peripheral resistance declines dramatically (−35%). This fall in peripheral resistance in the face of an increasing output is probably mediated by stretch receptors located in the cardiopulmonary region (84). However, regardless of the mechanism, the reduction in peripheral resistance is necessary for the full expression of the cardiac output reserve. As observed in exercising animals, if peripheral resistance were maintained, the peak change in cardiac output, stroke volume, and heart rate would be reduced.

There is little information on the effects of changes in the initial arterial pressure on the cardiac performance curve. One would predict that increasing arterial pressure would limit the cardiac output reserve while decreases might enhance the response. However, based upon animal studies, it is doubtful if changes in arterial pressure within the physiological range would have significant effects on either the maximum cardiac output or stroke volume.

VENTRICULAR PERFORMANCE DURING EXERCISE

In response to exercise stress, numerous adjustments take place in measureable cardiovascular parameters. The hemodynamic adaptations include changes in heart rate, stroke volume, preload, afterload, and myocardial contractile force. The net result is an increase in cardiac output which depends not only upon alterations in ventricular performance, but also upon alterations in the peripheral circulation (57, 101). Since the goal is to meet the increased oxygen demands of tissues active in exercise, both increases in output and redistribution of flow to active tissue help to satisfy these demands.

Normally, the heart utilizes increases in both heart rate and stroke volume to augment cardiac output during exercise. The relative contribution of each may vary according to the severity of the stress and the pre-exercise conditions. When exercise is performed in the supine position, there is little change in heart size and stroke volume, since both are near maximum at rest. In the erect position, the initial heart size is smaller since preload is less. Hence, stroke volume will be less initially, but can increase relatively more during exercise. Even in the latter case, the magnitude of the changes is dependent upon the severity of the exercise stress.

Heart Rate and Stroke Volume

In 1919, F.A. Bainbridge published a classical treatise entitled "The Physiology of Muscular Exercise," in which he attributed the increased cardiac output during exercise to increases in both heart rate and stroke volume (102). He postulated that the increase in stroke volume was due to increased venous return and the Frank-Starling mechanism. However, in 1956, Rushmer and his co-workers (9), on the basis of measurements of external left ventricular diameter, concluded that the entire increase in cardiac output was due to tachycardia, without associated increases in left ventricular end diastolic volume or stroke volume. Rushmer attributed the entire increase in cardiac output to autonomic, principally sympathetic, effects on heart rate (19). It is important to recognize that Rushmer and his co-workers studied only mild exercise loads, and did not measure stroke volume directly, but inferred changes on the basis of diameter measurements.

More recent studies provide evidence that Bainbridge was correct in implicating both stroke volume and heart rate. In 1971, Erickson et al. (10) observed that stroke volume measured with an ultrasonic flowmeter did increase in dogs performing extremely strenuous swimming exercise, even when heart rates exceeded 300 beats/min. Subsequently, Horwitz et al. (11) observed consistent increases in stroke volume of 14–20% in running dogs exposed to exercise stresses ranging from mild to severe loads. Associated with these modest increases in stroke volume were much larger increases in heart rate; the mean increase during near-maximal running exercise was 146%. If the heart rate is fixed, substantial increases in cardiac output can occur on the basis of increases in stroke volume alone. Warner and Toronto studied dogs in which the atrio-ventricular node region was ablated surgically (44). Stroke volume increased markedly during running, while heart rate was paced at a constant level.

Left Ventricular Dimensions and Exercise

Starling (103) and Bainbridge (102) invoked an increase in the end diastolic length of the ventricular muscle fibers as a likely event during exercise stress. According to this concept, the return of blood to the right side of the heart via the systemic veins is markedly enhanced by exercise. The consequent increase in preload, through the Frank-Starling mechanism, is associated with a greater force of contraction and a greater stroke volume. The measurements of Rushmer did not support this theory, in that, during mild exercise, he did not detect evidence of an increase in end diastolic left ventricular volume (9, 19). More recently, left ventricular internal diameter has been found to increase consistently at end diastole during near-maximal swimming or running exercise. However, the changes in preload are dependent upon the workload. With mild loads, there is no change in left ventricular internal diameter at end diastole, as compared to pre-exercise values (11).

Although most investigators concede that the Frank-Starling mechanism can contribute to the increase in stroke volume during exercise, its contribution relative to that of the increase in myocardial contractility is difficult to ascertain because of simultaneous changes in heart rate and the peripheral circulation. Indeed, when the initial heart rate is fixed by electrical pacing at the same high level encountered during very strenuous exercise, left ventricular end diastolic pressure and diameter and, therefore, stroke volume are initially quite low, but each increases progressively during running until the same maximum levels are reached which occur without pacing (104). Increased filling pressure due to increased venous return associated with exercise is, to some extent, analogous to a rapid intravenous infusion in a resting animal. Without this effect, end diastolic ventricular volume and stroke volume, rather than increasing, would not be maintained in the face of the tachycardia associated with exercise.

It is usual during exercise for left ventricular dimensions to decrease at end systole. This is the sole means by which stroke volume increases with mild effort. With severe exercise, most hearts continue to shorten to smaller end systolic dimensions than at rest, but others do not do so. The increased shortening is dependent upon increased myocardial contractile force due to sympathetic nervous system stimulation. It also is a function of afterload.

Increases in afterload during severe exercise may explain why some animals do not experience a decrease in end systolic left ventricular diameter.

Autonomic Control and Exercise

In 1914, Gasser and Meek (105) studied the effects of surgical cardiac denervation on exercise hemodynamics and concluded that the integrity of both sympathetic and parasympathetic systems was important for normal cardiac performance. More recently, surgical cardiac denervation did not appear to impair either running speed or heart rate in greyhounds (106). However, when propranolol, a β-adrenergic blocking agent, was given to the denervated dogs, both running speed and heart rate were reduced (107). Evidence has been presented that after surgical cardiac denervation the heart rate response to circulating catecholamines becomes exaggerated (108). Accordingly, to eliminate effectively β-adrenergic influences on cardiac performance, it is necessary to employ a pharmacological β-blocking agent.

Heart rate and stroke volume are reduced by propranolol at all levels of exercise (33, 104). The reductions in stroke volume are due to elevations in end systolic dimensions of the left ventricle. The increase in end systolic dimensions during severe effort is equivalent to approximately a 16% decrease in myocardial fiber shortening. Propranolol also reduces the rate of myocardial force development, as indicated by the maximum rate of increase in left ventricular pressure.

The initial onset of tachycardia of exercise at all levels is probably due to vagal inhibition (109). When vagal blockade is accomplished with atropine, heart rates are elevated at rest and during submaximal exercise. However, during near-maximal exercise, no significant difference in heart rate or other parameters occurs, whether or not atropine is given. Apparently, vagal withdrawal is complete during severe effort. At rest and during submaximal effort the high heart rate reduces preload and stroke volume, but does not significantly alter cardiac output or myocardial contractility. Therefore, the vagus appears to modulate heart rate primarily at low levels of exercise stress, whereas the sympathetic system modulates both heart rate and myocardial contractility (33, 104).

During exercise with β-blockade, marked reduction in the extent of myocardial fiber shortening occurs despite an elevated preload (33, 104). Data from these studies suggest that 50% of the heart rate response plus much of the increment in stroke volume are normally due to sympathetic stimulation. The Frank-Starling mechanism is obviously operative, yet the stroke volume is attenuated due to diminished myocardial contractility. It would appear that the Frank-Starling response is relatively ineffective in maintaining stroke volume in the absence of sympathetic innervation. This does not mean that the Frank-Starling mechanism can be minimized in the animal with intact autonomic control. Still, studies of both isolated hearts and intact animals have shown that the Frank-Starling response is significantly affected by the contractile state of the cardiac muscle (30, 93). Since β-blockade may depress the myocardium, responses in this state are not necessarily comparable to those when sympathetic innervation is intact. Furthermore, although end diastolic ventricular size may increase more after β-blockade during exercise, one cannot assume greater utilization of the Frank-Starling mechanism in view of differences in heart rates, and the accompanying effects on end diastolic size.

Peripheral Resistance during Exercise

When exercise begins, there is a prompt fall in mean systemic peripheral resistance. This is due to vasodilation in exercising skeletal muscle, in response to increased metabolic demands. When resistance is artificially controlled during exercise, sudden reductions are associated with reciprocal increases in heart rate, while elevations in resistance result in reductions in heart rate (110). On the basis of these findings, changes in peripheral resistance appear to modulate heart rate during strenous activity. It is also likely that stroke volume will be partially dependent upon aortic input inpedance. The extent to which impedance changes correlate with resistance changes is not known.

Dogs which have undergone surgical denervation of the carotid and aortic baroreceptors exhibit hemodynamic changes during exercise which are qualitatively similar to those of the intact dog but have significant quantitative differences (111). Heart rate is less during running, and the usual decline in systemic resistance is not sustained. The baroreceptors are active during exercise, although there is some controversy about their sensitivity. If a pressor agent is administered during exercise, heart rate is diminished (75, 109). Electrical or mechanical stimulation of the carotid sinus during running gives changes in peripheral resistance (112, 113). The absolute diminution in heart rate to a pressor challenge is the same at rest and during exercise. However, because of the higher heart rate during exercise, the percentage change and the relationship between the R-R interval and change in blood pressure are reduced (57).

Venous Return and Exercise

Exercise is associated with large increments in the quantity of blood entering the right side of the heart. Evidence has been presented for the role of a muscle pump action, whereby actively contracting muscles, and a one-way valve system in the systemic veins, push blood towards the heart (114). Vasodilatation occurs in small arteries and arterioles in exercising skeletal muscles, through local autoregulation, and this contributes to the increased venous return. An effect of sympathetic stimulation is to reduce ventricular pressure in early diastole, and this probably assists ventricular filling during this time (93).

Limitation of Cardiac Output in Exercise

The capacity of the cardiovascular system to deliver oxygen to active muscle determines the maximum oxygen consumption and, consequently, the level of activity which can be performed. Although it is appreciated that the capacity of the cardiovascular system is determined by numerous factors, including humoral and neural influences, the question as to whether the limitation of the maximum cardiac output during exercise resides in the heart or peripheral circulation is still unanswered. It seems likely that either can limit the capacity of the cardiovascular system to deliver oxygen in pathological states. When the heart itself is depressed or failing, the obvious limitation would be cardiac performance. In other circumstances, cardiac output might be limited through changes in venous return, as in hypovolemia, or peripheral resistance, as in early hypertension. It appears that the maximum output is dependent upon the coordinated responses between the heart and peripheral circulation.

There is evidence that under normal conditions the maximum cardiac output in man is determined by the heart. Robinson et al. (115) concluded that acute blood volume expansion did not increase cardiac output during exercise. Based upon the observation that the central blood volume was unaltered, these authors suggested that the upper limitation of the cardiac output was determined by the right ventricle. Paradoxically, in dogs, acute intravenous infusions during exercise increased the stroke volume, suggesting that the limitation of the cardiac output resides in the peripheral circulation (116). The differences observed in man and dog may be related to the experimental approach. In the human study by Robinson et al. (115), the expansion of the blood volume was slow and occurred prior to exercise, while in the studies of the dog by Keroes et al. (116), the expansion occurred during the exercise period. There was no evidence in either study that the maximum cardiac output or oxygen consumption was obtained.

It is of interest that the factor which determines the difference in the maximum oxygen consumption or cardiac output among individuals is the stroke volume (57). This is apparent among the general population as well as selected groups of physically conditioned or deconditioned individuals. The increase in maximum oxygen consumption resulting from physical conditioning is equally divided between an increase in stroke volume and an increased oxygen extraction. This could be accounted for by changes in venous compliance, but seems more likely to be due to the performance of the heart.

ACKNOWLEDGMENTS

The authors gratefully acknowledge the assistance of Linda Conn and Sue Newman for typing and Greg Riedel for proofreading the manuscript.

REFERENCES

1. Frank, O. (1895). Zur dyamik des herzmuskels. Z. schr. Biol. 32:370.
2. Patterson, S.W., and Starling, E.H. (1914). On the mechanical factors which determine the output of the ventricles. J. Physiol. 48:357.
3. Mommaerts, W.F.H.M., and Langer, G.A. (1963). Fundamental concepts of cardiac dynamics and energetics. Annu. Rev. Med. 14:261.
4. Gordon, A.M., Huxley, A.F., and Julian, F.J. (1966). The variation in isometric tension with sarcomere length in vertebrate muscle fibres. J. Physiol. 184:170.
5. Sonnenblick, E.H. (1968). Correlation of myocardial ultrastrusturcture and function. Circulation 38:29.
6. Sonnenblick, E.H., and Skelton, C.L. (1974). Reconsideration of the ultra-structural basis of cardiac length-tension relations. Circ. Res. 35:517.

7. Bishop, V.S., Horwitz, L.D., Stone, H.L., Stegall, H.F., and Engelken, E.J. (1969). Left ventricular internal diameter and cardiac function in conscious dogs. J. Appl. Physiol. 27:619.

8. Horwitz, L.D., and Bishop, V.S. (1972). Left ventricular pressure-dimension relationships in the conscious dog. Cardiovasc. Res. 6:163.

9. Rushmer, R.F., Franklin, D.L., and Ellis, R.M. (1956). Left ventricular dimensions recorded by sonocardiometry. Circ. Res. 4:684.

10. Erickson, H.H., Bishop, V.S., Kardon, M.B., and Horwitz, L.D. (1971). Left ventricular internal diameter and cardiac function during exercise. J. Appl. Physiol. 30:473.

11. Horwitz, L.D., Atkins, J. M., and Leshin, S.J. (1972). Role of the Frank-Starling mechanism in exercise. Circ. Res. 31:868.

12. Hamilton, W.F. (1955). Role of the Starling concept in regulation of normal circulation. Physiol. Rev. 35:161.

13. Imperial, E.S., Levy, M.N., and Zieske, H., Jr. (1961). Outflow resistance as an independent determinant of cardiac performance. Circ. Res. 9:1148.

14. Wilcken, D.E.L., Charlier, A.A., Hoffman, J.I.E., and Guz, A. (1964). Effects of alterations in aortic impedance on the performance of the ventricles. Circ. Res. 14:283.

15. Herndon, C.W., and Sagawa, K. (1969). Combined effects of aortic and right atrial pressures on aortic flow. Amer. J. Physiol. 217:65.

16. Sagawa, K. (1972). The use of control theory and systems analysis in cardiovascular dynamics. In D.H. Bergel (ed.), Cardiovascular Fluid Dynamics, Vol. 1, Chap. 5, p. 115. Academic Press, London.

17. O'Rourke, R.A., Pegram, B., and Bishop, V.S. (1972). Variable effect of angiotensin infusion on left ventricular function. Cardiovasc. Res. 6:240.

18. Langer, G.A. (1968). Ion fluxes in cardiac excitation and contraction and their relation to myocardial contractility. Physiol. Rev. 48:708.

19. Rushmer, R.F., and West, T.C. (1957). Role of autonomic hormones on left ventricular performance continuously analyzed by electronic computer. Circ. Res. 5:240.

20. Levine, H.J., and Britman, N.A. (1964). Force-velocity relationship in the intact dog heart. J. Clin. Invest. 43:1383.

21. Furnival, C.M., Linden, R.J., and Snow, H.M. (1973). Chronotropic and inotropic effects on the dog heart of stimulating the efferent cardiac sympathetic nerves. J. Physiol. 230 (1):137.

22. Gleason, W.L., and Braunwald, E. (1962). Studies on the first derivative of the ventricular pressure pulse in man. J. Clin. Invest. 41:80.

23. Levy, M.N., Ng, M., Martin, P., and Zieske, H. (1966). Sympathetic and parasympathetic interactions upon the left ventricle of the dog. Circ. Res. 19:5.

24. Sarnoff, S.J., and Berglund, E. (1954). Ventricular function. I. Starling's law of the heart studied by means of simultaneous right and left ventricular function curves in the dogs. Circulation 9:706.

25. Levy, M.N., Ng, M., Lipman, R.I., and Zieske, H. (1966). Vagus nerves and baroreceptor control of ventricular performance. Circ. Res. 18:101.

26. Martin, P.J., Levy, M.N., and Zieske, H. (1969). Bilateral carotid sinus control of ventricular performance in the dog. Circ. Res. 24:321.

27. Sarnoff, S.J., Gilmore, J.P., Brockman, S.K., Mitchell, J.H., and Linden, R.J. (1960). Regulation of ventricular contraction by the carotid sinus. Its effect on atrial and ventricular dynamics. Circ. Res. 8:1123.

28. Randall, W.C., Priola, D.V., and Ulmer, R.H. (1963). A functional study of distribution of cardiac sympathetic nerves. Amer. J. Physiol. 205:1227.

29. Vatner, S.F., Higgins, C.B., Franklin, D., and Braunwald, E. (1972). Extent of carotid sinus regulation of the myocardial contractile state in conscious dogs. J. Clin. Invest. 51:995.

30. Braunwald, E. (1971). On the difference between the heart's output and its contractile state. Circulation 43:171.

31. Noble, M.I.M., Trenchard, D., and Guz, A. (1966). Left ventricular ejection in conscious dogs. II. Determinants of stroke volume. Circ. Res. 19:148.

32. Bishop, V.S., and Horwitz, L.D. (1971). Effects of altered autonomic control on left ventricular function in conscious dogs. Amer. J. Physiol. 221:1278.

33. Horwitz, L.D., Atkins, J.M., and Leshin, S.J. (1974). Effect of β-adrenergic blockade on left ventricular function in exercise. Amer. J. Physiol. 227:839.
34. Suga, H., and Sagawa, K. (1974). Instantaneous pressure-volume relationships and their ratio in the excised, supported canine left ventricle. Circ. Res. 35:117.
35. Suga, H., Sagawa, K., and Shoukar, A.A. (1973). Load independence of the instantaneous pressure-volume ratio of the canine left ventricle and effects of epinephrine and heart rate on the ratio. Circ. Res. 32:314.
36. Miller, D.E., Gleason, W.L., Whalen, R.E., Morris, J.J., Jr., and McIntosh, H.D. (1962). Effects of ventricular rate on the cardiac output in dogs with chronic heart block. Circ. Res. 10:658.
37. Noble, M.I.M., Trenchard, D., and Guz, A. (1966). Effect of changing heart rate on cardiovascular function in the conscious dog. Circ. Res. 19:206.
38. Stone, H.L., and Bishop, V.S. (1968). Ventricular output in conscious dogs following acute vagal blockade. J. Appl. Physiol. 24:782.
39. Bishop, V.S., Stone, H.L., and Horwitz, L.D. (1971). Effects of tachycardia and ventricular filling pressure on stroke volume in the conscious dog. Amer. J. Physiol. 220:436.
40. Barnes, G.E., Bishop, V.S., Horwitz, L.D., and Kaspar, R.L. (1973). The maximum derivatives of left ventricular pressure and transverse internal diameter as indices of the inotropic state of the left ventricle in conscious dogs. J. Physiol. 235:571.
41. Bishop, V.S., and Horwitz, L.D. (1973). Influence of the cardiac sympathetics on left ventricular dynamics during tachycardia. Cardiology 58:326.
42. Sugimoto, T., Sagawa, K., and Guyton, A.C. (1966). Effect of tachycardia on cardiac output during normal and increased venous return. Amer. J. Physiol. 211:288.
43. Berglund, E., Borst, H.G., Duff, F., and Schreiner, G.L. (1958). Effect of heart rate on cardiac work, myocardial oxygen consumption and coronary blood flow in the dog. Acta Physiol. Scand. 42:185.
44. Warner, H.R., and Toronto, A.F. (1960). Regulation of cardiac output through stroke volume. Circ. Res. 8:549.
45. Folkow, B., and Neil, E. (1971). Circulation. Oxford University Press, New York.
46. Berne, R.M., and Levy, M.N. (1972). Cardiovascular Physiology. C.V. Mosby, St. Louis.
47. West, T.C. (1972). Electrical Phenomena in the Heart. W.C. DeMello (ed.). Academic Press, New York.
48. Levy, M.N., and Martin, P.J. (1974). Cardiac excitation and contraction. In A.C. Guyton and C.E. Jones (eds.), MTP International Review of Science. Cardiovascular Physiology, Vol. 1. Butterworths. London; University Park Press, Baltimore.
49. Jensen, D. (1971). Intrinsic Cardiac Rate Regulation. Appleton-Century-Crofts, New York.
50. Lange, G., Lu, H.H., Chang, A., and Brooks, C. McC. (1966). Effects of stretch on the isolated cat sinoatrial node. Amer. J. Physiol. 211:1192.
51. Brooks, C. McC., Lu, H.H., Lange, G., Mangi, R., Shaw, R.B., and Geoly, K. (1966). Effects of localized stretch of the sinoatrial node region of the dog heart. Amer. J. Physiol. 211:1197.
52. Horwitz, L.D., and Bishop, V.S. (1972). Effect of acute volume loading on heart rate in the conscious dog. Circ. Res. 30:316.
53. Knowlton, F.P., and Starling, E.H. (1912). The influence of variation in temperature and blood pressure of the performance of the isolated mammalian heart. J. Physiol. 44:206.
54. Jose, A.D., and Stitt, F. (1969). Effects of hypoxia and metabolic inhibitors on the intrinsic heart rate and myocardial contractility in dogs. Circ. Res. 25:53.
55. Garry, W.E., and Townsend, S.E. (1948). Neural responses and reactions of the heart of a human embryo. Amer. J. Physiol. 152:219.
56. Jose, A.D., Stitt, F., and Collison, D. (1970). The effects of exercise and changes in body temperature on the intrinsic heart rate in man. Amer. Heart J. 79:488.
57. Rowell, L.B. (1974). Human cardiovascular adjustments to exercise and thermal stress. Physiol. Rev. 54:75.
58. Carrier, G.O., and Bishop, V.S. (1972). The interaction of acetylcholine and norepinephrine on heart rate. J. Pharmacol. Exp. Ther. 180:31.

59. Trautwein, W., and Kassebaum, D.G. (1961). On the mechanism of spontaneous impulse generation in the pacemaker of the heart. J. Gen. Physiol. 45:317.

60. Wildenthal, K. (1972). Studies of isolated fetal mouse hearts in organ culture. Evidence for a direct effect triiodothyronine in enhancing cardiac responsiveness to norepinephrine. J. Clin. Invest. 51:2702.

61. Bishop, V.S., Stone, H.L., and Horwitz, L.D. (1971). Effects of tachycardia and ventricular filling pressure on stroke volume in the conscious dog. Amer. J. Physiol. 220:436.

62. Kardon, M.B., Peterson, D.F., and Bishop, V.S. (1974). Beat-to-beat regulation of heart rate by afferent stimulation of the aortic nerve. Amer. J. Physiol. 227:598.

63. Mizeres, N.J. (1958). Origin and course of the cardioaccelerator fibers in the dog. Anat. Rec. 132:261.

64. McKibben, J.S., and Getty, R. (1968). A comparative morphologic study of the cardiac innervation in domestic animals. I. The canine. Amer. J. Anat. 122:533.

65. Randall, W.C., Armour, J.A., Geis, W.P., and Lippincott, D.B. (1972). Regional cardiac distribution of the sympathetic nerves. Fed. Proc. 31:1199.

66. Kardon, M.B., Peterson, D.F., and Bishop, V.S. (1973). Reflex bradycardia due to aortic nerve stimulation in the rabbit. Amer. J. Physiol. 225:7.

67. Scher, A.M., Ohm, W.W., Bumgarner, K., Boynton, R., and Young, A.C. (1972). Sympathetic and parasympathetic control of heart rate in the dog, baboon and man. Fed. Proc. 31:1219.

68. Levy, M.N., Martin, P.J., Iano, T., and Zieske, H. (1969). Paradoxical effect of vagus nerve stimulation on heart rate in dogs. Circ. Res. 25:303.

69. Antonaccio, M.J. (1974). Paradoxical vagal tachycardia in ganglion-blocked anesthetized dogs. Eur. J. Pharmacol. 28:59.

70. Warner, H.R., and Russell, R.O., Jr. (1969). Effect of combined sympathetic and vagal stimulation on heart rate in the dog. Circ. Res. 24:567.

71. Levy, M.N., and Zieske, H. (1969). Autonomic control of cardiac pacemaker activity and atrioventricular transmission. J. Appl. Physiol. 27:465.

72. Levy, M.N. (1971). Sympathetic-parasympathetic interactions in the heart. Circ. Res. 29:437.

73. Grodner, A.S., Lahrtz, H.G., Pool, P.E., and Braunwald, E. (1970). Neurotransmitter control of sinoatrial pacemaker frequency in isolated rat atria and in intact rabbits. Circ. Res. 27:867.

74. Levy, M.N., and Zieske, H. (1969). Comparison of the cardiac effects of vagus nerve stimulation. Amer. J. Physiol. 216:890.

75. Pickering, T.G., Gribbin, B., Peterson, E.S., Cunningham, J.C., and Sleight, P. (1972). Effects of autonomic blockade on the baroreflex in man at rest and during exercise. Circ. Res. 30:177.

76. Higgins, C.B., Vatner, S.F., and Braunwald, E. (1973). Parasympathetic control of the heart. Pharmacol. Rev. 25:119.

77. Vatner, S.F., Franklin, D., and Braunwald, E. (1971). Effects of anesthesia and sleep on circulatory response to carotid sinus nerve stimulation. Amer. J. Physiol. 220:1249.

78. Sagawa, K., Kumada, M., and Schramm, L.P. (1974). Nervous control of the circulation. In A.C. Guyton and C.E. Jones (eds.), MTP International Review of Science. Cardiovascular Physiology, Vol. 1. Butterworths, London; University Park Press, Baltimore.

79. Levy, M.N., Martin, P.J., Iano, T., and Zieske, H. (1970). Effects of single vagal stimuli on heart rate and atrioventricular conduction. Amer. J. Physiol. 218:1256.

80. Levy, M.N., Iano, T., and Zieske, H. (1972). Effects of repetitive bursts of vagal activity on heart rate. Circ. Res. 30:186.

81. Cowley, A.W., Jr., Liard, J.R., and Guyton, A.C. (1973). Role of the baroreceptor reflex in daily control of arterial blood pressure and other variables in dogs. Circ. Res. 32:564.

82. Heymans, C., and Neil, E. (1958). Reflexogenic Areas of the Cardiovascular System. Churchill, London.

83. Ledsome, J.R., and Linden, R.J. (1964). Reflex increase in heart rate from distension of the pulmonary-vein-atrial junctions. J. Physiol. 170:456.

84. Edis, A.J., Donald, D.E., and Shepherd, J.T. (1970). Cardiovascular reflexes from stretch of pulmonary vein-atrial junctions in the dogs. Circ. Res. 27:1091.
85. Furnival, C.M., Linden, R.J., and Snow, H.M. (1971). Reflex effects on the heart of stimulating left atrial receptors. J. Physiol. 218:447.
86. Peterson, D.F., Kaspar, R.L., and Bishop, V.S. (1973). Reflex tachycardia due to temporary coronary occlusion in the conscious dog. Circ. Res. 32:652.
87. Daly, M. deB., and Scott, M.J. (1958). The effects of stimulation of the carotid body chemoreceptors on heart rate in the dog. J. Physiol. 144:148.
88. Rowell, L.B., Brengelmann, G.L., and Murray, J.A. (1969). Cardiovascular responses to sustained high skin temperature in resting man. J. Appl. Physiol. 27:673.
89. Johansson, B. (1962). Circulatory responses to stimulation of somatic afferents. Acta Physiol. Scand. 57, (Suppl.):198.
90. Guyton, A.C., Jones, C., and Coleman, T. (1973). Circulatory Physiology: Cardiac Output and Its Regulations. W.B. Saunders Co., Philadelphia.
91. Braunwald, E., Ross, J., Jr., and Sonnenblick, E.H. (1967). Mechanisms of Contraction of the Normal and Failing Heart. Little, Brown & Co., Boston.
92. Brady, A.J. (1968). Active state in cardiac muscle. Physiol. Rev. 48:570.
93. Sarnoff, S.J., and Mitchell, J.H. (1962). The control of the function of the heart. Handbook of Circulation, Sect. 2, Vol. 1. Williams & Wilkins Co., Baltimore.
94. Bishop, V.S., Stone, H.L., and Guyton, A.C. (1964). Cardiac function curves in conscious dogs. Amer. J. Physiol. 207:677.
95. Ross, J., Jr., and Braunwald, E. (1964). The study of left ventricular function in man by increasing resistance to ventricular ejection with angiotensin. Circulation 29:739.
96. Bishop, V.S., and Stone, H.L. (1967). Quantitative description of ventricular output curves in conscious dogs. Circ. Res. 20:581.
97. Stone, H.L., Bishop, V.S., and Guyton, A.C. (1966). Ventricular function following radiation damage of right ventricle. Amer. J. Physiol. 211:1209.
98. Stone, H.L., Bishop, V.S., and Dong, E. (1967). Ventricular function in cardiac-denervation and cardiac-sympathectomized conscious dogs. Circ. Res. 20:587.
99. Bishop, V.S., Kaspar, R.L., Barnes, G.E., and Kardon, M.B. (1974). Left ventricular function during acute regional myocardial ischemia in the conscious dog. J. Appl. Physiol. 37:785.
100. Diamond, G., and Forrester, J.S. (1972). Effect of coronary disease and acute myocardial infarction on left ventricular compliance in man. Circulation 45:11.
101. Bevegard, B.S., and Shepherd, J.T. (1967). Regulation of the circulation during exercise in man. Physiol. Rev. 47:178.
102. Bainbridge, F.A. (1919). The Physiology of Muscular Exercise. Longmann, Green & Co., London.
103. Starling, E.H. (1920). On the circulatory changes associated with exercise. J. R. Army Med. Corps 34:258.
104. Vatner, S.F., Franklin, D., Higgins, B., Patrick, T., and Braunwald, E. (1972). Left ventricular response to severe exertion in untethered dogs. J. Clin. Invest. 51:3052.
105. Gasser, H.S., and Meek, W.J. (1914). Study of the mechanisms by which muscular exercise produces acceleration of the heart. Amer. J. Physiol. 34:48.
106. Donald, D.E., Milburn, S.E., and Shepherd, J.T. (1964). Effect of cardiac denervation on the maximum capacity for exercise in the racing greyhound. J. Appl. Physiol. 19:849.
107. Donald, D.E., Ferguson, D.A., and Milburn, S.E. (1968). Effect of beta-adrenergic receptor blockade on racing performance of greyhounds with normal and with denervated hearts. Circ. Res. 22:127.
108. Donald, D.E., and Shepherd, J.T. (1965). Supersensitivity to l-norepinephrine of the denervated sinoatrial node. Amer. J. Physiol. 208:225.
109. Robinson, B.F., Epstein, S.E., Beisler, G.D., and Braunwald, E. (1966). Control of heart rate by the autonomic nervous system: studies in man on the inter-relation between baroreceptor mechanisms and exercise. Circ. Res. 19:400.
110. Warner, H.R., Topham, W.S., and Nicholes, K.K. (1964). The role of peripheral resistance in controlling cardiac output during exercise. Ann. N.Y. Acad. Sci. 115:669.
111. Krasney, J.A., Levityky, G., and Koehler, R.C. (1974). Sinoaortic contribution to the adjustment of systemic resistance in exercising dogs. J. Appl. Physiol. 36:679.

112. Epstein, S.E., Beiser, G.D., Goldstein, R.E., Stampfer, M., Wechsler, A.S., Glick, G., and Braunwald, E. (1969). Circulatory effects of electrical stimulation of the carotid sinus nerves in man. Circulation 49:269.
113. Bevegard, B.S., and Shepherd, J.T. (1966). Circulatory effects of stimulating the carotid arterial stretch receptors in man at rest and during exercise. J. Clin. Invest. 45:132.
114. Stegall, H.F. (1966). Muscle pumping in the dependent leg. Circ. Res. 19:180.
115. Robinson, B.F., Epstein, S.E., Kahler, R.L., and Braunwald, E. (1966). Circulatory effects of acute expansion of blood volume: studies during maximal exercise and at rest. Circ. Res. 19:26.
116. Keroes, J., Ecker, R.R., and Rapaport, E. (1969). Ventricular function curves in the exercising dog. Circ. Res. 25:557.

International Review of Physiology
Cardiovascular Physiology II, Volume 9
Edited by Arthur C. Guyton and Allen W. Cowley
Copyright 1976   University Park Press   Baltimore

7
The Physiology of Cyanotic Congenital Heart Disease

D. D. MAIR and D. G. RITTER

Mayo Clinic and Mayo Foundation

INTRODUCTION

Twenty years have passed since the advent of cardiopulmonary circulatory bypass techniques made the definitive repair of intracardiac congenital malformations feasible. Operations were first devised for the simpler deformities such as isolated septal defects and pulmonary stenosis. The techniques were improved

275

and applied to the more complex conditions, until at present only a few congenital cardiovascular malformations remain anatomically or physiologically uncorrectable. Although follow-up of patients who have undergone successful operations is limited to this maximum of 20 years, and in the more complex conditions to considerably shorter periods, the operations that have evolved have dramatically altered the natural history of many of these deformities (1) and have markedly improved the quality of life of the patients. Many patients have now reached adulthood, and evidence is accumulating that they do not have apparent limitations secondary to their prior cardiovascular deformity. Because follow-up on successfully operated patients is presently limited to 20 years in the earliest surgical cases and to increasingly shorter periods in later patients, many of whom initially had more complex anomalies, conclusive evidence as to whether these patients may anticipate a normal life-span will not be forthcoming for a number of years. However, there are reasons to be optimistic that many of these patients will enjoy normal longevity.

The development of potentially effective surgical therapy for congenital heart disease provided a tremendous impetus for the concomitant development of a better understanding of the pathophysiology of these lesions. Cardiac catheterization techniques, particularly, began to evolve rapidly as the need for accurate anatomical and physiological diagnosis of these lesions became of great practical importance; and during the past two decades, advances in cardiovascular laboratories have paralleled and complemented those in surgery. As Comroe and Dripps (2) have pointed out recently, the cardiovascular surgeon of 1975 can work his wonders only because of the many pathologists, physiologists, pharmacologists, physicists, chemists, cardiologists, radiologists, anesthesiologists, and hematologists who have contributed small pieces toward the solution of the puzzle and triggered the story of open-heart surgery as it has unfolded during the past 20 years as one of the truly great achievements of medical science.

Although great progress has been made in the understanding of the pathophysiology, in the diagnosis, and in the therapy of congenital cardiovascular disease, a disappointingly small amount has been learned about the etiology of these lesions. The cause of most of these deformities is still unknown, and their incidence has not decreased appreciably in recent years. Approximately six children per 1,000 live births will be born with a cardiovascular malformation (3), which means that in the United States nearly 20,000 children each year are born with congenital heart disease. Although many of the deformities are of relatively minor hemodynamic significance and do not require treatment, the number of patients needing accurate diagnosis and effective therapy remains substantial. With no effective means of preventing these malformations forthcoming, the continued thorough understanding of the anatomy and hemodynamics of these conditions will be necessary for accurate diagnosis and appropriate therapy.

This review will not examine the pathophysiology of all congenital cardiovascular malformations that have been described. Rather, it will concentrate on the hemodynamics of lesions that are characterized by desaturated systemic mixed venous blood recirculating directly to the aorta without passing through the lungs (that is, the presence of a "right-to-left shunt"). This group of malformations is commonly referred to as "cyanotic congenital heart disease," although when a large blood flow is also present through the lungs and the amount of desaturated systemic mixed venous blood recirculating directly to the aorta is relatively small in proportion to the amount of fully saturated pulmonary venous blood reaching the aorta, clinical cyanosis may not be actually apparent. In discussing the physiology of these malformations, the hemodynamics of complete transposition of the great arteries will be emphasized because this condition represents the most dramatic example of this group of deformities. The basic hemodynamic principles that characterize complete arterial transposition apply to the other malformations characterized by a "right-to-left shunt" as well; however, the differences are generally ones of degree. The important congenital cardiovascular malformations that are characterized by desaturated systemic mixed venous blood recirculating directly to the aorta without passing through the lungs are listed in Table 1.

MATERIAL AND METHODS

All hemodynamic data were obtained from cardiac catheterizations performed in the Mayo Clinic Cardiovascular Laboratory between 1962 and 1974 by methods previously described (4). Data from patients ages 1 year to 27 years are included. Physiological information from infants less than 1 year of age was not included, because such studies often were performed primarily for anatomical diagnosis in critically ill infants and complete hemodynamic data were not obtained. In most instances, the catheterization was performed in patients undergoing evaluation as to their suitability for surgical correction, and in patients in whom operation was appropriate and advisable the surgery was usually carried out within a few days

Table 1. Congenital cardiovascular malformations characterized by systemic arterial desaturation

---

Tetralogy of Fallot[a]
Severe pulmonary stenosis and atrial septal defect[a]
Pulmonary valve atresia[a]
Tricuspid valve atresia
Hypoplastic left heart syndrome (aortic and/or mitral valve atresia)
Complete transposition of great arteries[a]
Double-outlet right ventricle[a]
Common ventricle[a]
Common atrium[a]
Truncus arteriosus[a]
Total anomalous pulmonary venous connection[a]
Anomalous systemic venous connection[a]
Right pulmonary artery-to-left atrial fistula[a]

---

[a]Anatomical or physiological (or both) surgical correction now possible.

after the catheterization. In this review, only data obtained while the patients were breathing room air are included, although frequently additional determinations were carried out with the patients breathing various oxygen mixtures. Systemic arterial samples were obtained through a needle placed in the femoral artery or in a few instances the brachial artery. The hemoglobin level was measured for each patient at the beginning of the study, and, if significant blood was lost during the study, determinations were repeated. At all times during these studies, efforts were made to ensure that the patient maintained a satisfactory state of ventilation and that no serious disturbances in acid-base balance were present. Data from patients who did not maintain a relatively steady state insofar as these parameters are concerned were not included in this review. Flow data were calculated by use of the Fick equation, and in all instances total flow determined in this manner was then divided by the patient's surface area in square meters so that a flow index (liters min^{-1} m^{-2}) was obtained.

BASIC PHYSIOLOGICAL PRINCIPLES

The hemodynamics of the congenital cardiovascular abnormalities in which there are anatomical communications between the systemic and pulmonary circuits are characterized by the "shunting" of blood in either or both directions between the two circuits. The hemodynamics of these deformities can be understood better if the blood flow through each circuit is considered as effective and recirculated (or ineffective) flow.

Effective flow is the amount of systemic mixed venous blood per unit of time that enters the lungs, picks up oxygen, and then makes its way to the systemic capillaries where it delivers this oxygen to the tissues. The effective flow, therefore, is the volume of blood per unit time which flows in a sequential series manner between the systemic and pulmonary circuits and is responsible for picking up oxygen in the lungs and delivering it to the tissues and removing carbon dioxide. (The terms "effective pulmonary flow" and "effective systemic flow" are also frequently used to designate this blood which moves through the two circuits in a sequential series manner. These terms are synonymous with the term "effective flow," which will be used throughout the remainder of this review.) In the subsequent discussion, it is important to remember that this effective flow only takes part in the exchange of gases in the systemic and pulmonary capillary beds although such flow may comprise a relatively small amount of the total flow through these beds.

Recirculated systemic flow is the amount of systemic mixed venous blood per unit time that recirculates directly to the aorta without passing through the lungs. This recirculated systemic flow is usually referred to as the "right-to-left shunt." The term "right-to-left shunt," however, has anatomical connotations, implying that blood has physically moved from the right to the left side of the heart through one of the anatomical communications between the two sides of the heart such as a ventricular septal defect. In conditions such as complete transposition of the great arteries, however, this does not occur because the "right-to-left shunt" (that is, recirculated systemic mixed venous flow) moves from the right atrium to the right ventricle and then directly out into the aorta, which arises from the right ventricle. Therefore, to avoid unnecessary confusion in discussing the hemodynamics of the more complex anomalies such as complete transposition, it becomes advantageous to think in terms of recirculated systemic mixed venous flow and recirculated pulmonary venous flow rather than in terms of "right-to-left" and "left-to-right" shunts, respectively—terms that, although physiologically synonymous, carry anatomical implications that are not always accurate. For the remainder of this review, the term "recirculated systemic flow" will be used to denote relatively desaturated systemic mixed venous blood that recirculates directly to the aorta without passing through the pulmonary capillaries, and the term "recirculated pulmonary flow" will be used to denote fully saturated pulmonary venous blood that recirculates directly to the pulmonary artery without passing through the systemic capillaries. A schematic representation of the circulation in transposition of the great arteries using this terminology is seen in Figure 1.

The total volume of flow through either the systemic or the pulmonary circuit can be determined by use of the Fick principle, which states that flow is equal to oxygen consumption divided by the arteriovenous oxygen content difference across the appropriate capillary bed. The equations applied in these determinations are as follows:

Total systemic flow = oxygen consumption/[10 (systemic arterial oxygen content − systemic mixed venous oxygen content)]   (1)

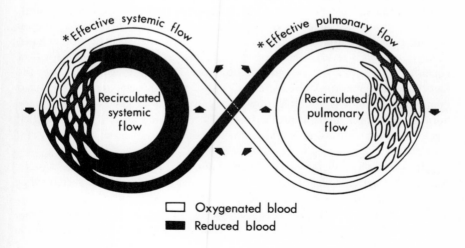

□ Oxygenated blood

■ Reduced blood

* Blood flow responsible for effective gas transport

Figure 1. Schematic representation of circulation in complete transposition of great arteries. (From Mair, D.D., Ritter, D.G., Ongley, P.A., and Helmholz, H.F., Jr. (1971). Hemodynamics and evaluation for surgery of patients with complete transposition of the great arteries and ventricular septal defect. Amer. J. Cardiol. 28:632. By permission of Dun·Donnelley Publishing Corporation.)

Total pulmonary flow = oxygen consumption/[10 (pulmonary venous oxygen content − pulmonary arterial oxygen content)]   (2)

The factor of 10 in equations 1 and 2 is necessary to obtain flow expressed as liters min^{-1} when oxygen contents are expressed in vol% (ml/100 ml).

As stated previously, effective flow is the amount of systemic mixed venous blood per unit time which enters the lungs, picks up oxygen in the pulmonary capillaries, and then makes its way in a sequential series manner to the systemic capillaries where it delivers this oxygen to the tissues. The formula for effective flow is as follows:

Effective flow = oxygen consumption/[10 (pulmonary venous oxygen content − systemic mixed venous oxygen content)]   (3)

When the total systemic and pulmonary flow and effective flow are calculated by use of these formulae, the recirculated (or ineffective) flow through each circuit can be calculated as follows:

Recirculated systemic flow = total systemic flow − effective flow   (4)

Recirculated pulmonary flow = total pulmonary flow − effective flow   (5)

In the normal circulation, where the systemic and pulmonary circuits are completely in series, with no abnormal communications between them, systemic mixed venous content equals pulmonary arterial content, and pulmonary venous content equals systemic arterial content. Thus, appropriate substitution in equations 1 through 3 reveals that total systemic flow equals total pulmonary flow equals effective flow for the normal situation. This relationship expresses the fact that in the normal circulation all blood flow through the systemic and pulmonary circuits is effective flow and participates in gas exchange. In the normal circulation, there is no ineffective (recirculated) flow, which is verified when the above expression is substituted into equations 4 and 5, and values of 0 are obtained.

However, when there are abnormal communications between the two circuits so that systemic venous blood may recirculate directly to the aorta or that pulmonary venous blood may recirculate directly to the pulmonary artery, or that both may occur (that is, the two circuits are no longer completely in series), the above equations become useful in understanding the hemodynamics involved.

The application of these equations in the delineation of the dynamics of a patient with congenital heart disease may be best demonstrated by the use of an example.

Consider the patient with transposition of the great arteries, from whom the following data were obtained at cardiac catheterization:

Oxygen consumption = 150 ml min^{-1}
Hemoglobin concentration = 14.9 g/100 ml
Oxygen capacity = 14.9 × 1.34 = 20 ml/100 ml of blood
Pulmonary venous saturation = 97.5%
Pulmonary artery saturation = 90%
Systemic artery saturation = 75%
Systemic mixed venous saturation = 60%

Then:

Pulmonary venous content = 20 × 0.975 = 19.5 vol%
Pulmonary artery content = 20 × 0.90 = 18.0 vol%
Systemic artery content = 20 × 0.75 = 15.0 vol%
Systemic mixed venous content = 20 × 0.60 = 12.0 vol%

Total systemic flow (Q_s) = 150 / [10(15.0 − 12.0)] = 150/30 = 5.0 liters min^{-1}

Total pulmonary flow (Q_p) = 150 / [10(19.5 − 18.0)] = 150/15 = 10.0 liters min^{-1}

Effective flow (Q_{eff}) = 150 / [10(19.5 − 12.0)] = 150/75 = 2.0 liters min^{-1}

Recirculated systemic flow = 5.0 − 2.0 = 3.0 liters min^{-1}
Recirculated pulmonary flow = 10.0 − 2.0 = 8.0 liters min^{-1}

From the analysis of these data, the following conclusions can be drawn concerning the hemodynamic status of this patient. He has an effective flow of 2.0 liters min^{-1}. This means that 2.0 liters min^{-1} of systemic mixed venous blood makes its way into the pulmonary artery, picks up oxygen in the pulmonary capillaries, and then moves into the pulmonary veins and later into the aorta, eventually reaching the systemic capillaries, where it delivers to the tissues the oxygen that it picked up in the lungs. This 2.0 liters min^{-1} is the only portion of the circulating blood in either circuit that is participating in the exchange of oxygen and carbon dioxide in the systemic and pulmonary capillary beds. However, even though this flow is the only portion of the flow through either circuit that participates in gas exchange, the amount of this flow is less than one-half the actual total flow through each circuit. The total flow through the systemic circuit is 5.0 liters min^{-1}, of which 5.0 − 2.0 = 3.0 liters min^{-1} is desaturated systemic mixed venous blood that recirculates through the systemic veins, the systemic arteries, and the systemic capillary bed. Conversely, the total flow through the pulmonary circuit is 10.0 liters min^{-1}, of which 10.0 − 2.0 = 8.0 liters min^{-1} is saturated pulmonary venous blood that recirculates through the pulmonary veins, the pulmonary artery, and the pulmonary capillary bed. Although the 3.0 liters min^{-1} of recirculated systemic flow and the 8.0 liters min^{-1} of recirculated pulmonary flow are completely "ineffective" and do not participate in gas exchange, they are pumped by the right and left ventricle, respectively, and constitute most of the volume load on these ventricles. In this specific example, in a patient with complete transposition of the great arteries, the left (pulmonary) ventricle is subjected to a significantly elevated volume load of 10.0 liters min^{-1}, although only 20% of this load (2.0 liters min^{-1}) is actually being used for effective gas transport. One of the physiological consequences seen in many of the congenital cardiac defects characterized by a large recirculated flow through either circuit or both circuits is a significant volume overload on one ventricle or both ventricles, often resulting in the clinical manifestation of congestive heart failure.

Although the recirculated systemic and pulmonary blood flows are completely ineffective in capillary gas exchange, the magnitude of their volume has a large role in determining the actual oxygen saturation and resultant oxygen tension of blood in the aorta and pulmonary artery, respectively. In the preceding example, the systemic arterial saturation of 75% is a result of 2.0 liters min^{-1} of 97.5% saturated pulmonary venous blood (the effective flow) reaching the aorta and mixing with the recirculated systemic flow of 3.0 liters min^{-1} of 60% saturated systemic mixed venous blood. Likewise, the pulmonary artery saturation of 90% is the result of 2.0 liters min^{-1} of 60% saturated systemic mixed venous blood combining with 8.0 liters min^{-1} of 97.5% saturated recirculating pulmonary venous blood. Thus, the "ineffective" or recirculated flow through each circuit present in many congenital heart lesions, although not having a role in gas exchange, nonetheless profoundly influences oxygen saturation in the aorta and pulmonary artery. The effect of recirculated systemic flow on systemic arterial oxygen saturation is particularly important because it helps determine the tension at which oxygen is available to the tissues in the systemic capillaries, a matter that will be investigated in detail in the following section.

This type of approach should aid in understanding the basic hemodynamics of the patient with a congenital heart disease that is characterized by blood flow recirculating through the systemic or pulmonary (or both) circuit. In such a patient:

1. Total blood flow through both systemic and pulmonary circuits should be determined. This should then be separated into effective flow and recirculated (or ineffective) flow through each circuit.

2. Uptake of oxygen from the lungs and delivery of oxygen to the tissues are achieved by the effective flow only.

3. Recirculated flow in each circuit does not enhance oxygen uptake and delivery in the pulmonary and systemic capillaries, although such flow may represent a substantial portion of the flow through each circuit and its magnitude profoundly influences pulmonary arterial and systemic arterial saturation.

4. Recirculated flow through each circuit, although completely ineffective in gas transport, must be pumped by the heart, and this often results in significant volume overloading of one ventricle or both ventricles.

FACTORS INFLUENCING SYSTEMIC ARTERIAL OXYGEN SATURATION

In a patient with a congenital cardiac deformity in which desaturated systemic mixed venous blood is recirculated directly to the aorta without passing through the lungs, the physiological consequence is systemic arterial oxygen desaturation. A significant reduction in systemic arterial saturation decreases arterial oxygen tension (5) and the "driving pressure" at which oxygen is available to the tissues in the systemic capillary bed. This decreased availability of oxygen to the tissues leads to tissue hypoxia and metabolic acidosis and results in the physical limitations seen in the severe forms of cyanotic congenital heart disease.

Systemic arterial blood in such patients is derived from two sources. A portion of the blood reaching the aorta is fully saturated pulmonary venous blood that represents the effective flow. The remainder is relatively desaturated systemic mixed venous blood that is being recirculated through the systemic circuit. Thus, systemic arterial oxygen saturation is dependent on the relative proportions from, and saturations of, these two sources.

With systemic arterial oxygen saturation partly dependent on the relative proportions of fully saturated pulmonary venous blood and relatively desaturated recirculating systemic venous blood reaching the aorta, the absolute magnitudes of the effective flow and the recirculated systemic flow are obviously important in its determination. The relative saturation levels of the pulmonary venous blood and systemic mixed venous blood are also important. In the absence of significant ventilatory problems, pulmonary venous blood in the person breathing room air, at or near sea level, has an oxygen saturation of approximately 97%. However, the saturation of systemic mixed venous blood is not constant but is dependent, at a given level of oxygen consumption, on the magnitude of the effective flow and the concentration of hemoglobin in the blood. A given level of oxygen uptake in the lungs demands that a given amount of reduced hemoglobin per unit time be delivered into the pulmonary capillaries. The only reduced hemoglobin that is delivered to the pulmonary capillaries is contained in systemic mixed venous blood that reaches the pulmonary artery (that is, the effective flow). Therefore, at a given level of oxygen uptake, the systemic mixed venous saturation (that is, the relative proportion of oxygenated and reduced hemoglobin in systemic mixed venous blood) must vary secondary to changes in the magnitude of effective flow or hemoglobin concentration, if a constant amount of reduced hemoglobin reaching the pulmonary capillaries is to be maintained. A decrease in either effective flow or hemoglobin concentration will necessitate a decrease in systemic mixed venous saturation, and this in turn will profoundly influence systemic arterial oxygen saturation if a significant proportion of the blood reaching the aorta is recirculating systemic mixed venous blood.

Therefore, at a given level of oxygen consumption, three potential variables influence systemic arterial oxygen saturation in these patients:

1. The magnitude of the effective flow to which systemic arterial saturation is directly related.

2. The magnitude of the recirculated systemic mixed venous flow to which systemic arterial saturation in inversely related.

3. The hemoglobin concentration to which systemic arterial saturation is directly related.

SYSTEMIC ARTERIAL OXYGEN SATURATION
IN COMPLETE TRANSPOSITION OF THE GREAT ARTERIES

The factors affecting systemic arterial oxygen saturation outlined above will be demonstrated using physiological data obtained from 100 patients with complete transposition of the great arteries. Significant systemic arterial desaturation is a consistent finding in such patients (6). Because of the parallel relationship of the systemic and pulmonary circuits in this condition, effective flow is accomplished only by blood that can move from the right to the left side of the heart and return through anatomical communications between the right and the left side of the heart, such as an atrial septal defect or a ventricular septal defect (or both) (7). The effective flow in this condition is often referred to as "intercirculatory mixing." A schematic representation of the circulation in transposition of the great arteries is seen in Figure 1.

The volume of the effective flow in transposition is limited, in part, by the number and size of the anatomical communications between the two circuits through which the blood must pass (7). The recirculated systemic and pulmonary flows, however, pass directly from right atrium to right ventricle to aorta for systemic flow and from left atrium to left ventricle to pulmonary artery for pulmonary flow. The unique parallel anatomical relationship of the two circuits that characterizes transposition, therefore, results in an effective flow which is relatively small and varies within a narrow range and recirculated systemic mixed venous and pulmonary venous flows which may be very large and can vary over wide ranges. The range and mean values for effective flow, recirculated systemic flow, and recirculated pulmonary flow for the 100 patients in our series are seen in Table 2. Because, as pointed out above, systemic arterial oxygen saturation in such patients is directly related to effective flow and inversely related to the volume of recirculated systemic flow, there is often severe systemic arterial desaturation in this condition which is characterized by a small effective flow and a large recirculated systemic mixed venous flow.

The systemic arterial oxygen saturations found in our series of patients are shown in Figure 2. The range was 40 to 86%, with 45 patients having saturations of 65% or less. Only two patients had an arterial oxygen saturation as high as 85%. The relationship between effective flow and systemic arterial oxygen saturation is apparent in Figure 3. Although the relationship is positive between

Table 2. Hemodynamic findings in 100 patients with complete transposition

Flow (liters min^{-1} m^{-2})	Mean	Range
Effective	1.7	0.8–2.8
Recirculated systemic venous	3.8	2.0–9.8
Recirculated pulmonary venous	7.3	1.5–14.1

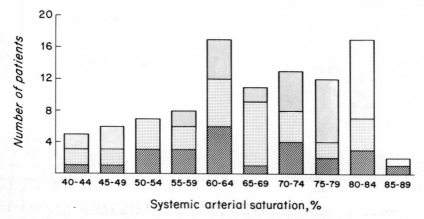

Figure 2. Systemic arterial oxygen saturation of 100 patients with transposition and breathing air. Patients grouped via anatomic intercirculatory communications present. Hatched segment represents atrial septal defect only (25 patients); light-dotted segment represents ventricular septal defect only (35 patients); dark-dotted segment represents patients with two or more communications (ASD + VSD, VSD + PDA, etc.) (40 patients). (From Mair, D.D., and Ritter, D.G. (1973). Factors influencing systemic arterial oxygen saturation in complete transposition of the great arteries. Amer. J. Cardiol. 31:742. By permission of Dun·Donnelley Publishing Corporation.)

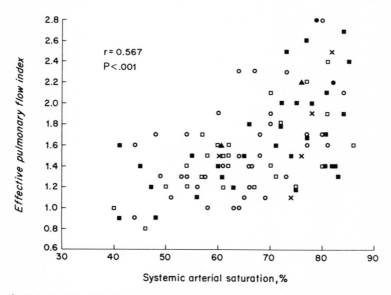

Figure 3. Relationship of systemic arterial oxygen saturation to effective flow index in 100 patients with transposition. □, atrial septal defect; ○, ventricular septal defect; ■, atrial septal defect and ventricular septal defect; ●, atrial septal defect, ventricular septal defect, and patent ductus arteriosus; X, ventricular septal defect and patent ductus arteriosus or Blalock-Taussig shunt; ▲, atrial septal defect and patent ductus arteriosus. (From Mair, D.D., and Ritter, D.G. (1973). Factors influencing systemic arterial oxygen saturation in complete transposition of the great arteries. Amer. J. Cardiol. 31:742. By permission of Dun· Donnelley Publishing Corporation.)

these two variables ($p < 0.001$), the correlation is not linear ($r = 0.567$) and demonstrates clearly that factors other than effective flow are important in the determination of systemic arterial oxygen saturation. Table 3 groups patients according to whether their effective flow index was 1.2 or less (23 patients), 1.3–1.9 (57 patients), or 2.0 or greater (20 patients). As expected, the mean arterial saturation is significantly greater in the group with the largest effective flow (best "intercirculatory mixing"), but there is overlap among values in all three groups. The finding that in one patient in the group with the lowest effective flow the systemic saturation was 75%, whereas it was only 64% in a patient in the group with the largest effective flow, further substantiates that factors in addition to effective flow are important in the determination of systemic arterial saturation. The usual benefits of a good effective flow are readily apparent, however. The saturation level was higher than 65% in only 26% of patients in the first group, whereas it reached this level in 95% of patients in the group with an effective flow index of 2.0 or greater. The highest level of saturation (75%) reached in the group with low effective flow occurred in the previously mentioned patient, whereas a systemic saturation of 80% or greater was achieved in 40% of patients in the group with greatest effective flow. Thus, although a relatively large effective flow does not guarantee a high systemic arterial oxygen saturation, the patient with transposition has little chance of obtaining a level higher than 75% unless reasonably "good mixing" is present.

Hemoglobin concentration was also significantly ($p \simeq 0.001$) related to systemic arterial saturation (Figure 4 and Table 4). The correlation was lower ($r = 0.330$) than that between effective flow index and systemic saturation, however. Although hemoglobin concentration directly affects systemic arterial oxygen saturation, the influence of hemoglobin should be less than the influence of the volume of effective flow because increased hemoglobin concentration serves only to increase the saturation of the systemic mixed venous blood that is

Table 3. Relationship of systemic arterial oxygen saturation to effective flow index in patients with transposition who had small ($\leqslant 1.2$), moderate (1.3–1.9), and large ($\geqslant 2.0$) effective flow indices

Effective flow index	No. of patients	Mean hemoglobin concentration (g/100 ml)	Systemic arterial oxygen saturation (range and mean), %	Systemic saturation (no. and % of patients)	
				>65%	$\geqslant 80\%$
$\leqslant 1.2$	23	19.8	40 to 75 (58)	6 (26%)	0 (0%)
1.3–1.9	57	17.5	41 to 86 (66)	30 (53%)	11 (19%)
$\geqslant 2.0$	20	16.1	64 to 85 (77)	19 (95%)	8 (40%)
Total	100	—	—	55 (55%)	19 (19%)

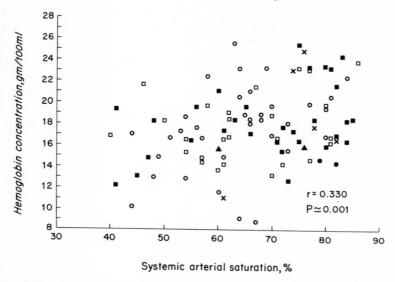

Figure 4. Relationship of systemic arterial oxygen saturation to hemoglobin concentration in 100 patients with transposition. Same symbolism as in Figure 3. (From Mair, D.D., and Ritter, D.G. (1973). Factors influencing systemic arterial oxygen saturation in complete transposition of the great arteries. Amer. J. Cardiol. 31:742. By permission of Dun· Donnelly Publishing Corporation.)

recirculating directly to the aorta, whereas an increased effective flow produces the dual result of increasing the systemic mixed venous saturation and also increasing the relative proportion of fully saturated pulmonary venous blood reaching the aorta.

Data in Tables 3 and 4 reveal that a relatively large effective flow and a high level of hemoglobin concentration are both beneficial in achieving a greater systemic arterial oxygen saturation, with the result that oxygen tension is increased, making oxygen more readily available to the tissues in the systemic capillary bed. Examining Tables 3 and 4 reveals, however, that effective flow and hemoglobin concentration tend to vary inversely. Patients with reduced effective flow have a significantly greater mean hemoglobin concentration than do patients with large effective flow (Table 3). Conversely, patients with increased hemoglobin concentration have a significantly less mean effective flow than do patients with low hemoglobin concentration (Table 4). The patient with a low systemic arterial oxygen saturation secondary to a poor effective flow tends to compensate for this by an increase in hemoglobin concentration and, by this method, raises the systemic saturation, and the resultant partial pressure at which oxygen is delivered to the tissues, to a more acceptable level.

The greater the proportion of desaturated systemic mixed venous blood that recirculates to the aorta, the lower the level of systemic arterial saturation. This effect of increased recirculated systemic flow is demonstrated in Figure 5, which shows a plot of systemic arterial saturation against a number determined by

Table 4. Relationship of systemic arterial oxygen saturation to hemoglobin concentration in patients with transposition who had low (≤15.0), moderate (15.1 to 19.9), and high (≥20.0) levels of hemoglobin concentration

Hemoglobin concentration (g/100 ml)	No. of patients	Effective flow index (mean)	Systemic arterial oxygen saturation (range and mean), %	Systemic saturation (no. and % of patients)	
				>65%	≥80%
≤15.0	22	1.8	41−82 (60)	7 (32%)	1 (5%)
15.1−19.9	55	1.6	40−85 (67)	31 (56%)	11 (20%)
≥20.0	23	1.4	46 to 86 (73)	17 (74%)	7 (30%)
Total	100	−	−	55 (55%)	19 (19%)

From Mair and Ritter (6). By permission of Dun·Donnelley Publishing Corporation.

multiplying the value for hemoglobin concentration times effective flow index and dividing this product by the systemic recirculated flow index. It will be recalled that systemic arterial saturation is directly related to hemoglobin concentration and effective flow and inversely related to systemic recirculated flow. The relationship between the variables plotted in Figure 5 is direct ($r = 0.933$). This positive relationship, which approaches linearity, demonstrates that the levels of effective flow, hemoglobin concentration, and systemic mixed venous recirculated flow are all important in determining the level of systemic arterial oxygen saturation achieved.

With establishment of the factors that influence systemic arterial oxygen saturation in patients in whom systemic mixed venous blood is recirculated directly to the aorta, it is of interest to see the degree to which systemic saturation is changed in a given patient if these factors are altered. This is best demonstrated by examining preoperative and postoperative catheterization data from a patient with transposition of the great arteries and a ventricular septal defect who underwent a palliative Mustard operation (8, 9) (Table 5). This patient had pulmonary vascular obstructive disease so severe that a definitive Mustard operation (10) involving an intra-atrial baffle and simultaneous ventricular septal defect closure was not possible. However, the palliative Mustard operation was performed in which an intra-atrial baffle is inserted but the ventricular septal defect is left open so that the pulmonary (left) ventricle has a means by which it can decompress itself. This operation has been effective in significantly increasing systemic arterial saturation and decreasing hemoglobin concentration (thus markedly reducing symptoms and increasing exercise tolerance) in patients with transposition of the great arteries and a ventricular septal defect in whom a complete separation of the two circuits via a definitive repair is not appropriate because of their severe pulmonary vascular obstructive disease. The immediate consequence of the palliative Mustard operation is a redirection

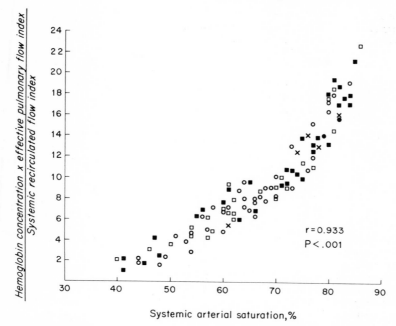

Figure 5. Relationship of systemic arterial oxygen saturation to a value obtained by multiplying hemoglobin concentration times effective flow index and dividing this product by systemic venous recirculated flow index. One hundred patients with transposition. Same symbolism as in Figure 3. (From Mair, D.D., and Ritter, D.G. (1973). Factors influencing systemic arterial oxygen saturation in complete transposition of the great arteries. Amer. J. Cardiol. 31:742. By permission of Dun·Donnelley Publishing Corporation.)

of the route of the flow of systemic mixed venous blood and pulmonary venous blood so that the systemic venous blood is directed across the mitral valve and the pulmonary venous blood is directed across the tricuspid valve. Anatomically, this procedure converts the systemic and pulmonary circulations from the parallel relationship, characteristic of transposition, to a series relationship resembling that of the patient with isolated ventricular septal defect. The effect of the procedure on the factors that influence systemic arterial saturation and the resultant effect on systemic saturation are demonstrated in Table 5.

Preoperatively, this patient had a systemic arterial saturation of 65% achieved only with a hemoglobin of 18.1 g/100 ml. The total flow through the systemic circuit of the patient was 6.0 liters min^{-1} m^{-2}, of which only 1.5 liters min^{-1} m^{-2} was effective flow. Thus, $6.0 - 1.5 = 4.5$ liters min^{-1} m^{-2} of systemic mixed venous blood was recirculating directly to the aorta. The arterial saturation of this patient of 65%, therefore, was the result of 1.5 liters min^{-1} m^{-2} of 97% saturated pulmonary venous blood reaching the aorta and mixing with 4.5 liters min^{-1} m^{-2} of 54% saturated recirculating systemic mixed venous blood.

The postoperative hemodynamic data reveal a different situation, however. **The** systemic arterial oxygen saturation is now 87%, as compared to the

Table 5. Catheterization data from a 5-year-old patient with transposition and ventricular septal defect 6 weeks after palliative Mustard operation

	Pressure (mm Hg)	Saturation (breathing room air), %
Preoperative		
Femoral artery	88/52 ($\bar{M}$ = 64)	65
Left atrium	12/4 ($\bar{M}$ = 6)	97
Pulmonary artery	84/52 ($\bar{M}$ = 62)	77
Right atrium	10/4 ($\bar{M}$ = 6)	54
Superior vena cava	–	54
$Hg = 18.1$ g/100 ml		
$Q_{s_I} = 6.0$ liters min^{-1} m^{-2}		
$Q_{p_I} = 3.3$ liters min^{-1} m^{-2}		
$R_s = 10.7$ units m^{-2}		
$R_{pa} = 17.0$ units m^{-2}		
$Q_{eff_I} = 1.5$ liters min^{-1} m^{-2}		
Postoperative		
Femoral artery	93/53 ($\bar{M}$ = 66)	87
Pulmonary artery wedge	13/6 ($\bar{M}$ = 8)	97
Pulmonary artery	91/51 ($\bar{M}$ = 64)	67
Systemic venous atrium	14/6 ($\bar{M}$ = 8)	67
Superior vena cava	14/7 ($\bar{M}$ = 8)	66
$Hg = 12.5$ g/100 ml		
$Q_{s_I} = 4.5$ liters min^{-1} m^{-2}		
$Q_{p_I} = 3.0$ liters min^{-1} m^{-2}		
$R_s = 14.7$ units m^{-2}		
$R_{pa} = 17.9$ units m^{-2}		
$Q_{eff_I} = 3.0$ liters min^{-1} m^{-2}		

preoperative level of 65%. This significant increase in systemic arterial oxygen saturation was accomplished concomitantly with a reduction of the patient's hemoglobin level from a preoperative polycythemic level of 18.1 g/100 ml to a normal level of 12.5 g/100 ml. The calculations reveal that the total systemic flow is now 4.5 liters min^{-1} m^{-2}. As a result of the intra-atrial baffle, his effective flow is now 3.0 liters min^{-1} m^{-2}. Thus, of the total of 4.5 liters min^{-1} m^{-2} of blood reaching the aorta, 3.0 liters min^{-1} m^{-2} is saturated pulmonary venous blood and only 4.5 − 3.0 = 1.5 liters min^{-1} m^{-2} is relatively desaturated recirculating systemic mixed venous blood. The systemic arterial saturation of 87% is the result of 3.0 liters min^{-1} m^{-2} of 97% saturated pulmonary venous blood reaching the aorta and mixing with 1.5 liters min^{-1} m^{-2} of 67% saturated recirculating systemic mixed venous blood.

Reviewing the data reveals that several changes have occurred as a result of the operation which converted the circulations from the parallel relationship characteristic of transposition to a series arrangement. First, the effective flow (amount of systemic mixed venous blood reaching the pulmonary artery, picking up oxygen in the pulmonary capillaries, and then moving on to the aorta and eventually reaching the systemic capillary bed) has doubled from 1.5 to 3.0 liters min^{-1} m^{-2}. Second, the amount of desaturated systemic mixed venous blood recirculating directly to the aorta was reduced to one-third its former value (4.5 to 1.5 liters min^{-1} m^{-2}). As a result of these changes, the proportions of saturated pulmonary venous blood and relatively desaturated recirculating systemic mixed venous blood reaching the aorta have changed markedly. Whereas, preoperatively, aortic blood was composed of one part pulmonary venous blood and three parts systemic mixed venous blood, postoperatively, aortic blood was composed of two parts pulmonary venous blood and one part systemic venous blood. The substantial increase in systemic arterial oxygen saturation gained from the operation is accomplished primarily because of this significant change in the proportions of pulmonary venous blood and systemic venous blood reaching the aorta. An additional factor that contributes to the improved systemic arterial saturation is that, as a result of the increased effective flow, the systemic mixed venous saturation increased from 54% preoperatively to 67% after operation although the patient no longer was polycythemic. As previously explained, a given oxygen uptake from the lungs demands that a given amount of reduced hemoglobin per unit time be delivered to the pulmonary capillaries. This reduced hemoglobin is contained in the systemic mixed venous blood, and with the effective flow now doubled (that is, the amount of systemic mixed venous blood per unit time reaching the pulmonary capillaries now doubled), the systemic mixed venous blood will contain relatively less reduced hemoglobin (that is, will be of higher saturation) despite a decrease in hemoglobin concentration. In the illustrated case, therefore, the improvement in systemic arterial saturation is due to the dual mechanism that a marked change (in fact, a near reversal) has occurred in the proportions of pulmonary venous blood and systemic venous blood reaching the aorta and also the systemic mixed venous blood, which continues to recirculate directly to the aorta through the still open ventricular septal defect, is not as desaturated, and so has a less detrimental effect on arterial oxygen saturation than it did before operation.

The preoperative and postoperative hemodynamic data in the illustrative case demonstrate conclusively the consequences that changes in the three variables (that is effective flow, hemoglobin concentration, and recirculated systemic mixed venous flow), which affect systemic arterial oxygen saturation in these patients, have on this saturation. In this instance, an increase in systemic arterial oxygen saturation from 65% to 87% resulted from an operation that doubled the effective flow (which has a direct relationship to systemic arterial saturation) and reduced the recirculated systemic mixed venous flow (which has an inverse relationship to systemic arterial saturation) to one-third its former value. This significant increase in arterial saturation is achieved despite a concomitant decrease in hemoglobin concentration (which has a direct relationship to systemic arterial saturation). The reduction in polycythemia seen in these patients is an important part of their palliation, because the severe polycythemia occurring in the severely cyanotic patients before palliation is often responsible for symptoms such as bleeding tendencies and recurrent headache and also predisposes them to possible thromboembolic complications. It is well recognized that the compensatory polycythemia that patients with cyanotic congenital heart disease develop in an attempt to maintain a satisfactory arterial oxygen tension is a two-edge sword in that the resulting increase in blood viscosity may lead to symptoms such as epistaxis and headache, and, on occasion, to a serious complication such as a cerebral thrombosis. Thus, effective palliation in these patients is aimed at accomplishing the twofold purpose of increasing arterial oxygen tension as well as reducing the degree of compensatory polycythemia, a goal that has been achieved in the case illustrated.

The previous discussion has concentrated on the findings in patients with complete transposition of the great arteries. However, the same hemodynamic principles apply, and methods of analysis should be used in patients with other lesions (Table 1) in which systemic mixed venous blood is recirculated directly to the aorta without passing through the lungs. As mentioned previously, this review concentrates on complete transposition because transposition represents the most dramatic example of this group of patients. Because of the anatomical parallel relationship of the circuits in transposition, the effective flow is limited to blood that can pass in both directions through anatomical communications between the right and the left side of the heart. Thus, the effective flow—which, as we have observed, directly affects systemic arterial oxygen saturation—is characteristically small in this condition, with resulting significant systemic arterial desaturation. The anatomy of complete transposition, with the aorta arising from the right ventricle, also predisposes to a large systemic mixed venous recirculated flow, and this further aggravates the systemic arterial desaturation. However, the differences in the hemodynamics of complete transposition and the other lesions (Table 1) are ones of degree, and the same principles and equations apply in the analysis of the dynamics of these lesions.

Patients with conditions such as total anomalous pulmonary venous connection, truncus arteriosus, and common ventricle often have an effective flow that is near normal (3.0–3.5 liters min^{-1} m^{-2}). When this is true, the dynamic findings are similar to those demonstrated in the preceding patient after his palliative Mustard operation, in that most of the blood reaching the aorta is fully saturated pulmonary venous blood and the systemic mixed venous blood that recirculates is of reasonably good saturation, thus causing a less devastating effect on systemic arterial saturation. Patients with such hemodynamics may have systemic arterial saturations greater than 90% (11, 12), with minimal compensatory polycythemia, and be clinically acyanotic. However, patients with lesions such as pulmonary atresia may have a very small effective flow and large recirculated systemic flow, similar to those characterizing complete transposition, with resultant similar very low systemic arterial saturations and marked compensatory polycythemia. The important fact to remember is that, at a given level of oxygen consumption, the same three variables (volume of effective flow, hemoglobin concentration, and volume of systemic recirculated mixed venous flow) determine the systemic arterial oxygen saturation in all these conditions because they are the only factors that influence the relative proportions of pulmonary venous blood and recirculating systemic mixed venous blood reaching the aorta and the saturation of this recirculating systemic mixed venous blood.

DISCUSSION

The concepts presented in the preceding section are relatively easily understood with one possible exception. This commonly misunderstood point deserves further elaboration, however, because it is critical to the understanding of the hemodynamics involved in patients with cyanotic congenital heart disease.

It has been pointed out above that the volume and saturation of recirculating systemic mixed venous blood are of great importance in the determination of systemic arterial oxygen saturation in these patients. A point that is often difficult to grasp is that the saturation of systemic venous blood, at a given level of oxygen consumption, is dependent only on effective flow and hemoglobin concentration and is not dependent on, nor does it vary with, changes in total flow through the systemic circuit, if the effective flow is unchanged. This concept is sometimes not well understood because, from the earliest days of medical school, students are taught in physiology that, if total flow through the systemic capillary bed increases, at a given level of oxygen consumption, systemic mixed venous saturation also must increase. This concept is true and valid in the normal person in whom the systemic and pulmonary circuits are completely in series. However, it is not true for the patient with congenital heart disease in whom there are communications between the two circuits through which systemic mixed venous blood may recirculate directly to the aorta without passing through the lungs. Its application in such a patient leads to confusion in understanding the effect on systemic arterial oxygen saturation of an increase in this recirculated systemic mixed venous flow. Using this concept, the student reasons falsely in the following manner: If recirculated systemic mixed venous flow increases, total systemic flow increases. This increase in total systemic flow means that systemic mixed venous saturation increases. Thus, although more systemic venous blood per unit time is now recirculating directly to the aorta, the blood is of higher saturation than previously and therefore has no great effect on systemic arterial oxygen saturation. The error in this reasoning can be demonstrated.

The equation for effective flow is as follows:

Effective flow = oxygen consumption / (pulmonary vein content −
systemic mixed venous content)   (6)

For the rest of this theoretical discussion, we shall assume that oxygen consumption is a constant, hemoglobin concentration is a constant (that is, oxygen capacity is a constant), and no ventilatory problems exist so that pulmonary venous saturation is also constant. Thus, equation 6 becomes:

Effective flow = K_1 / (K_2 − systemic mixed venous content)   (7)

Therefore, at a given level of oxygen consumption, there is a direct relationship between effective flow and systemic mixed venous content (or saturation). If the effective flow increases, the systemic mixed venous saturation also increases.

In the normal situation, the two circuits are entirely in series, and therefore effective flow equals total systemic flow equals total pulmonary flow. With all flow being effective, an increase in total systemic flow necessarily means an increase in effective flow. Because, as seen in the above formula, there is a direct relationship between effective flow and systemic mixed venous saturation, an increase in total systemic flow (which necessarily means an increase in effective flow) will lead to an increase in systemic mixed venous saturation. From our first days in physiology, therefore, we equate an increase in total systemic flow with an increase in systemic mixed venous saturation (if oxygen consumption is constant), and this is true in the normal (completely in series) situation.

However, when the two circuits are no longer completely in series and a "short circuit" around the lungs is possible, this direct relationship between total systemic flow and systemic mixed venous saturation no longer necessarily holds. In situations in which systemic mixed venous blood can be recirculated directly to the aorta without passing through the lungs, a change in total systemic flow may take place via a change in recirculated (or ineffective) flow without a change in effective flow. If effective flow does not change, systemic mixed venous saturation will not change because there is a direct relationship between these two variables.

This concept, which is vital to the understanding of the hemodynamics of cyanotic congenital heart disease, may then be summarized in the following manner. In any person (normal, transposition patient, tetralogy of Fallot patient, and so forth) at a given level of oxygen consumption, a certain amount of reduced hemoglobin per unit time must reach the lungs, pick up oxygen, and then deliver this oxygen to the tissues. This reduced hemoglobin is contained only in systemic mixed venous blood, and the amount of this systemic venous blood per unit time that reaches the lungs and is oxygenated, and then delivers oxygen to the tissues, is the effective flow. At a given level of oxygen consumption (and a given hemoglobin concentration), the systemic mixed venous saturation is directly related to, and determined by, the volume of effective flow as seen in the formula above.

In the normal arrangement (completely in series) of the systemic and pulmonary circuits, any increase in systemic flow must pass through the lungs (that is, all flow is effective). Because a direct relationship exists between systemic mixed venous saturation and effective flow, any increase in total systemic flow (which means a mandatory increase in effective flow) means an increase in systemic mixed venous saturation. We are taught this early in cardiovascular physiology, and although this is absolutely true in the normal series arrangement of the circulations, it is not necessarily true for patients with congenital cardiac lesions in whom systemic mixed venous blood may "short circuit" around the lungs. As seen by equation 7, systemic mixed venous saturation is determined by effective flow (not total systemic flow). In any condition in which there is a pathway around the lungs, total systemic flow may increase by virtue of an increase in the volume of systemic mixed venous blood per unit time that recirculates directly to the aorta. Thus, total systemic flow may increase without a mandatory increase in effective flow, and because systemic mixed venous saturation is determined by effective flow, total systemic flow can increase without an increase in systemic mixed venous saturation. In this situation, the decreased arteriovenous oxygen content difference across the systemic capillary beds seen with an increased total systemic flow results from a decrease in systemic arterial oxygen content rather than from an increase in systemic mixed venous oxygen content, as in the normal situation. The systemic arterial saturation (content) decreases as a result of a greater volume of systemic mixed venous blood (of constant saturation) per unit time reaching the aorta and mixing with a given amount of fully saturated pulmonary venous blood that is the result of a given effective flow. This change in proportions (as a result of an increase in recirculated systemic mixed venous flow) of saturated pulmonary venous blood and desaturated systemic venous blood reaching the aorta results in a change in systemic arterial saturation. Thus, in patients with cyanotic congenital heart disease, systemic arterial oxygen saturation may be decreased because of an increased volume of systemic mixed venous blood (of a given constant saturation determined by the effective flow) recirculating per unit time directly to the aorta. In such situations, the systemic mixed venous saturation does not result from a given total systemic flow or systemic arterial saturation but, at a given oxygen consumption, is a function of the effective flow only.

An additional fundamental point concerning the physiology of cyanotic congenital heart disease deserves clarification. The hallmark of these conditions, as previously reviewed in detail, is systemic arterial oxygen desaturation. The resulting lowered systemic arterial oxygen tension decreases the "driving pressure" of oxygen to the tissues in the systemic capillary bed and can lead to tissue hypoxia and acidosis of significant metabolic consequence.

A hypothesis for the physiological consequences of cyanotic congenital heart disease is that of "systemic oxygen transport" (13). Its application, however, can be misleading in the analysis of the severity of the physiological derangement present in these patients and has little usefulness. The hypothesis tends to obscure the important differences between arterial oxygen tension ($p_{a_{O_2}}$) and arterial oxygen content and hence leads to mistaken impressions as to the actual "availability" of oxygen to the tissues.

Systemic oxygen transport (SOT) has been defined as systemic arterial oxygen content multiplied by total systemic flow. Calculating the level for SOT by multiplying these two quantities has been used to arrive at a value for the amount of oxygen "potentially available" per unit time to the tissues in the systemic capillary bed. The manner in which use of the concept of SOT can be misleading in the analysis of the physiological status of the patient with cyanotic congenital heart disease can be illustrated best with an example. Consider the following two situations:

(a) Normal

Systemic arterial oxygen saturation = 97% (p_{aO_2} = 90 mm Hg)
Hemoglobin = 13.0 g/100 ml
Total systemic flow index = 3.0 liters min^{-1} m^{-2}
Arterial oxygen content = 13.0 × 1.34 × 0.97 = 16.9 ml O_2/100 ml
= 169 ml O_2 $liter^{-1}$
SOT = 169 ml × 3.0 = 507 ml O_2 min^{-1} m^{-2}

(b) Patient with transposition of the great arteries

Systemic arterial oxygen saturation = 65% (p_{aO_2} = 35 mm Hg)
Hemoglobin = 20.0 g/100 ml
Total systemic flow index = 5.0 liters min^{-1} m^{-2}
Arterial oxygen content = 20.0 × 1.34 × 0.65 = 17.4 ml O_2/100 ml
= 174 ml O_2 $liter^{-1}$
SOT = 174 × 5.0 = 870 ml O_2 min^{-1} m^{-2}

Review of the data from these two examples reveals that the actual volume of oxygen per unit time carried to the systemic capillary bed (SOT) by the blood is approximately 70% greater in the patient with transposition than in the normal person. Does this then imply that the patient with transposition is at a physiological advantage and has a greater capacity for exercise than does the normal?

Obviously it does not, as certainly the patient with transposition who has the type of hemodynamics illustrated would have significant symptoms and have a much reduced exercise tolerance as compared to the normal. The fallacy in equating SOT with the amount of oxygen "available" to the tissues in such patients is that, because of the low p_{aO_2} of the systemic arterial blood (35 mm Hg in the illustrated patient with transposition), and the resultant low "driving pressure" of oxygen to the tissues in the systemic capillary bed, much of the oxygen carried to the systemic capillaries does not reach the tissues. Hence, the described patient with transposition would exhibit severe tissue hypoxia despite the increased volume of oxygen per unit time actually passing through the systemic capillaries.

The real question is: Which is more important in delivery of oxygen to the tissues, arterial oxygen tension (p_{aO_2}) or arterial oxygen content? Comroe (14) has answered this question by stating, "Both are important: the tension must be high enough to permit diffusion from the capillary to the most remote cell; the content (number of molecules of oxygen per unit of blood) must be large enough so that the oxygen needs of all cells, near and remote, are met." As Comroe has pointed out, both oxygen tension and oxygen content are important. However, in the patient with cyanotic congenital heart disease, it is the arterial oxygen tension, rather than the oxygen content, that is inadequate, resulting in tissue hypoxia. Many patients with severe cyanotic congenital heart disease, such as the illustrated patient with transposition, actually will have increased volumes of oxygen per unit time passing through their systemic capillaries (increased SOT). This results from their significant compensatory polycythemia and the greater-than-normal total systemic flow that occurs because of systemic arteriolar dilatation (lowered systemic resistance) in response to low $P_{a_{O_2}}$. However, these patients still have severe tissue hypoxia because this oxygen "potentially available" is available only at a greatly reduced arterial oxygen tension and the resultant "driving pressure" of oxygen is inadequate to deliver the oxygen to the cells. Thus, the concept of SOT and the amount of oxygen "potentially available" to the cells, while perhaps useful in conditions such as anemia, is not helpful and indeed in often misleading when the patient has cyanotic congenital heart disease, because the patients with the largest SOT often will be the ones with extremely low arterial oxygen tensions and resultant severe tissue hypoxia.

EVALUATION OF SEVERITY OF
PULMONARY VASCULAR OBSTRUCTIVE DISEASE

The number of patients with congenital heart disease who are not candidates for definitive surgical repair, because of anatomically uncorrectable lesions, has decreased steadily during the past two decades. With the present state of the cardiovascular surgical art, only a few lesions, or combinations of lesions, remain anatomically or physiologically uncorrectable. However, the tendency for patients with some forms of congenital heart disease to develop significant occlusive changes (15) of the pulmonary arterioles as a consequence of their deformity has been well recognized for many years. If this process is severe, the patient with an anatomically correctable defect must be refused surgery, because experience has shown that he will not survive the procedure or, if he does survive, will not be benefited.

The cause of pulmonary vascular obstructive disease (PVOD) is still poorly understood, although increased pulmonary blood flow, increased pulmonary artery pressure, and elevated left atrial pressure all probably have some role. Some evidence (16) shows that patients who develop this problem have not had the normal involution of the pulmonary arterioles from the vessels with a thick medial layer seen during fetal life to the thin-walled vessels normally seen postnatally, and hence the process may have its origins very early in life. The unanswered question is: Why does one patient with a congenital cardiac lesion develop severe PVOD at an early age, whereas others with the same lesion show little evidence of this process after many years?

At particularly high risk for the development of significant PVOD are patients with some forms of cyanotic congenital heart disease. Patients with complete transposition of the great arteries (17, 18), double-outlet right ventricle, and single ventricle who do not have associated pulmonary stenosis often have severe occlusive changes, which preclude surgery, by 2 to 3 years of age. A significant proportion of the patients with truncus arteriosus studied in our laboratory also have had to be refused operation because of this problem (11). The accurate assessment in the cardiovascular laboratory of the degree of PVOD present in such patients is of critical importance in the appropriate selection of patients with such problems for surgical correction.

Hoffman (19) has pointed out the biophysical factors that must be considered in the assessment of PVOD and the inherent errors residing in methods presently available. Because techniques are not presently available to measure the resistance to pulmonary blood flow directly, reliance must be placed on a resistance that is calculated by dividing the measured mean driving pressure across the pulmonary arteriolar bed by the measured total pulmonary blood flow index. The expression for this is as follows: R_p (arteriolar) (units m^{-2}) = $\Delta P \; / \; Q_{p_I}$, in which R_p (arteriolar) = pulmonary arteriolar resistance, ΔP = mean pulmonary artery pressure (mm Hg) − mean left atrial pressure (mm Hg), and Q_{pI} = total pulmonary blood flow index (liters $min^{-1} \; m^{-2}$).

Hoffman has correctly pointed out the potential errors in applying an equation that was originally designed to measure resistance to steady flow through straight, nondistensible tubes to the measurement of resistance to pulsatile flow through the branching, distensible vessels of the human lung. However, despite the potential errors and limitations, the assessment of pulmonary resistance by this method, in the setting of the catheterization laboratory, has provided accurate assessment of the degree of PVOD and is the method now used to determine whether a patient with an anatomically correctable congenital cardiac lesion is a hemodynamically appropriate candidate for operation.

During the 1960s, the degree of PVOD in patients with congenital heart disease was assessed by the use of flow and resistance ratios (Q_p/Q_s and R_p/R_s) between the pulmonary and systemic circuits. These ratios were used because, at that time, oxygen consumption was difficult to measure accurately and reliable tables for assumed oxygen consumption were not available. The use of ratios of flows and resistances between the two circuits canceled out the factor of oxygen consumption; thus its determination was not necessary.

This method worked reasonably well in simple, acyanotic lesions such as isolated ventricular septal defect in which the systemic circuit flow and resistance usually were relatively stable. However, in the cyanotic lesions, and particularly in complete transposition, the use of these ratios was often extremely misleading in assessing the degree of PVOD. This was true because wide fluctuations in systemic flows and resistances often occurred, and these profoundly altered the value of the ratios but occurred independently of (and had little or no relationship to) the flow and resistance in the pulmonary circuit. The extreme degree to which the use of these flow and resistance ratios between the two circuits was misleading in assessing the severity of the PVOD in complete transposition was pointed out previously (4) in analyzing hemodynamic data from our laboratory.

With the development of simple and accurate means to measure oxygen consumption in the catheterization laboratory (20), the reason for the evolution of the use of flow and resistance ratios between the two circuits to assess the degree of PVOD ceased to exist. Therefore, in the cardiovascular laboratory of the 1970s, the assessment of PVOD in all congenital cardiac lesions should be made by the use of the absolute pulmonary arteriolar resistance, if the status of the pulmonary arteriolar bed is to be determined accurately.

A previous review from our institution indicated that patients with transposition and associated ventricular septal defect who had a pulmonary arteriolar resistance of greater than 10.0 units m^{-2} had severe occlusive disease of the pulmonary arterioles and were not likely to survive definitive surgical correction. However, it was emphasized that this value of 10.0 units m^{-2} could not be used as an absolute dividing line below which patients were operable and above which they were not. A number of patients would be "borderline" insofar as the likelihood of surviving and benefiting from a definitive operation was concerned. Subsequent to that initial review, a corrective Mustard operation was attempted in five additional such patients with transposition in whom the pulmonary arteriolar resistance was greater than 10.0 units m^{-2}. Our present total experience with attempted definitive correction in this group is shown in Table 6. Corrective operation was undertaken in the last five patients with the realization by the parents and physicians that the risks were substantial. However, these children were extremely hypoxemic and polycythemic and had associated severe disability, and under these circumstances an attempt at surgical intervention seemed justified.

Seven of the eight patients died during the immediate postoperative period. Of the five patients in whom intraoperative pressures were obtained after repair, all had pulmonary artery peak systolic pressures that were three-quarters or more of aortic systolic pressures. In every case in which autopsy was performed, the associated ventricular septal defect was found to be completely and accurately closed. In all autopsy cases, microscopic examination of the lungs revealed extensive Heath-Edwards grade 4 histologic changes in the pulmonary arterioles. With these findings, the failure of these operative patients to survive could be

Table 6. Results from definitive Mustard operation in eight patients with transposition who had pulmonary arteriolar resistance greater than 10.0 units m^{-2}

Case	Age at surgery (yrs)	Pulmonary arteriolar resistance (units m^{-2})	Ratio of pulmonary artery pressure to aorta pressure (postop), %	Outcome	Autopsy findings
1[a]	5	12.0	—	Died	Heath-Edwards grade 4
2[a]	6	11.8	—	Died	Heath-Edwards grade 4
3[a]	4	10.8	175	Died	Heath-Edwards grade 4
4	2½	10.8	80	Died	Heath-Edwards grade 4
5	6	12.9	95	Died	Heath-Edwards grade 4
6	5	11.1	75	Died	No autopsy
7	10/12	10.7	120	Died	Heath-Edwards grade 4
8	1 4/12	12.1[b]	68	Survived	—

[a]Previously reported (4).
[b]7.7 units m^{-2} breathing 100% oxygen.

attributed primarily to severe PVOD and the postoperative pulmonary hypertension secondary to this process. The only surgical survivor (case 8) was a 16-month-old child who had a pulmonary arteriolar resistance (while breathing room air) of 12.1 units m^{-2}. However, this resistance decreased to 7.7 units m^{-2} while breathing 100% oxygen (a study that is always performed in our laboratory in patients on whom room air data indicate a significant elevation in pulmonary arteriolar resistance). This patient was the only one of the eight who had a significant decrease in pulmonary arteriolar resistance while breathing 100% oxygen. The decrease in resistance to 7.7 units m^{-2} while this patient was breathing oxygen indicated that her elevated pulmonary arteriolar resistance on room air was partially due to constriction of hypertrophied arteriolar smooth muscle in response to hypoxia and acidosis and that actual occlusive changes of the intima were not as severe as the value of 12.1 units m^{-2} with room air would ordinarily indicate. Although the immediate postoperative course of this child was stormy and she required assisted ventilation and very close monitoring of acid-base balance for several days, she is now doing well and is taking no medications nearly 2 years after operation.

These data indicate that the risk of an attempt at definitive correction is prohibitively high for the patient with transposition who has a pulmonary arteriolar resistance of greater than 10.0 units m^{-2} while breathing room air and whose resistance does not decrease significantly in response to 100% oxygen.

A recent review (11) of 40 patients with truncus arteriosus who underwent cardiac catheterization and subsequent surgical correction at the Mayo Clinic further substantiated the conclusion drawn from the transposition group: Patients with a total pulmonary resistance of greater than 12.0 units m^{-2} or a pulmonary arteriolar resistance of greater than 10.0 units m^{-2} which does not decrease with breathing of 100% oxygen have severe occlusive pulmonary arteriolar disease that precludes definitive surgical correction. This degree of PVOD after operation will leave them with a significant elevation of pulmonary artery pressure (greater than three-quarters systemic), which greatly increases the risk of operation.

Our study of patients with truncus arteriosus also revealed that, if a patient with congenital heart disease has unilateral absence of a pulmonary artery (an occurrence seen in 16% of our patients with truncus arteriosus), different hemodynamic criteria must be used to assess the severity of the PVOD if pulmonary flow and pulmonary resistance are calculated in the usual manner and if these parameters are correlated with the degree of PVOD. For a given degree of histological PVOD, the patient with only one pulmonary arteriolar bed that is exposed to a given driving pressure would have approximately one-half the pulmonary flow and hence have twice the calculated pulmonary resistance as would the patient with comparable PVOD and the given driving pressure in two lungs. Because there is often bronchial collateral supply to the lung to which the pulmonary artery is absent, this explanation is an oversimplification. However, there is no doubt that different criteria for operability are necessary. This reasoning suggests that, in the patient with only one lung perfused by a high-pressure pulmonary artery, pulmonary resistance, once calculated in the usual way, must be halved if it is to be correlated with the degree of histologic PVOD present in the lung supplied by the pulmonary artery, in an attempt to assess operability. In one series (11), this approximation was reliable in selecting for surgery the appropriate patients who had a single pulmonary artery.

These data from our patients with transposition and truncus arteriosus indicate that patients with congenital heart disease who have pulmonary arteriolar resistances of greater than 10.0 units m^{-2} have significant PVOD and that the risk of definitive surgical repair is high. Even if such patients survive operation, the pulmonary artery pressure will remain significantly elevated, and they will likely have signs of congestive heart failure and a significantly decreased exercise tolerance. In addition, in patients with advanced PVOD, even if they survive operation, there is significant doubt that the complete separation of the two circulations achieved by definitive corrective surgery will have any influence on the natural history of the pulmonary vascular disease. Thus, we now do not offer definitive surgical correction of congenital cardiac lesions to patients with severe PVOD who have pulmonary arteriolar resistances greater than 10.0 units m^{-2}. However, in the patient with transposition and associated ventricular septal defect and severe PVOD, the "palliative" Mustard operation (intra-atrial baffle while the ventricular septal defect is left opened) may improve his hemodynamic status significantly and may reduce his symptoms greatly. Moreover, patients with truncus arteriosus or isolated ventricular septal defect may have PVOD severe enough to preclude an attempt at definitive repair and yet be able to lead relatively useful and productive lives, although obviously somewhat limited in regard to activities requiring strenuous physical exertion. Such patients often have systemic arterial oxygen saturations greater than 80%, achieved with modest degrees of polycythemia, and exhibit minimal clinical cyanosis. A review (21) of patients with inoperable isolated ventricular septal defect who had severe PVOD and were seen at the Mayo Clinic several years ago revealed that many were young adults leading useful lives. Although these patients eventually do die of PVOD, survival into the fourth decade has been reported.

CONCLUSIONS

Congenital cardiac lesions are malformations for which the definitive therapy, if necessary, is surgical. The advent of cardiopulmonary bypass techniques 20 years ago, which made possible the effective repair of intracardiac defects, provided a great impetus for the understanding of the pathophysiology of these deformities. Advances in surgical techniques during the past two decades have been accompanied by parallel achievements in the cardiac catheterization laboratory that were necessary if patients were to be appropriately selected for surgery. Present angiocardiographic methods make possible the delineation of cardiovascular anatomy in great detail, and accurate pressure recording devices, oximetric techniques, and dye dilution techniques have provided the means for assessing the hemodynamic status of patients with these lesions. Some relatively new techniques using ultrasound and radioactive isotopes, as well as those of roentgen videometry (22) and three-dimensional radiographic computer reconstruction of the heart (23), should further improve our abilities to delineate the pathophysiology of even the most complex deformity. If this new methodology, in combination with older established techniques, is to be effective in the accurate assessment of congenital heart disease, however, the physician must have a comprehensive understanding of the basic physiological consequences of these malformations.

The hemodynamics of the group of patients with cyanotic congenital cardiac lesions outlined in the present paper represent the most complex of the congenital cardiac deformities. The understanding of these hemodynamic principles and their applications to data obtained by the methods presently (and in the future) available should enable us to continue to add to our understanding of these deformities and thus provide an ever-increasing fund of knowledge on which to base the important decisions regarding clinical management of patients with these malformations.

REFERENCES

1. Liebman, J., Collum, L., and Belloc, N.B. (1969). Natural history of transposition of the great arteries: anatomy and birth and death characteristics. Circulation 40:237.
2. Comroe, J.H., Jr., and Dripps, R.D. (1974). Ben Franklin and open heart surgery. Circ. Res. 35:661.
3. Nadas, A.S., and Fyler, D.C. (1972). Pediatric Cardiology, 3rd ed., 749 pp. W.B. Saunders Co., Philadelphia.
4. Mair, D.D., Ritter, D.G., Ongley, P.A., and Helmholz, H.F., Jr. (1971). Hemodynamics and evaluation for surgery of patients with complete transposition of the great arteries and ventricular septal defect. Amer. J. Cardiol. 28:632.
5. Comroe, J.H., Jr. (1965). Physiology of Respiration. An Introductory Text, p. 61. Year Book Medical Publishers, Chicago.
6. Mair, D.D., and Ritter, D.G. (1973). Factors influencing systemic arterial oxygen saturation in complete transposition of the great arteries. Amer. J. Cardiol. 31:742.
7. Mair, D.D., and Ritter, D.G. (1972). Factors influencing intercirculatory mixing in patients with complete transposition of the great arteries. Amer. J. Cardiol. 30:653.
8. Lindesmith, G.G., Stiles, Q.R., Tucker, B.L., Gallaher, M.E., Stanton, R.E., and Meyer, B.W. (1972). The Mustard operation as a palliative procedure. J. Thorac. Cardiovasc. Surg. 63:75.
9. Mair, D.D., Ritter, D.G., Danielson, G.K., Wallace, R.B., and McGoon, D.C. (1975). The palliative Mustard operation. Rationale and results. Amer. J. Cardiol. In press.
10. Mustard, W.T., Keith, J.D., Trusler, G.A., Fowler, R., and Kidd, L. (1964). The surgical management of transposition of the great vessels. J. Thorac. Cardiovasc. Surg. 48:953.
11. Mair, D.D., Ritter, D.G., Davis, G.D., Wallace, R.B., Danielson, G.D., and McGoon, D.C. (1974). Selection of patients with truncus arteriosus for surgical correction. Anatomic and hemodynamic considerations. Circulation 49:144.
12. Ritter, D.G. (1975). Hemodynamics of common ventricle. In preparation.
13. Murray, J.F., Gold, P., and Johnson, B.L., Jr. (1962). Systemic oxygen transport in induced normovolemic anemia and polycythemia. Amer. J. Physiol. 203:720.
14. Comroe, J.H., Jr (1965). Physiology of Respiration. An Introductory Text, p. 169. Year Book Medical Publishers, Chicago.
15. Heath, D., and Edwards, J.E. (1958). The pathology of hypertensive pulmonary vascular disease. A description of six grades of structural changes in the pulmonary arteries with special reference to congenital cardiac septal defects. Circulation 18:533.

16. Rudolph, A.M. (1970). The changes in the circulation after birth. Their importance in congenital heart disease. Circulation 41:343.
17. Ferencz, C. (1966). Transposition of the great vessels. Pathophysiologic considerations based upon a study of the lungs. Circulation 33:232.
18. Viles, P.H., Ongley, P.A., and Titus, J.L. (1969). The spectrum of pulmonary vascular disease in transposition of the great arteries. Circulation 40:31.
19. Hoffman, J.I.E. (1972). Diagnosis and treatment of pulmonary vascular disease. Birth Defects 8:9.
20. Kappagoda, C.T., and Linden, R.J. (1972). A critical assessment of an open circuit technique for measuring oxygen consumption. Cardiovasc. Res. 6:589.
21. Clarkson, P.M., Frye, R.L., DuShane, J.W., Burchell, H.B., Wood, E.H., and Weidman, W.H. (1968). Prognosis for patients with ventricular septal defect and severe pulmonary vascular obstructive disease. Circulation 38:129.
22. Ritman, E.L., Sturm, R.E., and Wood, E.H. (1973). Biplane roentgen videometric system for dynamic (60/sec) studies of the shape and size of circulatory structures, particularly the left ventricle. Amer. J. Cardiol. 32:180.
23. Johnson, S.A., Robb, R.A., Greenleaf, J.F., Ritman, E.L., Gilbert, B.K., Storma, M.T., Sjostrand, J.D., Donald, D.E., Herman, G.T., Sturm, R.E., and Wood, E.H. (1974). Dynamic three-dimensional reconstruction of beating heart and lungs from multiplanar roentgen-television images. Mayo Clin. Proc. 49:958.

International Review of Physiology
Cardiovascular Physiology II, Volume 9
Edited by Arthur C. Guyton and Allen W. Cowley
Copyright 1976   University Park Press   Baltimore

8
The Evolution
of Myocardial Infarction:
Physiological Basis
for Clinical Intervention

R. A. GUYTON and W. M. DAGGETT
National Heart and Lung Institute and Massachusetts General Hospital

INTRODUCTION   306

INITIATION OF INFARCTION   307
  Role of Pathological Increases in Coronary Resistance   307
  Hemodynamic Determinants of Oxygen Supply   307
  Special Vulnerability of the Subendocardium   308
  Determinants of Myocardial Oxygen Consumption   309

EVOLUTION OF INFARCTION:
 ROLE OF POSITIVE FEEDBACK MECHANISMS   311
  Power Failure (Cardiogenic Shock)   311
  Pain   311
  Local Change in Coronary Resistance   311
  Competitive Redistribution of Coronary Flow   315
    Hemodynamic Considerations   315
    *Evidence for Coronary Steal
    After Acute Coronary Occlusion in Normal Dogs*   317
    *Evidence for Coronary Steal
    After Chronic Coronary Occlusion in Dogs*   319
    Definition of "Competitive Redistribution"   320

END OF INFARCTION   321
  Increased Oxygen Supply   321
  Decreased Oxygen Demand   323
  Necrosis of a Portion of the Jeopardized Myocardium   324

INTRODUCTION

Coronary artery occlusion has dramatic sequelae. Experimentally, within a few seconds myocardial contraction in the affected area deteriorates and systolic bulging of the ischemic area ensues (1–4). After only 20 min of ischemia, a number of the deprived cells are damaged beyond repair (5). But, even with permanent occlusion, a variable portion of the ischemic cells recover. Myocardial infarction, even in the simple dog model, is a complex function of the multiple determinants of myocardial oxygen consumption and oxygen supply.

In man, infarction pathophysiology is further complicated: coronary occlusion does not seem to be a prerequisite for infarction (6, 7), collateral coronary flow may be well developed, and multiple coronary lesions are more common than not. But there is a starting point from which this complexity may be attacked. Myocardial infarction, as it is clinically recognized, is a discrete event. It has a beginning, which implies that precipitating factors are involved. But just as importantly, presuming a nonfatal outcome, it has an end. Infarction ceases when the mechanisms that tend to cause self-propagation of cellular necrosis (positive feedback mechanisms) are outweighed by the sequelae of infarction that tend to limit additional necrosis (negative feedback mechanisms). Recent advances in the identification of these precipitating factors and in the understanding of potential positive and negative feedback mechanisms have opened new frontiers in clinical intervention.

Several concepts to receive special consideration in this review are: 1) acute infarction is often not associated with acute thrombosis; 2) coronary reserve (i.e., potential oxygen supply/oxygen demand) is not homogeneous in the left ventricle. The subendocardium is particularly vulnerable to oxygen supply/demand deficiency; 3) a time-related local increase in coronary resistance in ischemic muscle leads to the demarcation of an ischemic region into a necrotic subregion and a subregion with adequate flow for survival; 4) a reciprocal relationship may exist between changes in coronary flow to a normal region of the myocardium and changes in flow from that region to an adjacent ischemic region. For example, an increase in flow to a nonischemic area may decrease collateral flow to a nearby ischemic area. This relationship may multiply the effects of induced alterations in myocardial oxygen consumption after myocardial infarction; 5) there is substantial experimental support for various pharmacological and surgical interventions in the evolution of myocardial infarction.

INITIATION OF INFARCTION

Role of Pathological Increases in Coronary Resistance

The discrepancy between oxygen supply and delivery that leads to infarction is uncommon in the absence of coronary artery disease (6–8). But coronary occlusion is certainly not a prerequisite for infarction. Indeed, in autopsy series, fatal infarction has been found to occur with no evidence of an acute increase in proximal coronary resistance in as many as 37–62% of specimens (6, 7). There is a low incidence of thrombosis in patients dying suddenly from circulatory collapse and in patients with fatal infarction limited to the subendocardium. But an increasing incidence of thrombosis is found in large transmural infarctions, particularly those leading to cardiogenic shock. It has been suggested that thrombosis is caused by coronary stasis during shock and that those patients dying suddenly do not have time to develop thrombi (6, 7). This hypothesis, that thrombosis is the consequence rather than the cause of transmural infarction, is not widely accepted (8).

Development of collateral vessels after chronic coronary occlusion in animals ordinarily permits normal flow to the myocardium in the resting state. There is no experimental evidence that "chronic ischemia" of viable myocardium is a stable condition (9–13). But proximal coronary stenosis can limit coronary reserve. The response of coronary flow to stress (tachycardia, isoproterenol infusion) has been shown to be impaired by coronary disease (10, 14, 15). The inability to adequately decrease coronary resistance in response to increased oxygen demand is the presumptive basis for acute infarction without acute occlusion and may account for an unknown portion of fatal infarctions and a majority of nonfatal subendocardial infarctions (6–8).

Hemodynamic Determinants of Oxygen Supply

Both collateral flow and blood flow across a proximal coronary stenosis are linearly dependent upon the effective pressure gradient (10, 16–19). Collateral flow, estimated by retrograde coronary flow from the acutely divided artery of the dog, has been shown to vary directly with diastolic arterial pressure and to be relatively independent of systolic pressure. This observation suggests that collateral flow may occur primarily during diastole (18). The possible diastolic nature of collateral flow has not been studied in animals with chronic coronary occlusion, whose collateral system is anatomically quite different from that in animals with acute occlusion (12). The duration of diastole and heart rate obviously could mechanically limit the time available for collateral flow, if flow is indeed diastolic in the chronic as well as in the acute situation. After acute coronary occlusion in dogs, bradycardia seems to increase collateral flow, but tachycardia does not alter flow to the ischemic area (by xenon-133 washout techniques (17)).

Special Vulnerability of the Subendocardium

In the resting state, coronary flow to the left ventricle is relatively homogeneous (10, 13, 20, 21). Moreover, in response to moderate stress, coronary flow tends to increase in a homogeneous manner (10, 13, 21). But, if coronary flow is inadequate to meet metabolic demands, oxygen supply is then distributed throughout the myocardium in a distinctly *nonhomogeneous* manner. This maldistribution of flow has been demonstrated during coronary occlusion (22, 23), coronary stenosis (20, 24, 25), severe left ventricular pressure overload (21, 26), severe left ventricular volume overload (26), catecholamine infusion (26), and severe anemia (21). The inner layers of the ventricle, in these situations, are invariably more severely compromised than the outer layers. Coronary flow may normally be homogeneous in the left ventricle, but *coronary reserve,* i.e., potential oxygen supply/oxygen demand, is clearly *heterogeneous* in the left ventricle.

Local coronary reserve in the subendocardium is limited by the mechanics of subendocardial blood flow. Intramyocardial pressure during systole in the inner layers of the heart always equals or exceeds aortic pressure (27, 28). Blood flow, therefore, occurs primarily during diastole in the subendocardium (26, 27), implying that, since normal coronary flow is homogeneous, coronary resistance during diastole must be lower in the subendocardium than in the subepicardium. Studies of capillary density have indeed revealed a greater density of open capillaries in the subendocardium (27, 29). But suppose coronary flow is limited by proximal resistance, and distal coronary pressure falls. To maintain equal regional flows, coronary resistance in the subendocardium would need to decrease proportionally *more* than regional resistance in the subepicardium (since the coronary to right atrial diastolic pressure gradient falls proportionally more than the systolic pressure gradient). But this does not occur. Measurement of regional coronary flow distal to a coronary stenosis reveals that autoregulation maintains subepicardial flow over a wide range of distal coronary artery pressures, while endocardial flow decreases in a linear manner as mean distal coronary pressure is decreased below about 70 mm Hg (Figure 1) (25). The subendocardial vascular system is not normally fully dilated. Subendocardial flow can be increased up to three times normal with vasodilators or by reactive hyperemia (20, 25), and it also increases appropriately in response to metabolic demands (13, 21, 26). Asphyxia increases the small vessel blood content of the subendocardium, suggesting a reserve of unopened capillaries (29). But the dilatory capacity of the subendocardial oxygen delivery system, hampered as it is by the effects of systolic compression, is inadequate to successfully compete with the subepicardial system for a limited coronary blood supply. Indeed, as coronary flow is decreased, subepicardial flow is maintained with remarkable effectiveness *at the expense of the subendocardium* (Figure 1) (25).

A subendocardial deficiency in coronary flow represents a deficiency in local coronary reserve since oxygen consumption in the inner layers of the heart equals or exceeds that of the outer layers (30, 31). Intramyocardial oxygen tension is normally lower in the deeper layers than in the outer layers of the left ventricle (27, 30, 32). If local coronary pressure is decreased in a region of the myocardium, intramyocardial p_{O_2} decreases in a stepwise manner in the sub- endocardium, but tends to be maintained in the subepicardium (30). Studies of high energy phosphate depletion (31), lactate accumulation (31), creatine phosphokinase depletion (33), and the histologic distribution of infarction (5) further verify the transmural gradient of oxygen supply/demand deficiency during limited coronary inflow.

Subendocardial S-T segment elevation, which is associated with subendo- cardial ischemia and eventual subendocardial infarction, increases progressively as distal coronary pressure is decreased by stenosis. Epicardial S-T segments are progressively depressed by increasing stenosis until rapid elevation of the epi- cardial S-T segments occurs below a critical level of distal coronary pressure (Figure 1) (25). These S-T segment alterations closely parallel the changes observed in both regional flow (Figure 1) and intramyocardial p_{O_2} (30) as distal coronary pressure is reduced. Proximal coronary stenosis leads to a marked heterogeneity of coronary reserve which is manifest in infarction by *conservation of the subepicardium and sacrifice of the subendocardium.*

Determinants of Myocardial Oxygen Consumption

Since infarction seems to occur often without postmortem evidence of acutely increased coronary resistance, it is reasonable to believe that alterations in oxygen utilization may precipitate positive feedback mechanisms that lead to necrosis. The three major determinants of oxygen consumption are myocardial wall tension, the contractile state of the heart, and heart rate. The development of wall tension can be estimated by assuming a spherical ventricle and by utilizing the Laplace relation: tension = (pressure) X (radius) / 2 (wall thickness). Changes in left ventricular size clearly alter wall tension and therefore alter oxygen consumption. A great variety of interventions that affect the contractile state of the heart have been shown to change energy utilization, but the mechanism of this effect is not clear. Changes in heart rate lead to changes in both tension development and in the contractile state of the heart. An increased heart rate may or may not lead to increased oxygen consumption per beat, but, generally, oxygen consumption per minute is elevated. Volume work is a relatively less important determinant of energy utilization, and the basal oxygen

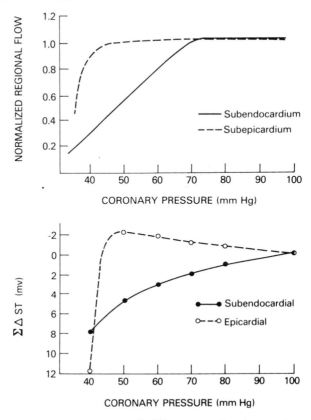

Figure 1. Changes in regional flow and S-T segment elevation as circumflex coronary pressure is decreased at constant aortic pressure, heart rate, and cardiac output. Normalized regional flow is flow in the inner or outer ¼ of the circumflex area divided by flow in the full thickness of the anterior descending area. Flow curves from 35 to 70 mm Hg are based on 43 determinations of regional flow, and curves above 70 mm Hg are based on observed equal regional flows at control pressure in this preparation. S-T segment changes represent the sum of changes in six pairs of electrodes in a typical experiment conducted under identical hemodynamic conditions (25).

requirement necessary for maintenance of cellular viability is a relatively constant background upon which alterations in energy utilizations are superimposed (34, 35).

One of the critical questions that remains unanswered is the degree to which the energy requirement for maintenance of cellular viability can be dissociated from the energy requirement for contraction. If the myocardial cell could neglect contraction and concentrate on viability, then it might be able to tolerate as much as an 80% reduction in oxygen supply (34). McDonald and MacLeod have shown that contraction and transmembrane electrical activity in the guinea pig myocardium are *not* equally dependent upon aerobic metabolism.

During twelve hours of anoxic incubation in a solution containing 50 mM glucose, rapid depletion of ATP and marked decline in developed tension occur, but transmembrane electrical activity remains essentially unchanged (36).

EVOLUTION OF INFARCTION:
ROLE OF POSITIVE FEEDBACK MECHANISMS

After irreversible cellular damage is precipitated by inadequate coronary reserve, the cardiac and systemic consequences of infarction may lead to a further decrease in coronary reserve and self-propagation of infarction. At least four positive feedback mechanisms have been implicated in the propagation of infarction: 1) power failure (cardiogenic shock); 2) pain; 3) local changes in coronary resistance; 4) competitive redistribution of coronary blood flow (coronary steal).

Power Failure (Cardiogenic Shock)

Left ventricular power failure with subsequent cardiogenic shock has been related to the proportion of the myocardium found to be infarcted at autopsy (6, 37). In one study, 19 of 20 patients succumbing to power failure after infarction had lost 40–70% of the contractile mass of the left ventricle, while 12 of 14 patients who died of other causes suffered less than 40% myocardial loss. But more importantly, the patients who died of cardiogenic shock almost universally had postmortem evidence of recent extension of the infarction, both in regions adjacent to the original infarction and in regions remote from the original infarction (37). Thus cardiogenic shock is initiated by loss of left ventricular mass which in turn leads to loss of additional myocardial mass. The mechanisms responsible for this vicious cycle are numerous and beyond the scope of this review (38), but a superficial summary of three major mechanisms is illustrated in Figure 2. The results of therapeutic intervention in cardiogenic shock are dismal, with mortality usually exceeding 80% (38).

Pain

The origin of either angina pectoris or the pain of infarction is obscure, but the sequelae of pain lead to increased oxygen demand and decreased coronary reserve as illustrated in Figure 3. Pharmacological intervention in this cycle is effective and has been for years a mainstay of medical treatment of infarction.

Local Change in Coronary Resistance

After 10 min of ischemia in the rabbit brain, or after 60 min of ischemia in the rat kidney, attempted reperfusion of the organ is hampered by increased vascular resistance (39–41). In the brain, the increase in resistance is associated with swelling of the perivascular glial cells (42). In the kidney, the mechanism of increased resistance is less clear. The infusion of a hypertonic mannitol solution does seem to diminish the resistance elevation, and again cellular swelling has

POWER FAILURE
(CARDIOGENIC SHOCK)

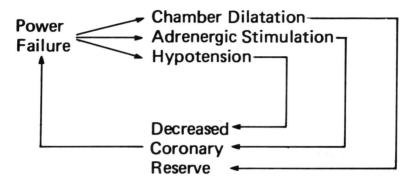

Figure 2. Three important positive feedback mechanisms in cardiogenic shock.

been implicated (41, 43). This observation of increased vascular resistance after ischemia has been termed the no-reflow phenomenon (39), and it has also been investigated in the heart.

Reperfusion after coronary artery occlusion in the cat, as measured by vital dye staining, has been found to be retarded after a 120-min occlusion followed by 1–6 hours of reperfusion. Thirty- and 60-min occlusions produced lesser derangements in vital dye staining, generally worse in the subendocardium than the subepicardium, that were reversed by 60 min of reperfusion (44). Radioactive microsphere estimation of regional flow during reperfusion has demonstrated no impairment of epicardial reperfusion after 1, 2, or 6 hours of coronary occlusion. But endocardial reperfusion is variably diminished after 2 hours of occlusion and uniformly diminished after 6 hours (45).

Peak reactive hyperemia has been used as an estimate of the dilatory capacity or minimum resistance of a coronary vascular bed (46). After 2 hours of left circumflex occlusion in a series of conscious dogs, peak reactive hyperemia was unchanged from control levels. But after 6 hours of coronary occlusion, peak reactive hyperemia was reduced to 62% of control levels, and returned to normal only after 3 days of reperfusion. Longer occlusions further reduced the peak reactive hyperemia response and lengthened the reperfusion time necessary for return to normal. Caution must be exercised in interpreting these results, since the development of collateral flow after longer coronary occlusions may blunt the reactive hyperemia response, but it appears that the dilatory capacity of the ischemic vascular bed may be reduced in a time-related manner after coronary occlusion. Edema in the ischemic area has been proposed as a possible mechanism for this reduction (46). In this regard, it must be pointed out that reperfusion has been shown to accelerate dramatically electrolyte

PAIN

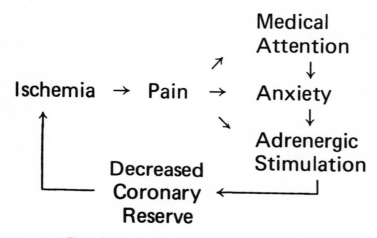

Figure 3. The sequelae of pain tend to propagate ischemia.

disorders and the accumulation of water in ischemic areas (47, 48). Moreover, reperfusion after 3–5 hours of coronary occlusion may, in some instances, actually cause extension of infarction which is generally associated with hemorrhage into the necrotic area (49, 50). A direct extrapolation, therefore, of resistance derangements found upon reperfusion cannot be made to the situation of permanent coronary occlusion or stenosis. A study of this latter situation has been conducted in our laboratory.

Temporal changes in regional coronary resistance during partial-thickness infarction were measured using radioactive microsphere techniques (the density of trapped radioactive microspheres is proportional to blood flow). An open-chest canine preparation was used which allowed maintenance of constant blood pressure, cardiac output, left atrial pressure, heart rate, and adrenergic tone. A variable resistance placed on the proximal circumflex coronary artery (Figure 4) allowed control of distal coronary pressure (P_{Cx}). This pressure was maintained at a constant level (about 40 mm Hg) which created a partial thickness infarction as determined by subendocardial S-T segment elevation and no elevation in epicardial S-T segments. During 3 hours of constant pressure circumflex ischemia, regional flow to the inner one-quarter of the myocardium decreased from 0.22 to 0.09 ml/g min ($p < 0.02$) while no significant change in flow was observed in the outer 1/4 of the myocardium. Upon reperfusion, flow in both areas returned to normal. Resistance in the more ischemic subendocardium more than doubled over the 3-hour period of stenosis while resistance in the subepicardium was unchanged. Intermediate layers of the heart showed intermediate changes in resistance (51). This time-related increase in resistance in an ischemia coronary vascular system was anticipated from reperfusion studies, but both the

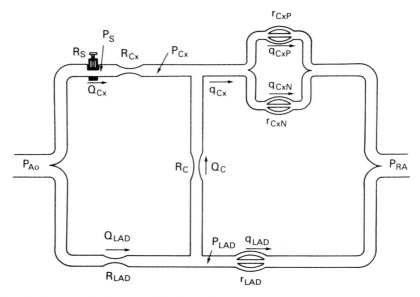

Figure 4. A schematic diagram of single vessel coronary stenosis with collateral flow. Upper case R and Q represent resistance and flow proximal to the collateral network and lower case r and q represent resistances and flows distal to the collateral network. Subscripts S, C, *LAD*, and *Cx* denote the imposed stenosis, collateral, anterior descending, and circumflex, respectively. *CxN* refers to the endocardial portion of the distal circumflex vascular bed and *CxP* refers to the epicardial portion of the distal circumflex vasculature. P_{Ao} and P_{RA} are pressures in the aorta and right atrium. P_{LAD} is a hypothetical pressure representing the average pressure at the junction of the collateral system with the anterior descending system. P_{Cx} is a similar pressure in the circumflex system. P_S is the pressure in the circumflex system just distal to the imposed stenosis. P_S is the pressure usually measured in experimental animals and may or may not equal P_{Cx}.

magnitude of the increase and the speed with which it developed were greater than expected.

The etiology of this local resistance increase is not clear. The increase occurred too quickly to be attributed entirely to cellular swelling (43) since changes in water or sodium content of the central ischemic area of an infarct are small for at least the first 10 hours after coronary occlusion (47). The reversal of regional flow abnormalities by 30 min of reperfusion after 3 hours of ischemia in the study discussed above does not imply that this resistance change is completely reversible. The dilatory capacity of the area after 30 min of reperfusion was not measured and may have been impaired although resting flows were normal. Indeed, since reperfusion after this degree of ischemia leads to at least partial infarction (25) one would expect some impairment in the dilatory capacity of the region.

The implications of a time-related increase in local resistance during myocardial ischemia are profound. If ischemia increases local resistance, then ischemia causes more ischemia, and a powerful positive feedback system is possible (Figure 5). Rees and Redding found no xenon uptake in the most

LOCAL CHANGES IN CORONARY RESISTANCE

Local Ischemia   →   Time-related Increase
in Local Coronary
↑   Resistance

Decreased Local ↙
Coronary Flow

Figure 5. The effect on ischemia of local resistance increase.

severely damaged area of a dog heart after 72 hours of coronary occlusion (52). Grayson et al. often measured a progressive decline in regional blood flow after coronary occlusion to near zero levels using a thermocouple embedded within the myocardium (53). Although these thermocouple measurements cannot be extended to generalizations about collateral flow, the regional nature of these flow measurements and the thermocouple location in the deeper, more vulnerable layers of the myocardium make this observation of a progressive decline in flow consistent with the hypothesis under consideration.

If the ischemia-induced time-dependent resistance change mechanism is operative in infarction evolution, it may be a major factor in the demarcation of infarction and could well explain the observation that, even in infarction that is not massive, cellular death and cellular survival occur not in scattered cells but instead in circumscribed foci (7, 37, 54, 55).

Competitive Redistribution of Coronary Flow

Hemodynamic Considerations   The possibility that changes in blood flow in the nonischemic myocardium might alter collateral flow to the ischemic area has been the subject of much investigation. A consideration of the hemodynamics of coronary occlusion suggests the probable existence of "coronary steal," i.e., a reciprocal relationship between flow to the donor area and flow to the recipient area (51). Figure 4 is a schematic representation of the dog left coronary arterial, capillary, and venous system upon which an acute increase in proximal circumflex resistance (R_s) has been imposed. A resistance proximal to the collateral system is identified for both the stenosed circumflex vessel ($R_s + R_{Cx}$) and for the normal anterior descending artery (R_{LAD}). Similarly, a resistance in each system distal to the collateral system is designated by a lower case r. To simplify the system for initial consideration, let us consider complete acute occlusion of the circumflex artery ($R_s = \infty$) in the normal dog. Ischemia is profound in this situation, with estimates of collateral flow varying from 8% of normal (retrograde flow method), to 15–20% of normal (radioactive micro-sphere method), to 15–30% of normal (radioactive isotope washout method) (12, 52). Infarction of at least the endocardial layers invariably occurs (5, 12). Since flows to both the endocardial and epicardial layers are reduced, the autoregulatory vessels in the circumflex distribution may be considered to be maximally dilated. Collateral vessels in this situation are small, averaging about 40 μm in diameter (12). Their walls are deficient in muscular components, and they are usually considered to act as passively distensible conduits (11, 12). Collateral resistance (R_C) and distal circumflex area resistance (r_{Cx}) in this simplified system are functions only of transmural pressure, increasing as trans-mural pressure decreases. These resistances will remain constant if intraluminal pressure and intramyocardial pressure are constant. Collateral flow is also con-stant if distal anterior descending pressure (P_{LAD}), right atrial pressure, and intramyocardial pressure are fixed. But suppose P_{LAD} decreases. Collateral resistance may increase since transmural pressure in the collateral system will fall. Distal circumflex pressure (P_{Cx}) will tend to fall as collateral resistance is increased, and distal circumflex resistance (r_{Cx}) will tend to rise as P_{Cx} falls. Therefore, any intervention that changes distal anterior descending pressure (P_{LAD}) may alter not only the pressure gradient along the collateral vessels, but also the resistance in the distal circumflex system. The effect of an alteration in distal anterior descending artery pressure (P_{LAD}) on collateral flow may be magnified by these resistance considerations.

A decrease in coronary perfusion pressure in dogs has been shown to lead to an increase in resistance both in large vessels and in small distal vasculature (56). This indicates that pressure induced distension of coronary vessels may signifi-cantly alter coronary resistance. But this conclusion is disputed by studies of pressure-flow relationships in coronary vasculature maximally dilated by reactive hyperemia. Pressure and flow (coronary flowmeter) during maximal dilation were linearly related over low pressure ranges (20–60 mm Hg), and alterations in resistance were small (57, 58). But pressure-flow studies at very low pressures using these techniques are difficult to interpret because collateral flow is neglected by the flowmeter.

Consideration of data from radioactive microsphere studies suggests that pressure-induced changes in coronary vascular diameter (and therefore in resis-tance) may indeed be important. According to Schaper's data (12), acute occlusion of a coronary artery in the dog leads to a pressure distal to the occlusion of about 15–20 mm Hg. Blood flow in the ischemic area is about 15% of nonischemic area flow. Since a pressure gradient of approximately 15% of normal in this situation leads to a regional blood flow of 15%, regional resistance is approximately normal (15%/15%). But these vessels are in a cardiac area of profound ischemia. Why has resistance not fallen with autoregulatory dilation? The decrease in vascular diameter consequent to low perfusion pressure must have counterbalanced autoregulatory dilation so that resistance did not decrease. Similar results have been observed when coronary stenosis was used to decrease distal coronary pressure from 100 to 40 mm Hg. Resistance decreased from 122 to 105 mm Hg/ml/min (51). This resistance change is small when one considers that exercise can quadruple coronary flow at normal coronary pressures (a decrease in coronary resistance to 1/4 normal). Again, a transmural pressure related decrease in vascular diameter seems to have limited the efficacy of autoregulatory dilation at low distal coronary pressure.

The pressure gradient across the collateral/recipient vascular system is a critical determinant of collateral flow. If changes in flow to the donor area (q_{LAD} in Figure 4) are to influence collateral flow, distal anterior descending pressure (P_{LAD}) must be altered by changes in distal anterior descending resistance (r_{LAD}), and proximal resistance (R_{LAD}) must not be negligible compared to distal resistance (r_{LAD}). Winbury, Howe, and Hefner, in attempting to explain the mechanism of action of nitroglycerin, measured pressure gradients between the aorta and a small (0.5 mm) coronary artery under different flow conditions. They found that large vessel resistance was about 5% of total resistance under resting conditions and increased to about 20% of total resistance during reactive hyperemia (56). Since these distal coronary pressures were measured in 500-μm vessels, and collaterals average 40 μm in diameter in these hearts (12), P_{LAD} in Figure 4 is actually a pressure that is more distal in the vascular system than the pressure measured and must therefore be lower. It follows that R_{LAD} must be larger than the proximal resistance estimated by Winbury et al., and r_{LAD} must be smaller than the estimated distal resistance. Proximal resistance, then, is not negligible relative to distal resistance particularly when distal resistance becomes small in high flow situations. Therefore, P_{LAD} *can* be altered by changes in flow if the ratio between proximal and distal resistance is altered.

Evidence for Coronary Steal After Acute Coronary Occlusion in Normal Dogs   In 1959, Kattus and Gregg, using retrograde flow as an indicator of collateral flow, considered the possibility that coronary inflow alters collateral flow (16). Seeking to determine the participation of collateral vessels in reactive hyperemia, they found that, during reactive hyperemia in the donor coronary vasculature, retrograde flow measured from the recipient (ischemic) area was *decreased* in 21 of 24 dogs. They concluded that reactive hyperemia does not lead to dilation of collateral channels. But one cannot rule out a *reciprocal* relationship between metabolically determined flow to the donor area and collateral flow from Kattus and Gregg's study. Indeed, such a relationship is suggested by these data.

Vasodilators have been used to study the relationship between donor area flow and collateral flow. Dipiramidole, administered after acute coronary occlusion, causes no increase in collateral flow (using retrograde flow to estimate collateral flow) at the same time that donor area flow is greatly increased (59). Flameng, Winston, and Schaper injected dipiramidole after mild circumflex stenosis (circumflex/aortic diastolic pressure = 0.9). Dipiramidole led to non-homogeneous blood flow in the circumflex region (using radioactive microspheres to estimate regional flow) and total recipient area flow was *increased.* But after severe stenosis (circumflex/aortic diastolic pressure = 0.56) dipiramidole not only exacerbated the nonhomogeneity of flow to the circumflex region but also *decreased* total recipient area flow. Interpretation of these results is complicated by the fact that aortic pressure decreased after dipiramidole, and the authors concluded that this decrease in aortic pressure was the cause of the observed decrease in recipient area flow (20). But an increase in R_{LAD}/r_{LAD} (Figure 4) with a consequent decrease in P_{LAD} and collateral flow may also have been a contributing factor.

Our own studies relating to coronary steal also utilized stenosis of the circumflex coronary artery to produce nonhomogeneous flow in the circumflex distribution (51). In an open-chest canine preparation, aortic pressure, heart rate, and adrenergic tone were maintained at constant levels. Left atrial pressure and cardiac output were adjusted to effect a minimum change in these two variables as the evolution of infarction altered contractility. Figure 4 is a schematic representation of the model used. The imposed resistance R_S was adjusted to decrease mean circumflex pressure to about 40 mm Hg and to effect S-T segment elevation in both subendocardial and epicardial electrograms. Mean aortic pressure and heart rate were maintained at 100 mm Hg and 150 beats/min respectively. The stenosis R_S was not changed over a 3-hour period, and the evolution of the infarction was observed with respect to epicardial and subendocardial electrograms, distal coronary pressure (P_S), regional coronary blood flow (estimated by radioactive microspheres), and ventricular performance.

Collateral flow increased initially in all the dogs studied. Between 5 and 30 min, flow to the circumflex area increased, distal coronary pressure increased, epicardial S-T elevation disappeared, endocardial S-T elevation persisted, and left ventricular function improved. But after 30 min, five dogs continued to improve while seven dogs behaved quite differently. The five dogs that improved all had 5-min circumflex regional flows greater than 0.4 ml g^{-1} min^{-1} while the seven dogs that failed to improve had 5-min flows less than 0.4 ml g^{-1} min^{-1}. Other differences between the two groups at 5 and 30 min were small. After 30 min, regional blood flow in the less ischemic group was unchanged over the ensuing 150 min. But more ischemic hearts suffered a decrease in circumflex flow and pressure and an increase in donor area flow compared to the less ischemic group ($p<0.05$). Left ventricular function in this group tended to deteriorate slightly, but changes in atrial pressure, cardiac output, and maximum DP/DT were small. The most important relationship observed was that in the more ischemic group changes in flow to the donor area were reciprocally related to changes in flow to the recipient area (using linear regression analysis, $r = 0.80$, $p<0.001$) and also to changes in distal coronary pressure (P_S) ($r = 0.79$, $p<0.001$). The failure of the reciprocal relationship in the less ischemic group to achieve statistical significance might have been predicted. Recipient area dilatory capacity in this group was not depleted (19, 25) and changes in flow to the recipient area may have been attenuated by autoregulation. But can the observation that a change in flow to the donor area reciprocally alters flow and pressure in the recipient area be extended to the conclusion that *collateral* flow is reciprocally altered? The only assumption necessary to make this extension is that flow across the stenosis is increased by an elevation of the pressure gradient across the stenosis ($P_{Ao} - P_S$ in Figure 4). Collateral flow is equal to recipient area flow (q_{Cx}) minus flow across the stenosis (Q_{Cx}), i.e., $Q_C = q_{Cx} - Q_{Cx}$. Then coincident decreases in both P_S and q_{Cx} necessarily imply an increase in pressure gradient across the stenosis ($P_{Ao} - P_S$), a consequent increase in flow across the stenosis (Q_{Cx}), and a decrease in collateral flow (= $q_{Cx} - Q_{Cx}$). A reciprocal relationship between donor area flow and *either* P_S or q_{Cx} could be explained by a metabolic mechanism without a change in collateral flow (e.g., decreased power from ischemic area increased power demand from nonischemic area increased flow to nonischemic area). But the fact that changes in donor area flow are reciprocally related to *both* pressure and flow in the recipient area can only be explained by a reciprocal relationship between changes in donor area flow and changes in collateral flow. These data, then, suggest that coronary steal may be a major mechanism in the temporal evolution of infarction in acute coronary stenosis in normal dogs. It should be noted that pharmacological intervention was not utilized in this study and that the changes in left ventricular function involved were sufficiently small to be below the level of clinical detection without special monitoring equipment.

Evidence for Coronary Steal After Chronic Coronary Occlusion in Dogs Chronic single vessel coronary occlusion with extensive collateral development presents an entirely different hemodynamic situation than acute occlusion. Collaterals are fewer, larger vessels (12), P_{LAD} (Figure 4) is more proximal in the vascular system, and proximal resistance (R_{LAD}) is probably smaller relative to distal resistance (r_{LAD}) than in the normal dog. P_{LAD}, then, may change less with changes in donor area flow than in the situation previously considered, tending to decrease the effect of coronary steal after chronic coronary occlusion. Collateral resistance, on the other hand, may be quite low, with as little as a 10–20 mm Hg pressure gradient across R_c, allowing normal flow to the recipient area. This low resistance might exacerbate coronary steal. A third factor, which would tend to decrease coronary steal, is that coronary pressure in the recipient area is relatively high. This implies that some dilatory reserve is available in the recipient area, and that autoregulation might attenuate the effects of coronary steal.

The relationship of donor area flow to collateral flow has been studied in dogs after ameroid occlusion of the circumflex coronary artery. Fam and McGregor found that dipiramidole led to a reduction in retrograde flow from the circumflex artery, but aortic pressure decreased as well (59). Flameng, Schaper, and Lewis found that flow in the circumflex region (by radioactive microsphere measurements), after ameroid constriction of the circumflex and right coronary arteries, was able to respond normally to an increase in heart rate from 80 to 160 beats/min. But infusion of a potent vasodilator (e.g., dipiramidole) in these animals caused a smaller increase in flow in the circumflex region than in the anterior descending region. If rapid atrial pacing and the coronary vasodilator were combined, the increase in collateral flow was further retarded relative to the increase in the normal area, and marked nonhomogeneity of flow in the recipient area led to a fall in flow to the subendocardium (13). *Total* collateral flow has *not* been shown to be decreased by vasodilators or pacing in these microsphere studies, but the coronary reserve of the recipient area is sufficiently limited that vasodilators may lead to a fall in recipient area coronary pressure with consequent "steal" of blood from the subendocardium by the subepicardium. True coronary steal (i.e., a decrease in total collateral flow with an increase in donor area flow) has not been convincingly demonstrated after chronic coronary occlusion. But vasodilators, the intervention usually used to increase donor area flow, tend to decrease both proximal resistance and distal resistance (R_{LAD} and r_{LAD}, Figure 4). A metabolically mediated increase in donor area flow might decrease distal resistance more than proximal resistance and have a greater effect on P_{LAD} (and thus on collateral flow) than a vasodilator-mediated increase.

Human coronary disease differs from these experimental situations in that multiple vessel disease often exists at the time of infarction. The proximal resistance in the donor area (R_{LAD} in Figure 4) is usually larger relative to distal resistance (r_{LAD}) than in the animal models just discussed. This difference would tend to make changes in the proximal/distal resistance ratio greater as distal resistance is altered and thereby increase the effect of changes in donor area flow on collateral flow.

Definition of "Competitive Redistribution" "Coronary steal" usually refers to a decrease in collateral flow in association with an increase in donor area flow. Since subendocardial to subepicardial flow redistribution distal to a coronary stenosis may be sufficient to lead to infarction (25), this form of "steal" within the recipient area represents a second type of flow redistribution that must be considered as well. A third form of redistribution may occur when two stenoses are in series in a vascular system. Consider, for example, a proximal anterior descending stenosis combined with stenosis of one of its diagonal branches. An increase in total anterior descending flow in response to metabolic demands in this situation would increase the pressure gradient across the proximal stenosis and decrease the gradient across the stenosed diagonal vascular system, thus causing at least a relative decrease in flow to the diagonal system if not an absolute decrease in flow. *"Competitive redistribution of flow"* is a broader term than "coronary steal" and is used here to refer to *any impairment (relative or absolute) of local coronary flow in response to an increase in coronary flow in another region.*

Figure 6 illustrates a positive feedback mechanism involving competitive redistribution which may participate in infarct evolution. This mechanism involves compensatory increases in power supply from the nonischemic regions as power becomes less available from the ischemic area. Such a compensation has been demonstrated (2, 60). The changes in ventricular function which are associated with competitive redistribution may be small (51). If systolic bulging of the ischemic area occurs, that area not only may fail to make a power contribution, but may be a power liability and further increase the power demand from the remainder of the heart.

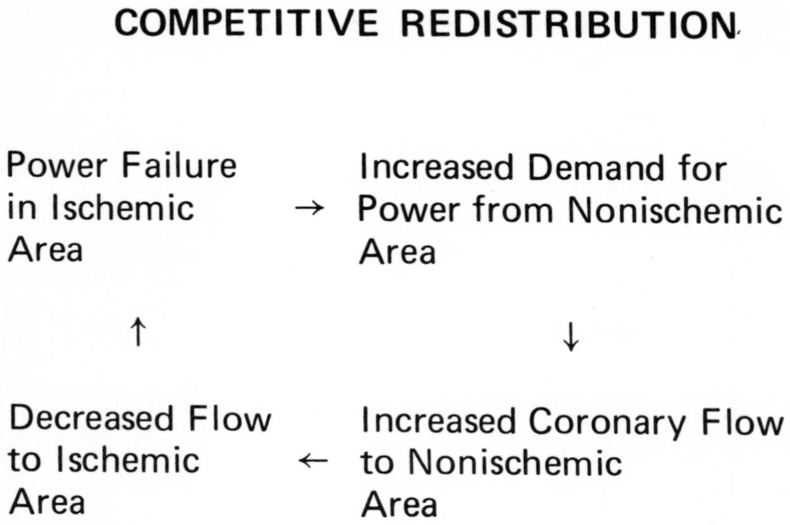

COMPETITIVE REDISTRIBUTION

Power Failure   Increased Demand for
in Ischemic   →   Power from Nonischemic
Area   Area

↑   ↓

Decreased Flow   Increased Coronary Flow
to Ischemic   ←   to Nonischemic
Area   Area

Figure 6. Competitive redistribution may lead to self-propagation of ischemia.

END OF INFARCTION

Infarction ends when the factors that tend to prevent further necrosis over-whelm the mechanisms that tend to propagate infarction. If inadequate coronary reserve implies a sufficient oxygen supply/demand imbalance to cause necrosis, then the *end of infarction necessarily means the restoration of adequate local coronary reserve* in the areas of jeopardized myocardium that have not been irreversibly damaged. Coronary reserve may be restored by increased oxygen supply, decreased oxygen demand, or necrosis of a portion of the jeopardized myocardium (Figure 7).

Increased Oxygen Supply

The most important single factor determining infarct size after coronary occlu-sion is the state of the collateral circulation at the time of occlusion. Abrupt occlusion of the circumflex coronary artery in the dog leads to infarction in 100% of the dogs and to death in more than 80% (12). If gradual occlusion is effected by an ameroid constrictor (complete occlusion in 50% of dogs by 2½ weeks), no infarction at all occurs in greater than 40% of dogs and mortality is reduced to 11% (12). Elliot et al. observed infarction on only one of five dogs in which occlusion was produced within 7 days (61). Collaterals develop rapidly in response to reduction in coronary reserve, and the rate of development is associated with the degree of ischemia. If the circumflex artery is constricted in the dog, a small pressure gradient (10 mm Hg) is sufficient to stimulate collateral development even though coronary inflow remains normal. Larger pressure gradients with impairment of arterial inflow lead to more rapid collateral development (11).

CORONARY RESERVE
MAY BE RESTORED BY

1. Increased Collateral Flow

↑ Competitive Redistribution

2. Decreased Oxygen Demand

3. Necrosis of a Portion of the Jeopardized Myocardium

Figure 7. Factors which tend to terminate infarction. A decrease in total myocardial oxygen demand may lead to increased flow to the ischemic area by competitive redistribution.

The changes in collateral flow that occur within the first few days after abrupt occlusion of the canine circumflex coronary artery are not clear. Peripheral coronary pressure (P_S in Figure 4) rises rapidly after occlusion (11), but the validity of this index of collateral development has been questioned (62). Certainly, if infarction occurs, the time-related increase in resistance that occurs in the ischemic vascular bed will tend to elevate peripheral coronary pressure (51). Measurements of collateral flow by precordial counting of xenon-133 washout in conscious dogs have demonstrated gradual increase in flow over the first day in one study (11) and no increase in flow for the first 3 days in a second study (52). Two problems greatly compromise the reliability of this technique for estimating collateral flow. A saline flush may not completely remove the injected xenon from the occluded artery. This results in an intra-arterial depot of radioactivity which is variably washed out by collateral flow (12). Second, regional resistance is markedly heterogeneous in severely ischemic myocardium (23, 51, 63). Xenon may therefore be selectively injected into low resistance (and high flow) subregions, leading to an overestimation of collateral flow. Microsphere estimation of collateral flow in conscious dogs revealed no increase in collateral flow in the first 6 hours after occlusion, but a substantial increase between 6 and 24 hours (63).

In our own studies, using radioactive microspheres in open-chest dogs, changes in collateral flow within 3 hours after coronary stenosis were reciprocally related to changes in flow to the donor area (51). There is general agreement that collateral blood supply is increased within less than a week. By this time there is histological and biochemical evidence for actual growth of collateral vessels (12). Collateral development will, therefore, eventually restore adequate coronary reserve to jeopardized muscle, but the time course of this restoration is not fully known for single vessel occlusion in the dog, much less for multiple vessel stenoses in man.

A second mechanism that augments oxygen supply during infarction is increased oxygen extraction. Mean coronary A-V O_2 difference may increase by 2–3 ml/100 ml in acute coronary occlusion (64). Since resting pO_2 is considerably higher in the subepicardium than in the subendocardium (27, 30, 32), increased oxygen extraction could potentially make a greater contribution to the outer layers of the heart than to the inner layers. A decreased hemoglobin affinity for oxygen has been demonstrated during myocardial infarction and this may help to increase oxygen extraction by the ischemic myocardium (65).

Decreased Oxygen Demand

In the critical first few days after infarction, myocardial oxygen demand offers an attractive target for pharmacologic intervention. A decrease in oxygen demand, if there is not a concomitant decrease in oxygen supply, obviously increases coronary reserve. But it is not clear how the determinants of oxygen demand, which have been defined in normal muscle, affect severely ischemic muscle. Suppose the contraction of an ischemic segment of myocardium is so limited by energy deficit that systolic expansion occurs. Changing the systolic intraventricular pressure will *not* necessarily alter the tension *developed* in the ischemic segment. It is therefore not obvious that increasing aortic pressure, against which the segment is failing to contract, will increase the energy utilized for contraction in the segment and thereby decrease the energy available for preservation of viability. Increases in heart rate, however, would increase the number of contractions per minute in the ischemic segment and probably would increase the oxygen supply/demand deficit. Changes in the contractile state of the heart have been shown to alter contractility in ischemic as well as normal segments (2, 66), and consequent changes in oxygen utilization in ischemic segments are probably qualitatively similar to those in normal segments. In the margins of the ischemic zone, one would expect that the application of the determinants of oxygen utilization would be quantitatively more accurate as one approaches more normal muscle.

One consequence of necrosis which may lead to a time-related decrease in overall myocardial oxygen demand is a progressive increase in wall stiffness in the infarcted area. This increase may diminish systolic bulging of the ischemic zone, and thus progressively decrease the power liability of such aneurysmal bulging (67).

Competitive redistribution relates, in a reciprocal manner, changes in myocardial oxygen demand in nonischemic areas to changes in blood flow in ischemic areas. Thus if overall power demand is decreased (e.g., by analgesia and sedation) after infarction, a consequent decrease in flow to nonischemic areas may lead to an *increase* in flow to ischemic areas. This is particularly true if the infarction was initiated by competitive redistribution. Consider, for example, a subendocardial infarction distal to a proximal coronary stenosis. Ischemia sufficient to cause infarction may be present in the subendocardium while *normal* flow is maintained in the subendocardium (25). If overall power demand is decreased, oxygen utilization in the subepicardium may decrease and lead to an autoregulatory increase in local coronary resistance (r_{CxP} in Figure 4). If r_{CxP} increases, distal coronary pressure (P_{Cx}) will increase and blood flow to the subendocardium (q_{CxN}) will tend to increase (following the pressure-flow relationship in Figure 1) even though total circumflex flow (q_{Cx}) may remain constant or even decrease. This reversal of the endocardial to epicardial coronary steal phenomenon leads to an increase in coronary reserve (supply/demand) in the subendocardium that is independent of any effect that the decrease in overall power demand may have had on local oxygen utilization. *Competitive redistribution, then, magnifies the effects of alterations in myocardial oxygen demand.* Moreover, the feedback loop in Figure 6 may be reversed as decreased power demand from the nonischemic area leads to increased flow and increased power contribution in the ischemic areas which in turn lead to decreased power demand from the nonischemic area, etc. The quantitative importance of competitive redistribution has been defined in only a few experimental situations, but in those situations it has been found to be functionally significant (10, 51). One must, therefore, not neglect its possible impact on the termination of infarction, and it is appropriate to include competitive redistribution in the list of factors tending to restore coronary reserve shown in Figure 7.

Necrosis of a Portion of the Jeopardized Myocardium

The increase in resistance that occurs in ischemic areas of the coronary vasculature has a second consequence that is as important as local self-propagation of ischemia. The resistance increase tends to restore coronary reserve in less ischemic portions of the jeopardized region. Consider again the partial thickness infarction distal to a proximal stenosis illustrated in Figure 4. As resistance in the infarcting subendocardium (r_{CxN}) increases, distal circumflex pressure (P_{Cx}) will rise, and flow to the less ischemic subepicardium (q_{CxP}) will tend to increase. Thus *ischemia in a region of the myocardium leads to demarcation of the tissue into a necrotic subregion and a region with adequate coronary reserve for survival.* Figure 8 summarizes this mechanism.

The existence of this ischemia-mediated mechanism for increasing coronary reserve in residual myocardium is undeniable. Intramural vasculature in necrotic areas is itself a victim of coagulation necrosis (54), and avascular areas after infarction are demonstrable by angiography (12) or at postmortem examination (54). There is abundant evidence that collateral flow does not decrease with time (11, 12, 63). Flow per unit mass in the residual myocardium must therefore increase as necrotic areas become avascular.

But is the time course of this mechanism compatible with the termination of infarction? Resistance in a severely ischemic area rises dramatically within 3 hours (51), and by 3 days there is no flow to the necrotic area (52). The kinetics of serum creatine phosphokinase concentration have been studied to assess infarct size. Creatine phosphokinase release seems to be minimal after about 14 hours in the dog (68), and after about 30 hours in uncomplicated human

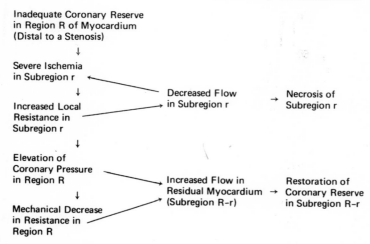

Figure 8. A possible mechanism for the demarcation of an ischemic area into a necrotic subregion and a subregion with adequate coronary reserve for survival.

infarction (69). Since extension of infarction is associated with additional creatine phosphokinase release (70), it seems that *extension of necrosis in uncomplicated human infarction has essentially ended in less than 30 hours.* Adequate coronary reserve, then, has been restored in the residual myocardium within about one day. Since large resistance changes in ischemic coronary vascular areas are demonstrable in 3 hours, this mechanism is certainly fast enough to participate in the termination of infarction. But all three of the mechanisms illustrated in Figure 7 probably combine to effect the restoration of coronary reserve that ends infarction.

Cox and his co-workers studied regional coronary flow (radioactive microspheres) in conscious dogs after anterior descending occlusion (Figure 9) (63). Total collateral flow increased over the first 24 hours. Flow in the subepicardium of the central zone and in both layers of the marginal area returned to near-normal levels. But similar return of flow failed to occur in the subendocardium in the central zone of the infarct. This exclusion of flow from the most ischemic portion of the infarct clearly directs flow to other areas. *Demarcation in this animal model seems to be nearly complete in 24 hours, with flow restored to normal in the areas destined to survive and flow excluded from the area destined to die.* Both an increase in total collateral flow and elevated resistance in the most ischemic subregion contributed to the restoration of coronary reserve in the residual myocardium. Once again, *sacrifice of the subendocardium contributes to survival of the subepicardium.*

In humans, the steady state established after the initial infarction event is a precarious one. Using precordial electrocardiographic monitoring, extension of necrosis was documented in at least 8 of 14 patients with transmural infarction. This extension occurred 2–15 days (average 5.8 days) after hospital admission

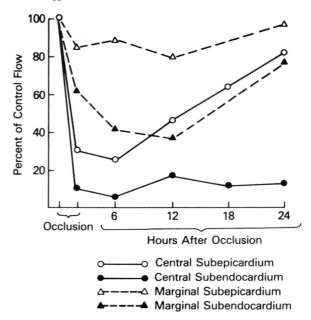

Figure 9. Temporal changes in regional coronary flow (radioactive microspheres) after occlusion of the anterior descending artery in conscious dogs. Reproduced from Cox et al. with kind permission from the authors and the American College of Surgery (63).

(55). It is tempting to speculate that ischemia-mediated demarcation initially leads to just enough necrosis to restore a barely adequate coronary reserve in the residual myocardium. Extension of necrosis may then occur with relatively small increases in oxygen demand or additional small decreases in oxygen supply during a *vulnerable period*. The duration of this vulnerable period is determined by the rate of anatomic growth of collateral vessels, which leads to sufficient excess coronary reserve to allow temporary increases in oxygen demand.

Prevention of ischemia-mediated resistance increase has been suggested as a goal of pharmacological intervention (43). But such prevention, if not accompanied by an augmentation of total flow to the region in jeopardy, could retard demarcation, expose mildly ischemic areas to a prolonged period of mild ischemia, and actually increase the eventual total area of necrosis. Studies of interventions in this mechanism should aim at reduction of eventual infarct size as well as diminution of early indicators of the severity of ischemia (such as S-T segment elevation).

EXPERIMENTAL INTERVENTION IN EVOLUTION OF NECROSIS

Methods for Estimating Infarct Size

Techniques for estimation of infarct size without altering the evolution of infarction have allowed correlation of various therapeutic interventions with the actual target variable. The first such technique to be widely applied was *epicardial S-T segment elevation* (71). Epicardial S-T segment elevation in the 15 min after occlusion is closely associated with local concomitant depression of intramyocardial pO_2 (4), local lactate accumulation and high energy phosphate depletion at 17 min (72), local decrease in subepicardial blood flow at 1 day (33), myocardial creatine phosphokinase depletion 1 day later (33, 73), and histological evidence of necrosis at 1 and 7 days after occlusion (71, 74, 75). But epicardial S-T segment elevations reflect subsequent subepicardial damage better than they reflect subendocardial damage (33), and the absence of epicardial S-T segment elevation in no way implies the absence of severe ischemia in deeper layers of the myocardium (25). Subendocardial S-T segment elevation may be produced without epicardial S-T segment elevation, and regional flow measurements in this situation reveal near-normal flow in the subepicardium despite severe ischemia in the subendocardium (25).

Epicardial electrograms (71, 73), intramyocardial recordings (76, 77), and the combination of epicardial and subendocardial electrograms (78, 79) have been used to estimate infarct size in dogs. Extension of the epicardial S-T segment technique to precordial S-T segment mapping may provide an atraumatic method for estimating infarct size in humans (80).

A second noninvasive technique for estimation of infarct size is derived from analysis of serial changes in serum creatine phosphokinase concentration. The total amount of creatine phosphokinase released has been found to be linearly related ($r = 0.96$) to subsequent myocardial creatine phosphokinase depletion in dogs (68). Myocardial creatine phosphokinase depletion is in turn associated with histological evidence of necrosis (73). Infarct size estimated by this technique in humans has been shown to be related to the incidence of power failure after infarction (69).

Alteration of Hemodynamic Parameters

Elevation of arterial pressure by methoxamine in open-chest dogs decreases S-T segment elevation after coronary occlusion (73). Hypotension produced by hemorrhage in the same experimental model leads to an increase in epicardial S-T segment elevation (73). Similar observations have been made using intramyocardial electrodes in sedated, closed-chest dogs (76) and using paired epicardial and subendocardial electrodes in open-chest dogs after ablation of autonomic reflexes (78). Changes in arterial pressure seem to alter epicardial S-T elevation more than subendocardial S-T segment elevation (78). Arterial pressure is a major determinant of both collateral flow and oxygen demand. *Elevation of arterial pressure presumably increases oxygen supply more than it elevates oxygen demand in the ischemic area.* In the heart whose function is acutely depressed by pharmacological agents, however, elevation of arterial pressure tends to cause left ventricular failure and leads to an *increase* in epicardial S-T segment elevation (81). In this situation, an increase in flow to the nonischemic myocardium may limit, by competitive redistribution, the pressure-induced increase in flow to the ischemic area. Johansson, Linder, and Seeman observed a decrease in collateral flow (krypton-85 washout) as aortic pressure was increased in only 1 of 17 dogs after acute coronary occlusion. Left ventricular end diastolic pressure at high aortic pressure in this animal was elevated to 32 mm Hg (17). Aortic pressure as an isolated variable seems reciprocally related to infarct size, but if ventricular function is impaired the deleterious effects of hypertension may lead to infarct extension.

Heart rate has been shown to be directly related to infarct size in open-chest dogs by epicardial S-T elevation (73), in closed-chest dogs by intramyocardial S-T elevation (76), and in hemodynamically controlled open-chest dogs after autonomic reflex ablation by epicardial and subendocardial S-T elevation (78). Even at low heart rates, infarct size may be reduced by further decreases in rate (76). Elevation of heart rate after initial demarcation of infarction may lead to substantial extension of necrosis as evidenced by renewed creatine phosphokinase release (70). Tachycardia does not seem to change collateral flow, but bradycardia may increase collateral flow (krypton-85 washout estimation) (17). When circumflex coronary flow in the dog is limited by proximal stenosis, tachycardia potentiates the redistribution of flow from the endocardium to the epicardium and may cause subendocardial ischemia (82). *An increase in heart rate elevates oxygen demand while not altering or, if anything, diminishing oxygen supply.*

Elevation of preload by volume expansion leads to augmentation of epicardial and subendocardial S-T segment elevation after coronary occlusion in open-chest dogs at constant aortic pressure and heart rate (78). A decrease in collateral flow subsequent to volume expansion has been demonstrated using radioactive microspheres (83). The subendocardium is especially compromised by elevation of preload. Elevation of end diastolic pressure leads to a decrease in subendocardial perfusion (estimated by dye injection) in dogs subject to coronary arterial hypotension (84). Kjekshus elevated preload by blood infusion after ligation of a branch of the left anterior descending artery. Augmented preload caused an increase in epicardial and endocardial flow in the nonischemic muscle, no change in epicardial flow in the ischemic area, and a 30% decrease in endocardial flow in the ischemic area (23). An elevated diastolic intramyocardial pressure may cause part of this effect by enhancement of endocardial to epicardial competitive redistribution. And increased flow to the surrounding normal muscle could theoretically cause total collateral flow to decrease, again by competitive redistribution. But elevated preload could also lead to an increase in collateral resistance or ischemic area vascular resistance which may diminish collateral flow independent of any "coronary steal" mechanism. *Elevated preload tends to increase oxygen demand and decrease oxygen supply in an ischemic area, with particularly deleterious effects upon the subendocardium.*

Inotropic Interventions

After coronary occlusion, isoproterenol, digitalis, and glucagon all lead to an increase in the magnitude and extent of epicardial S-T segment elevation in the nonfailing heart (71, 73, 79). Isoproterenol infusion leads to extension of necrosis as determined by creatine phosphokinase release after infarction (70) and by myocardial creatine phosphokinase depletion hours after infarction (71). Isoproterenol has been shown to increase contractile force in the center of ischemic areas while increasing the extent of S-T segment elevation and of contractile abnormality (66). The deleterious effects of positive inotropic agents after infarction have been attributed primarily to a local increase in oxygen demand (71, 73). Microsphere studies of the effect of isoproterenol on ischemic area flow have had variable results. In open-chest dogs after occlusion of multiple anterior descending branches, isoproterenol infusion led to a small increase in ischemic area flow (85). In conscious dogs with anterior descending occlusion, isoproterenol caused a decrease in ischemic area flow both 4 hours and 1 week after infarction (86). In open-chest dogs with circumflex artery stenosis, isoproterenol caused either an increase or a decrease in ischemic area flow, depending upon its hemodynamic effects (87).

In the heart whose function has been acutely depressed by drugs, digitalis has been shown to improve left ventricular function sufficiently to lead to a *reduction* of epicardial S-T segment elevation. In this situation, changes in ventricular size seem to reduce oxygen demand more than demand is increased by the direct effects of digitalis. Again, both decreased demand in the ischemic area and alterations in collateral flow secondary to changes in oxygen demand and coronary flow in the nonischemic area may contribute to this improvement. Isoproterenol, when administered to this acutely failing infarct model, has variable effects on infarct size (81).

Propranolol competitively blocks β-adrenergic receptors. It causes a decrease in the three major hemodynamic determinants of oxygen consumption and a decrease in oxygen consumption and coronary flow in animals and man (88). The effects of propranolol on preload vary with dosage (88). Propranolol decreases S-T elevation after acute coronary occlusion and diminishes the area of myocardial creatine phosphokinase depletion present 24 hours later (71, 73, 79). Gross and histological estimation of the extent of necrosis 2–5 days after a 40-min coronary occlusion demonstrated a dramatic reduction in necrosis in dogs pretreated with propranolol (89). In humans, propranolol administered 6–8 hours following hospital admission for acute transmural infarction seems to improve abnormal myocardial oxygenation (90). Propranolol clearly decreases total myocardial oxygen demand, although its effect on oxygen consumption in the ischemic area is not known (89). Data on the effect of propranolol on collateral flow, like that concerning isoproterenol, are variable, but generally ischemic area flow remains constant or decreases slightly while flow to the nonischemic area decreases markedly (91, 92). β-Adrenergic blockade by propranolol seems beneficial in human and auimal hearts after infarction, providing ventricular failure is absent.

Maroko, Libby, and Braunwald (93) studied the relation between reduction in left main coronary flow and left ventricular failure in open-chest dogs. Isoproterenol infusion led to severe left ventricular failure (left atrial pressure greater than 20 mm Hg) at a *higher* coronary flow than prior to infusion. Propranolol, on the other hand, allowed coronary flow *lower* than control values before failure occurred (84). Thus, as far as power production is concerned, the left ventricle becomes progressively less tolerant of reductions in coronary flow as inotropic stimulation is increased.

Vasodilator Therapy

Peripheral vasodilation may decrease left ventricular wall tension and thus decrease myocardial oxygen demand after myocardial infarction. In patients with elevated ventricular filling pressure (greater than 15 mm Hg) after infarction, treatment with phentolamine or nitroprusside led to a fall in filling pressure, improvement in cardiac output, and only a small decrease in aortic pressure. Coronary sinus flow did not rise, indicating no metabolic cost for this improvement in ventricular function (94). Infarct size, estimated by serial creatine phosphokinase levels in a group of hypertensive patients, was reduced by infusion of trimethaphan. Moreover, mortality in the treated group was lower than that in a control group of nonhypertensive patients with matched predicted infarct size (95).

Nitroglycerin

In the open chest, hemodynamically controlled dogs, nitroglycerin infusion does not decrease S-T segment elevation after coronary occlusion (96). In a series of experiments on conscious dogs, however, Epstein and co-workers found a significant reduction of S-T segment elevation after nitroglycerin, and this reduction was augmented if arterial pressure was maintained by adding methoxamine infusion (77). This combination of nitroglycerin and methoxamine was also shown to reduce myocardial damage (determined by myocardial creatine phosphokinase depletion at 24 hours) consequent to 5 hours of coronary occlusion in conscious dogs. In dogs with chronic multivessel occlusions, nitroglycerin and methoxamine led to a diminution of S-T segment elevation after additional acute coronary occlusion (77). The discrepancy between the negative results of the hemodynamically controlled open-chest study and the encouraging results found in Epstein's closed-chest dogs suggests that the noncardiac effects of nitroglycerin may be important. Although methoxamine reversed the hypotension and tachycardia consequent to nitroglycerin in closed-chest dogs, a change in ventricular filling pressure persisted; but, unfortunately, the contractile state of the heart was not assessed (77).

Nitroglycerin infusion in patients with left ventricular failure after infarction led to marked improvement in ventricular filling pressure with little change in aortic pressure or stroke work. A decrease in precordial S-T segment elevation was also demonstrated (97). The effects of nitroglycerin upon systemic hemodynamics in this study were so great, however, that no statement can be made about the direct effect of the nitroglycerin on the heart.

The mechanism of action of nitroglycerin is not clear. Its systemic effects reduce the hemodynamic determinants of oxygen consumption. But this effect is common with other vasodilators which are relatively ineffective in the treatment of angina. It seems reasonable, therefore, to expect that nitroglycerin has some beneficial effect on oxygen supply independent of its effect on oxygen demand. After chronic coronary occlusion in dogs, retrograde flow increases after nitroglycerin if aortic pressure is held constant (59). If aortic pressure is allowed to fall, retrograde flow does not significantly decrease (59), nor does collateral flow estimated by microsphere techniques (98). In humans with coronary disease, measurements made at the time of coronary bypass operation indicate a similar increase in collateral flow estimated by xenon-133 washout (99) or retrograde flow (100).

In normal dogs, whose collateral circulation is not nearly as well developed as in those with chronic occlusion, nitroglycerin after acute coronary occlusion has variable effects on collateral flow. Retrograde flow has been shown to fall after nitroglycerin in one study (59), while another observer, who maintained aortic pressure, observed a small increase (101). Peripheral coronary pressure was unchanged by nitroglycerin in the 24 hours following acute coronary occlusion (102).

Nitroglycerin seems to improve selectively endocardial oxygenation (estimated by intramyocardial pO_2) in both normal and ischemic areas after coronary occlusion (30). Regional coronary flow studies have demonstrated an increase (103), no change (98), and a decrease (104) in the subendocardial/subepicardial flow ratio if nitroglycerin is administered while coronary inflow is limited. But differences in initial conditions may explain this apparent discrepancy. Autoregulation may persist in the epicardial region even at low coronary pressures while the endocardium seems to maximally dilate after only a moderate reduction in coronary pressure (19, 25). Relative regional perfusion and oxygenation before and after an intervention will therefore depend on the degree of ischemia initially present. For example, in the situation illustrated in Figure 1, an elevation of coronary pressure from 35 to 45 mm Hg will have very different relative effects on the subendocardium and subepicardium than will an elevation from 45 to 55 mm Hg.

An interesting mechanism has been suggested for the efficacy of nitroglycerin relative to other vasodilators in regard to collateral flow (29, 30, 56, 105, 106). Nitroglycerin seems to dilate selectively large coronary vessels, while other vasodilators act primarily on autoregulatory vessels (105). The dilatory effect of nitroglycerin might be easily overcome by autoregulatory mechanisms while other vasodilators lead to dilation that is resistant to autoregulatory constriction (105, 106). Nitroglycerin, then, could improve collateral flow by decreasing the proximal/distal resistance ratio (R_{LAD}/r_{LAD} in Figure 4), or by decreasing collateral resistance (R_C).

Hyaluronidase

Hyaluronidase may facilitate the transport of substrate through the extracellular space by its ability to depolymerize hyaluronic acid. It has been found to diminish substantially both S-T segment elevation and subsequent myocardial creatine phosphokinase depletion after coronary occlusion in the dog (71).

Glucose-Potassium-Insulin

In an attempt to augment energy production by anaerobic metabolism, glucose-potassium-insulin infusion has been utilized after infarction. In dogs, a clear reduction in subsequent myocardial creatine phosphokinase depletion has been observed, but clinical studies have not demonstrated uniform benefit (71).

Steroid Treatment

Hydrocortisone, administered 30 min after coronary occlusion, diminishes subsequent infarct size (by creatine phosphokinase depression and histology). The membrane and lysozomal stabilizing properties of corticosteroids may be the source of this effect. Clinical trials are again inconclusive (71).

Mannitol

Hypertonic mannitol, infused before and during coronary occlusion, decreases S-T segment elevation in the area of occlusion (107) and increases collateral flow to the area (krypton-86 washout and radioactive microspheres). Augmentation of left ventricular function and of total coronary flow was observed as it has been after mannitol infusion in normal hearts (107, 108). A reversal of cellular swelling by mannitol has been suggested as the primary mechanism for this improvement (43), but the demonstration of improvement after only fifteen minutes of ischemia (107) makes this hypothesis unlikely.

Counterpulsation

Counterpulsation is the removal of arterial blood during systole and subsequent infusion during diastole (109). One counterpulsation device, the intra-aortic balloon pump, has had wide clinical application. After acute coronary occlusion in dogs, counterpulsation leads to a decrease in S-T segment elevation, even when instituted as long as three hours after occlusion (110). An increase in the developed tension in the ischemic area has been demonstrated, concomitant with a decrease in the magnitude and extent of S-T segment elevation (66). The extent of necrosis in dogs (histologic determination) after acute coronary occlusion is diminished by counterpulsation (111). Intra-aortic balloon pumping in patients can alleviate crescendo angina that is refractory to medical therapy (112). Cardiogenic shock after infarction can be reversed by balloon pumping in 80% of patients, but only a small portion of these patients can subsequently be withdrawn from counterpulsation therapy (113). Counterpulsation affects both oxygen supply and oxygen demand. Systolic unloading of the left ventricle decreases myocardial oxygen demand (114). In the failing heart both afterload and preload are decreased (114). Retrograde flow in open-chest dogs after acute coronary occlusion is linearly related to the elevation of diastolic pressure caused by intra-aortic balloon pumping (18). Collateral flow measured by radioactive microspheres is increased by counterpulsation, but a delay between coronary occlusion and the beginning of counterpulsation may limit this effect (115–117).

Coronary Revascularization

Coronary artery bypass surgery may be performed in elective patients with a mortality of less than 5% (118, 119) and similar mortality figures have been achieved in a group of patients with impending or established infarction (120, 121). The benefit of early reperfusion on infarct size is critically dependent upon very early intervention. Reperfusion within 3 hours of occlusion in dogs results in an overall decrease in infarct size (by histology and myocardial creatine phosphokinase depletion) (74, 75). Reperfusion at intervals up to 6 hours leads to progressively decreasing improvement in electrophysiological derangements consequent to infarction in dogs (122). Reperfusion after 3–5 hours of occlusion not infrequently leads to extension of infarction associated with hemorrhage into the infarcted area (49, 50). Local contractile impairment *subsequent* to reperfusion has been documented after as little as 1–2 hours of occlusion (123). The rapid demarcation of uncomplicated infarction (in less than 30 hours in humans) with increasing resistance to coronary inflow in ischemic areas suggests that salvage of jeopardized myocardium will be minimal if revascularization cannot be accomplished within 1 day after infarction. Pharmacological interventions and counterpulsation are beneficial if instituted within 3–6 hours after uncomplicated infarction (71, 73, 110), and these interventions may extend the period of potential benefit from revascularization. Berg and his co-workers have advocated emergency revascularization after uncomplicated infarction. Revascularization within 6–12 hours has a mortality similar to that of medical management, but infarct size has not been assessed (121).

Revascularization may have a greater role in the treatment of cardiogenic shock, since infarction in this situation is progressive. Fifty-two patients in cardiogenic shock, refractory to medical treatment and stabilized by intra-aortic balloon counterpulsation, would not tolerate withdrawal from counterpulsation and underwent coronary angiography. Of 24 patients who were selected for operation, nine survived. Mortality in this group of 24 patients was reduced from 100 to 63% by this aggressive management (113). The use of counterpulsation and emergency revascularization earlier in the course of developing ischemia (prior to the onset of cardiogenic shock) may significantly limit the progression of necrosis.

CONCLUSION

Myocardial infarction is the consequence of oxygen supply/demand discrepancy and may be precipitated by alteration in either supply or demand. If the cumulative myocardial loss from chronic and acute infarction is large, cardiogenic shock may lead to progressive extension of infarction and death. If the residual myocardium is adequate to maintain hemodynamic stability, the heterogeneous distribution of myocardial ischemia and a time-related increase in coronary resistance in severely ischemic areas lead to rapid demarcation of an ischemic region into a progressively more ischemic subregion and a progressively less ischemic subregion. Human monitoring techniques are available that are directly related to infarct size, and numerous therapeutic interventions experimentally reduce the extent of necrosis after infarction. But the rapidity with which demarcation occurs in uncomplicated infarction demands that therapy be instituted as soon as possible (certainly within 1 day after infarction) to achieve significant benefit. The high frequency of discrete infarction extension within the first few weeks after the initial event suggests that these interventions may have a role in the prevention of such extension even after initial demarcation has occurred.

ACKNOWLEDGMENT

The authors wish to acknowledge the excellent secretarial assistance of Mrs. Linda Johnson in the preparation of this manuscript.

REFERENCES

1. Tennant, R., and Wiggers, C.J. (1935). Effect of coronary occlusion on myocardial contraction. Amer. J. Physiol. 112:351.
2. Theroux, P., Franklin, D., Ross, J., and Kemper, W.S. (1974). Regional myocardial function during acute coronary occlusion and its modification by pharmacologic agents in the dog. Circ. Res. 35:896.
3. Banka, V.S., and Helfant, R.H. (1974). Temporal sequence of dynamic contractile characteristics in ischemic and nonischemic myocardium after acute coronary ligation. Amer. J. Cardiol. 34:158.
4. Sayen, J.J., Warner, F.S., Peirce, C., and Kuo, P.T. (1958). Polarographic oxygen, the epicardial electrocardiogram and muscle contraction in experimental acute regional ischemia of the left ventricle. Circ. Res. 6:779.
5. Jennings, R.B., and Ganote, C.E. (1974). Structural changes in myocardium during acute ischemia. Circ. Res. 34–35 (Suppl. III):156.
6. Buja, L.M., and Roberts, W.C. (1974). The coronary arteries and myocardium in acute myocardial infarction and shock. In R.M. Gunnar, H.S. Loeb, and S.H. Rahimtoola (eds.), Shock in Myocardial Infarction, p. 1. Grune and Stratton, New York.
7. Baroldi, G. (1972). Viewpoint of a human pathologist on the current coronary problem. In C.M. Bloor and R.A. Olsson (eds.), Vol. 39, p. 141, Current Topics in Coronary Research. Advances in Experimental Medicine and Biology, Plenum Press, New York.
8. Chandler, A.B., Chapman, I., Erhardt, L.R., Roberts, W.C., Schwartz, C.J., Sinapius, D., Spain, D.M., Sherry, S., Ness, P.M., and Simon, T.L. (1974). Coronary thrombosis in myocardial infarction. Report of a workshop on the role of coronary thrombosis in the pathogenesis of acute myocardial infarction. Amer. J. Cardiol. 34:823.
9. Becker, L.C., and Pitt, B. (1971). Coronary blood flow in conscious dogs with chronic coronary occlusion. Amer. J. Physiol. 221:1507.
10. Flameng, W., Wüsten, B., Winkler, B., Pasyk, S., and Schaper, W. (1975). Influence of perfusion pressure and heart rate on local myocardial flow in the collateralized heart with chronic coronary occlusion. Amer. Heart J. 89:51.
11. Gregg, D.E. (1974). The natural history of coronary collateral development. Circ. Res. 35:335.
12. Schaper, W. (1971). The Collateral Circulation of the Heart, p. 276. North-Holland, Amsterdam/London.
13. Flameng, W., and Schaper, W. (1973). Multiple coronary occlusion without infarction. Effects of heart rate and vasodilation. Amer. Heart J. 85:767.
14. Knoebel, S.B., McHenry, P.L., Bonner, A.J., and Phillips, J.F. (1973). Myocardial blood flow in coronary artery disease. Effect of right atrial pacing and nitroglycerin. Circulation 47:690.

15. Knoebel, S.B., Elliott, W.C., McHenry, P.L., and Ross, E. (1971). Myocardial blood flow in coronary artery disease. Amer. J. Cardiol. 27:51.
16. Kattus, A.A., and Gregg, D.E. (1959). Some determinants of coronary collateral blood flow in the open-chest dog. Circ. Res 7:628.
17. Johansson, B., Linder, E., and Seeman, T. (1966). Effects of heart rate and arterial blood pressure on coronary collateral flow in dogs. Acta Physiol. Scand. 68 (Suppl. 272):33.
18. Brown, B.G., Gundel, W.D., Gott, V.L., and Covell, J.W. (1974). Coronary collateral flow following acute coronary occlusion: A diastolic phenomenon. Cardiovasc. Res. 8:621.
19. Gould, K.L., Lipscomb, K., and Calvert, C. (1975). Compensatory changes of the distal coronary vascular bed during progressive coronary constriction. Circulation 51:1085.
20. Flameng, W., Wüsten, B., and Schaper, W. (1974). On the distribution of myocardial flow: Effects of arterial stenosis and vasodilation. Basic Res. Cardiol. 69:435.
21. Brazier, J., Cooper, N., and Buckberg, G. (1974). The adequacy of subendocardial oxygen delivery. The interaction of determinants of flow, arterial oxygen content, and myocardial oxygen need. Circulation 49:968.
22. Becker, L.D., Ferreira, R., and Thomas, M. (1973). Mapping of left ventricular blood flow with radioactive microspheres in experimental coronary artery occlusion. Cardiovasc. Res. 7:391.
23. Kjekshus, J.K. (1973). Mechanism for flow distribution in normal and ischemic myocardium during increased ventricular preload in the dog. Circ. Res. 33:489.
24. Timogiannakis, G., Amende, I., Martinez, E., and Thomas, M. (1974). ST segment deviation and regional myocardial blood flow during experimental partial coronary occlusion. Cardiovasc. Res. 8:469.
25. Guyton, R.A., Andrews, M.J., Ferrans, V.J., McClenathan, J.H., Newman, G.E., and Michaelis, L.L. Experimental subendocardial infarction I. S-T segment elevation in subendocardial electrograms associated with epicardial S-T segment depression, local ischemia, and subsequent necrosis. In preparation.
26. Hoffman, J.I.E., and Buckberg, G.D. (1974). Regional myocardial ischemia—causes, prediction and prevention. Vasc. Surg. 8:115.
27. Honig, C.R., Kirk, E.S., and Myers, W.W. (1966). Transmural distributions of blood flow, oxygen tension, and metabolism in myocardium: mechanism and adaptations. In International Symposium on the Coronary Circulation and the Energetics of the Myocardium, p. 31. Karger 1967, Basel/New York.
28. Armour, J.A., and Randall, W.C. (1971). Canine left ventricular intramyocardial pressures. Amer. J. Physiol. 220:1833.
29. Weiss, H.R., and Winbury, M.M. (1974). Nitroglycerin and chromonar on small vessel blood content of the ventricular walls. Amer. J. Physiol 226:838.
30. Winbury, M.M., Howe, B.B., and Weiss, H.R. (1971). Effect of nitroglycerin and dipyridamole on epicardial and endocardial oxygen tension—further evidence for redistribution of myocardial blood flow. J. Pharmacol. Exp. Ther. 176:184.
31. Griggs, D.M., Tchokoev, V.V., and Chen, C.C. (1972). Transmural differences in ventricular tissue substrate levels due to coronary constriction. Amer. J. Physiol. 222:705.
32. Moss, A.J. (1968). Intramyocardial oxygen tension. Cardiovasc. Res. 2:314.
33. Kjekshus, J.K., Maroko, P.R., and Sobel, B.E. (1972). Distribution of myocardial injury and its relation to epicardial ST segment changes after coronary artery occlusion in the dog. Cardiovasc. Res. 6, 490.
34. Braunwald, E. (1971). Control of myocardial oxygen consumption: Physiologic and clinical considerations. Amer. J. Cardiol. 27:416.
35. Guyton, R.A., Andrews, M.J., Ferrans, V.J., McClenathan, J.H., Newman, G.E., and D.N. Ghista, and H. Sandler, (eds.), Cardiac Mechanics: Physiological, Clinical, and Mathematical Considerations, p. 113. Wiley, New York.
36. McDonald, T.F., and MacLeod, D.P. (1973). Metabolism and the electrical activity of anoxic ventricular muscle. J. Physiol. (Lond.) 229:559.
37. Page, D.L., Caulfield, J.B., Kaston, J.A., DeSanctis, R.W., and Sanders, C.A. (1971). Myocardial changes associated with cardiogenic shock. New Eng. J. Med 285:133.
38. Kones, R.J. (1974). Cardiogenic Shock. Mechanism and Management. 386 pp. Futura, Mount Kisco, N.Y.

39. Ames, A., Wright, R.L., Kowada, M., Thurston, J.M., and Majno, G. (1968). Cerebral ischemic renal damage and the protective effect of hypertonic solute. J. Clin. Invest.
40. Summers, W.K., and Jamison, R.L. (1971). The no-reflow phenomenon in renal ischemia. Lab Invest. 25:635.
41. Flores, J., DiBona, D.R., Beck, C.H., and Leaf, A. (1972). The role of cell swelling in ischemic renal damage and the protective effect of hypertonic solute. J. Clin. Invest. 51:118.
42. Chang, J., Kowada, M., Ames, A., Wright, R.L., and Majno, G. (1968). Cerebral ischemia. III. Vascular changes. Amer. J. Pathol. 52:455.
43. Leaf, A. (1973). Cell swelling. A factor in ischemic tissue injury. Circulation 48:455.
44. Krug, A., de Rochemont, W., and Korb, G. (1966). Blood supply of the myocardium after temporary coronary occlusion. Circ. Res. 19:57.
45. Beller, G.A., Smith, T.W., and Hood, W.B. (1974). Altered regional myocardial blood flow following coronary reperfusion in acute myocardial ischemia. Clin. Res. 22:262A.
46. Bloor, C.M. (1972). Coronary Artery Reperfusion: Early Effects on Coronary Hemodynamics. In C.M. Bloor and R.A. Olsson (eds.), Current Topics in Coronary Research. Advances in Experimental Medicine and Biology, Vol. 39, 279. Plenum Press, New York.
47. Jennings, R.B., Sommers, H.M., Kaltenbach, J.P., and West, J.J. (1963). Electrolyte alterations in acute myocardial ischemic injury. Circ. Res. 14:260.
48. Shen, A.C., and Jennings, R.B. (1972). Myocardial calcium and magnesium in acute ischaemic injury. Amer. J. Pathol. 67:417.
49. Bresnahan, G.F., Roberts, R., Shell, W.E., Ross, J., and Sobel, B.E. (1974). Deleterious effects due to hemorrhage after myocardial reperfusion. Amer. J. Cardiol. 33:82.
50. Lang, T.W., Corday, E., Gold, H., Merrbaum, S., Rubins, S., Costantini, C., Hirose, S., Osher, J., and Rosen, V. (1974). Consequences of reperfusion after coronary occlusion. Effects on hemodynamic and regional myocardial metabolic function. Amer. J. Cardiol. 33:69.
51. Guyton, R.A., McClenathan, J.H., and Michaelis, L.L. Partial thickness infarction: Evidence for ischemia-induced increase in local resistance and for a coronary steal phenomenon. In preparation.
52. Rees, J.R., and Redding, V.J. (1967). Anastomotic blood flow in experimental myocardial infarction. A new method, using [133]Xenon clearance, for repeated measurements during recovery. Cardiovasc. Res. 1:169.
53. Grayson, J., and Lapin, B.A. (1966). Observations on the mechanisms of infarction in the dog after experimental occlusion of the coronary artery. Lancet 1284.
54. Baroldi, G., and Scomazzoni, G. (1965). Coronary Circulation in the Normal and the Pathologic Heart, p. 173. U.S. Government Printing Office, Washington.
55. Reid, P.R., Taylor, D.R., Kelley, D.T., Weisfeldt, M.L., Humphries, J.O., Ross, R.S., and Pitt, B. (1974). Myocardial infarct extension detected by precordial ST-segment mapping. New Eng. J. Med. 290:123.
56. Winbury, W.M., Howe, B.B., and Hefner, M.A. (1969). Effect of nitrates and other coronary dilators on large and small coronary vessels: an hypothesis for the mechanism of action of nitrates. J. Pharmacol. Exp. Ther. 168:70.
57. Fam, W.M., and McGregor, M. (1969). Pressure-flow relationships in the coronary circulation. Circ. Res. 25:293.
58. Mosher, P., Ross, J., McFate, P.A., and Shaw, R.F. (1964). Control of coronary blood flow by an autoregulatory mechanism. Circ. Res. 14:250.
59. Fam, W.M., and McGregor, M. (1964). Effects of coronary vasodilator drugs on retrograde flow in areas of chronic myocardial ischemia. Circ. Res. 15:355.
60. Hood, W.B., Jr. (1970). Experimental myocardial infarction. III. Recovery of left ventricular function in the healing phase: Contribution of increased fiber shortening in noninfarcted myocardium. Amer. Heart J. 79:531.
61. Elliot, E.C., Bloor, C.M., Jones, E.L., Mitchell, W.J., and Gregg, D.E. (1971). Effect of controlled coronary occlusion on collateral circulation in conscious dogs. Amer. J. Physiol. 220:857.
62. Bloor, C.M., and White, F.C. (1972). Functional development of the coronary collateral circulation during coronary artery occlusion in the conscious dog. Amer. J. Pathol. 67:483.

63. Cox, J.L., Wechsler, A.S., Oldham, H.N., and Sabiston, D.C. (1973). Evolution and transmural distribution of collateral blood flow in acute myocardial infarction. Surg. Forum 24:154.

64. Meerbaum, S., Lang, T.W., Corday, E., Rubins, S., Hirose, S., Costantini, C., Gold, H., and Dalmastro, M. (1974). Progressive alterations of cardiac hemodynamics and regional metabolic function after acute coronary occlusion. Amer. J. Cardiol. 33:60.

65. Kostuk, W.J., Suwa, K., Bernstein, E.F., and Sobel, B.E. (1973). Altered hemoglobin affinity in patients with acute myocardial infarction. Amer. J. Cardiol. 31:295.

66. Schelbert, H.R., Covell, J.W., Burns, J.W., Maroko, P.R., and Ross, J. (1971). Observations on factors affecting local forces in the left ventricular wall during acute myocardial ischemia. Circ. Res. 39:306.

67. Hood, W.B., Jr., Bianco, J.A., Kumar, R., and Whiting, R.B. (1970). Experimental myocardial infarction. IV. Reduction of left ventricular compliance in the healing phase. J. Clin. Invest. 49:1316.

68. Shell, W.E., Kjekshus, J.K., and Sobel, B.E. (1971). Quantitative assessment of the extent of myocardial infarction in the conscious dog by means of analysis of serial changes in serum creatine phosphokinase activity. J. Clin. Invest. 50:2614.

69. Sobel, B.E., Bresnahan, G.F., Shell, W.E., and Yoder, R.D. (1972). Estimation of infarct size in man and its relation to prognosis. Circulation 46:640.

70. Shell, W.E., and Sobel, B.E. (1973). Deleterious effects of increased heart rate on infarct size in the conscious dog. Amer. J. Cardiol. 31:474.

71. Maroko, P.R., and Braunwald, E. (1973). Modification of myocardial infarction size after coronary occlusion. Ann. Inter. Med. 79:720.

72. Karlsson, J., Templeton, G.H., and Willerson, J.T. (1973). Relationship between epicardial S-T segment changes and myocardial metabolism during coronary insufficiency. Circ. Res. 32:725.

73. Maroko, P.R., Kjekshus, J.K., Sobel, B.E., Watanabe, T., Covell, J.W., Ross, J., and Braunwald, E. (1971). Factors influencing infarct size following experimental coronary artery occlusion. Circulation 43:67.

74. Maroko, P.R., Libby, P., Ginks, W.R., Bloor, C.M., Shell, W.E., Sobel, B.E., and Ross, J. (1972). Coronary artery reperfusion. I. Early effects on local myocardial function and the extent of myocardial necrosis. J. Clin. Invest. 51:2710.

75. Ginks, W.R., Sybers, H.D., Maroko, P.R., Covell, J.W., Sobel, B.E., and Ross, J. (1972). Coronary artery reperfusion. II. Reduction of myocardial infarct size at one week after the coronary occlusion. J. Clin. Invest. 51:2717.

76. Redwood, D.R., Smith, E.R., and Epstein, S.E. (1972). Coronary artery occlusion in the conscious dog. Effects of alterations in heart rate and arterial pressure on the degree of myocardial ischemia. Circulation 46:323.

77. Epstein, S.E., Kent, K.M., Goldstein, R.E., Borer, J.S., and Redwood, D.R. (1975). Reduction of ischemic injury by nitroglycerin during acute myocardial infarction. New Eng. J. Med. 292:29.

78. Guyton, R.A. (1975). Subendocardial S-T segment changes during acute coronary occlusion. Ann. Thorac. Surg. 20:55.

79. Wendt, R.L., Canavan, R.C., and Michalak, R.J. (1974). Effects of various agents on regional ischemic myocardial injury: Electrocardiographic analysis. Amer. Heart J. 87:468.

80. Maroko, P.R., Libby, P., Covell, J.W., Sobel, B.E., Ross, I., and Braunwald, E. (1972). Precordial S-T segment elevation mapping: An atraumatic method for assessing alterations in the extent of myocardial ischemic injury. Amer. J. Cardiol. 29:223.

81. Watanabe, T., Covell, J.W., Maroko, P.R., Braunwald, E., and Ross, J. (1972). Effects of increased arterial pressure and positive inotropic agents on the severity of myocardial ischemia in the acutely depressed heart. Amer. J. Cardiol. 30:371.

82. Neill, W.A., Oxendine, J., Phelps, N., and Anderson, R.P. (1975). Subendocardial ischemia provoked by tachycardia in conscious dogs with coronary stenosis. Amer. J. Cardiol. 35:30.

83. Greene, L., Kelly, D., and Pitt, B. (1972). Changes in regional myocardial blood flow in acute myocardial infarction with plasma volume expansion. Circulation 46 (Suppl. 2):100.

84. Salisbury, P.F., Cross, C.E., and Rieben, P.A. (1963). Acute ischemia of inner layers of ventricular wall. Amer. Heart J. 66:650.

85. Becker, L.C., Ferreira, R., and Thomas, M. (1975). Effect of propranolol and isoprenaline on regional left ventricular blood flow in experimental myocardial ischaemia. Cardiovasc. Res. 9:178.

86. Sharma, G.V.R.K., Kumar, R., Molokhia, F., and Messer, J.V. (1971). "Coronary steal": Myocardial blood flow during isoproterenol infusion in acute and healing myocardial infarction. Clin. Res. 19:339.

87. McClenathan, J.H., Guyton, R.A., and Michaelis, L.L. The effect of isoproterenol on regional coronary blood flow in dogs after circumflex artery stenosis. In preparation.

88. Wolfson, S., and Gorlin, R. (1969). Cardiovascular pharmacology of propranolol in man. Circulation 40:501.

89. Reimer, K.A., Rasmussen, M.M., and Jennings, R.B. (1973). Reduction by propranolol of myocardial necrosis following temporary coronary artery occlusion in dogs. Circ. Res. 33:353.

90. Mueller, H.S., Ayres, S.M., Religa, A., and Evans, R.G. (1974). Propranolol in the treatment of acute myocardial infarction. Circulation 49:1078.

91. Pitt, B., and Craven, P. (1970). Effect of propranolol on regional myocardial blood flow in acute ischemia. Cardiovasc. Res. 4:176.

92. Becker, L.C., Fortuin, N.J., and Pitt, B. (1971). Effect of ischemia and antianginal drugs on the distribution of radioactive microspheres in the canine left ventricle. Circ. Res. 28:263.

93. Maroko, P.R., Libby, P., and Braunwald, E. (1973). Effect of pharmacologic agents on the function of the ischemic heart. Amer. J. Cardiol. 32:930.

94. Chatterjee, K., Parmley, W.W., Ganz, W., Forrester, J., Walinsky, M.D., Crexells, C., and Swan, H.J.C. (1973). Hemodynamic and metabolic responses to vasodilator therapy in acute myocardial infarction. Circulation 48:1183.

95. Shell, W.E., and Sobel, B.E. (1974). Protection of jeopardized ischemic myocardium by reduction of ventricular afterload. New Eng. J. Med. 291:481.

96. Bleifeld, W., Wende, W. Bussmann, W.D., and Meyer, J. (1973). Influence of nitroglycerin on the size of experimental myocardial infarction. Naunyn-Schmiedeberg's Arch. Pharmacol. 277:387.

97. Flaherty, J.T., Reid, P.R., Kelly, D.T., Taylor, D.R., Weisfeldt, M.L., and Pitt, B. (1975). Intravenous nitroglycerin in acute myocardial infarction. Circulation 51:132.

98. Schaper, W., Lewi, P., Flameng, W., and Gijpen, L. (1973). Myocardial steal produced by coronary vasodilation in chronic coronary artery occlusion. Basic Res. Cardiol. 58:3.

99. Horwitz, L.D., Gorlin, R., Taylor, W.J., and Kemp, H.G. (1971). Effects of nitroglycerin on regional myocardial blood flow in coronary artery disease. J. Clin. Invest. 50:1578.

100. Goldstein, R.E., Stinson, E.B., Scherer, J.L., Seningen, R.P., Grehl, T.M., and Epstein, S.E. (1974). Intraoperative coronary collateral function in patients with coronary occlusive disease. Nitroglycerin responsiveness and angiographic correlations. Circulation 49:298.

101. Leighninger, D.S., Rueger, R., and Beck, C.S. (1959). Effect of glyceryl trinitrate (nitroglycerin) on arterial blood supply to ischaemic myocardium. Amer. J. Cardiol. 3:638.

102. Gregg, D.E. (1972). Coronary vasodilator effects of nitroglycerin during coronary insufficiency. In G.G. Gensini (ed.), The Study of the Systemic, Coronary and Myocardial Effects of the Nitrates, p. 292. Charles C Thomas, Springfield, Ill.

103. Mathes, P., and Rival, J. (1971). Effect of nitroglycerin on total and regional coronary blood flow in the normal and ischaemic canine myocardium. Cardiovasc. Res. 5:54.

104. Forman, R., Kirk, E.S., Downey, J.M., and Sonnenblick, E.H. (1973). Nitroglycerin and heterogeneity of myocardial blood flow: Reduced subendocardial blood flow and ventricular contractile force. J. Clin. Invest. 52:905.

105. Fam, W.M., and McGregor, M. (1968). Effect of nitroglycerin and dipyridamole on regional coronary resistance. Circ. Res. 22:649.

106. McGregor, M., and Fam, W. (1972). On the site of vasomotion in the coronary vascular bed. In G.G. Gensini (ed.), The Study of the Systemic, Coronary and Mycardial Effects of the Nitrates, p. 323. Charles C Thomas, Springfield, Ill.

107. Willerson, J.T., Powell, W.J., Guiney, T.E., Stark, J.J., Sanders, C.A., and Leaf, A. (1972). Improvement in myocardial function and coronary blood flow in ischemic myocardium after mannitol. J. Clin. Invest. 51:2989.

108. Hutton, I., Marynick, S.P., Fixler, D.E., Templeton, G.H., and Willerson, J.T. (1975). Changes in regional coronary blood flow with hypertonic mannitol in conscious dogs. Cardiovasc. Res. 9:47.

109. Clauss, R.H., Birtwell, W.C., Albertal, G., Lunzer, S., Taylor, W.J., Fosberg, A.M., and Harken, D.E. (1961). Assisted circulation. I. The arterial counterpulsator. J. Thorac. Cardiovasc. Surg. 41:447.

110. Maroko, P.R., Bernstein, E.F., Libby, P., DeLaria, G.A., Covell, J.W., Ross, J., and Braunwald, E. (1972). Effects of intra-aortic balloon counterpulsation on the severity of myocardial ischemic injury following acute coronary occlusion: Counterpulsation and myocardial injury. Circulation 45:1150.

111. Goldfarb, D., Friesinger, G.C., Conti, C.R., Brown, B.G., and Gott, V.L. (1968). Preservation of myocardial viability by diastolic augmentation after ligation of the coronary artery in dogs. Surgery 63:320.

112. Gold, H.K., Leinbach, R.C., Sanders, C.A., Buckley, M.J., Mundth, E.D., and Austen, W.G. (1973). Intra-aortic balloon pumping for control of recurrent myocardial ischemia. Circulation 47:1197.

113. Leinbach, R.C., Gold, H.K., Dinsmore, R.E., Mundth, E.D., Buckley, M.J., Austen, W.G., and Sanders, C.A. (1973). The role of angiography in cardiogenic shock. Circulation 47–48 (Suppl. III):95.

114. Powell, W.J., Daggett, W.M., Magro, A.E., Bianco, J.A., Buckley, M.J., Sanders, C.A., Kantrowitz, A.R., and Austen, W.G. (1970). Effects of intra-aortic balloon counterpulsation on cardiac performance, oxygen consumption, and coronary blood flow in dogs. Circ. Res. 26:753.

115. Gill, C.C., Wechsler, A.S., Newman, G.E., and Oldham, H.N. (1973). Augmentation and redistribution of myocardial blood flow during acute ischemia by intra-aortic balloon pumping. Ann. Thorac. Surg. 16:445.

116. Watson, J.T., Willerson, J.T., Fixler, D.E., and Sugg, W.L. (1974). Temporal changes in collateral coronary blood flow in ischemic myocardium during intra-aortic balloon pumping. Circulation 49–50 (Suppl. II):249.

117. Reneman, R.S., Jageneau, A.H.M., Schaper, W.K.A., Brouwer, F.A.S., and Van Gerven, W. (1972). Influence of counterpulsation on collateral circulation after acute occlusion of the left anterior descending coronary artery in dogs. Cardiovasc. Res. 6:45.

118. Sheldon, W.C., Rincon, G., Effler, D.B., Proudfit, W.L., and Sones, F.M. (1973). Vein graft surgery for coronary artery disease. Survival and angiographic results in 1000 patients. Circulation 47–48 (Suppl. III):184.

119. Anderson, R.P., Rahimtoola, S.H., Bonchek, L.I., and Starr, A. (1974). The prognosis of patients with coronary artery disease after coronary bypass operations. Time related progress of 532 patients with disabling angina pectoris. Circulation 50:274.

120. Cheanvechai, C., Effler, D.B., Loop, F.D., Groves, L.K., Sheldon, W.C., Razavi, M., and Sones, F.M. (1973). Emergency myocardial revascularization. Amer. J. Cardiol. 31:125.

121. Berg, R., Rudy, L. W., Ganji, J.H., Kendall, R.W., Everhart, F.J., and Duvoisin (1975). Acute myocardial infarction: A surgical emergency. Presented at the 55th Annual Meeting of the American Association for Thoracic Surgery, April 14–16, New York.

122. Cox, J.L., Daniel, T.M., and Boineau, J.P. (1973). The electrophysiologic time-course of acute myocardial ischemia and the effects of early coronary artery reperfusion. Circulation 48:971.

123. Banka, V.S., Chadda, K.D., and Helfant, R.H. (1974). Limitations of myocardial revascularization in restoration of regional contractile abnormalities produced by coronary occlusion. Amer. J. Cardiol. 34:164.

International Review of Physiology
Cardiovascular Physiology II, Volume 9
Edited by Arthur C. Guyton and Allen W. Cowley
Copyright 1976   University Park Press   Baltimore

9
Integration and Control
of Circulatory Function

A. C. GUYTON, A. W. COWLEY, JR., D. B. YOUNG, T. G. COLEMAN, J. E. HALL, and J. W. DeCLUE
University of Mississippi Medical Center

INTRODUCTION

Too often after reading highly specialized reviews in subspecialty fields of the circulation the physiologist comes away with a highly distorted view of overall circulatory function. Therefore, it is important occasionally to discuss the way in which the parts of the circulatory system function together and how overall function is controlled.

For the past 25 years, the goal of the research in our laboratory has been to develop better understanding of the integration and control of the circulation. Therefore, the present review is based mainly on these studies but also on many related studies from other laboratories as well. Much of our work has been presented previously in the form of quantitative systems analyses of circulatory function (1–5). This chapter presents primarily the logic of the overall system.

BASIC GOAL OF CIRCULATORY CONTROL

The basic goal of the circulation is to provide blood flow to each tissue as needed to meet its requirements. This is achieved by a combination of local reactions of the blood vessels in the tissues as well as simultaneous gross adjustments of large portions of the circulation. For instance, during heavy exercise, local conditions in the muscles cause intense vasodilatation (6), thus allowing rapid blood flow in the muscles. Unfortunately, this also decreases the peripheral resistance, which leads to an immediate tendency for the arterial pressure to decrease (7); yet, normally this is more than offset by a host of nervous reactions that rapidly elevate the pressure (8). And, if for some reason the peripheral vasodilatation continues for days at a time, still other adjustments take place, such as retention of fluid by the kidneys to increase the blood volume, secretion of aldosterone to enhance the retention of salt, secretion of ADH to enhance the retention of water, and even hypertrophy of the heart if the stimulus lasts for weeks. Thus, all of the usual factors of circulatory control—control of cardiac output, control of arterial pressure, control of blood volume, and others—are geared toward the same end of providing appropriate blood flow to each tissue of the body according to its need.

Basic Organization of Circulatory Control—Intrinsic Versus Extrinsic Controls

Generally, the overall system for circulatory control can be divided into two different parts: 1) those controls that are intrinsic in the hemodynamic complex itself, and 2) those that are extrinsic to the hemodynamic components.

The intrinsic controls include automatic control of the heart by the Frank-Starling law of the heart, automatic equilibration of pressures on the two sides of the capillary membrane in accordance with the Starling principles of capillary exchange, automatic control of blood volume by the kidneys in response to changes in arterial pressure, automatic control of local blood flow in response to tissue metabolic demands, as well as others.

The extrinsic controls include the nervous reflexes—that are especially important in arterial pressure control—and the hormonal feedback systems—that are especially important in the control of body fluid composition.

In the subsequent discussions of this chapter, we shall see that the basic hemodynamic system with its own intrinsic controls is capable of functioning with a high degree of long-term stability even without the extrinsic controls. But, to provide the very rapid circulatory changes required during acute stress such as heavy exercise, hemorrhage, or even adjustments to myocardial damage, the rapidly acting circulatory reflexes are indispensable. And to provide fine degrees of control of sodium ion concentration, potassium ion concentration, and body fluid osmolality, the hormonal systems are equally necessary.

Now, let us look in much greater detail especially at the quantitative aspects of both the intrinsic and the extrinsic controls.

LESSONS TO BE LEARNED FROM CIRCULATORY FUNCTION OF THE HEADLESS OR OTHERWISE AREFLEXIC ANIMAL

One of the best ways to study the importance of both the intrinsic controls of the circulation and the extrinsic circulatory reflexes is to compare circulatory function in animals whose nervous reflexes are normal with that in animals having no reflexes. An expeditious way to prepare an animal without reflexes for acute studies lasting up to several days is to remove its head without loss of blood volume and to destroy its spinal cord by injecting alcohol into the spinal cord (9–11). Or a way to eliminate the circulatory reflexes for 30 min to 3 hours is to inject an anesthetic into the spinal canal, then run the anesthetic all the way up the canal into the subarachnoid spaces surrounding the brain (12–13). If a sufficient quantity of anesthetic is administered, all of the outflow nerves, both the spinal nerves and the cranial nerves, become anesthetized, and all reflexes are lost. Obviously, in both these instances, respiration is also lost so that the animal must be maintained on artificial respiration.

The most important acute effect of paralyzing the nervous outflow to the body is an acute fall in arterial pressure, caused principally by decreased peripheral arterial resistance. In the acute headless preparations, the arterial pressure can be returned artificially to normal by infusing a vasoconstrictor drug or a combination of drugs such as norepinephrine, epinephrine, vasopressin, or angiotensin. If the rate of infusion is maintained at an appropriate and constant level, normal vasomotor tone and normal arterial pressure will return, but neither of these can any longer respond to reflexes. Thus, the animal is areflexic.

It must also be noted that when human beings suffer brain death, either as a result of pathology in the brain or as a result of trauma, many of these persons are kept alive for days or even weeks with a circulatory system that is functioning in an areflexic state.

General Characteristics of Circulatory Function in the Areflexic State

The most striking characteristic of circulatory function in the areflexic state is its almost complete normalcy under quiet resting conditions. On the other hand, disturbances to the animal, such as rolling the animal from side to side, holding its head upright, removing only a few milliliters of blood, pressing on the abdomen, or changing the body temperature, all have profound effects on the circulation, much more so than in the animal with normal reflexes (9–13).

Capability of the Areflex Circulation to
Control Local Blood Flow and Cardiac Output—Whole Body Autoregulation

In the animal without circulatory reflexes, the local factors in the tissues that control the degree of vasoconstriction are uninhibited in their function. The blood vessels dilate in response to oxygen deficiency (14, 15), they dilate in response to muscle activity (16), and they dilate in response to tissue ischemia (17).

Especially striking has been the demonstration that whole body "autoregulation" occurs in the areflexic circulation and that this is an automatic mechanism for controlling cardiac output in the entire animal (14, 15, 18–20). Let us explain. When an areflexic animal's arterial pressure is artificially decreased by removing blood from the circulation, the cardiac output at first falls by almost exactly the same per cent that the arterial pressure decreases. However, within 10–30 min the blood vessels throughout the entire body dilate and the cardiac output returns about three-quarters of the way back toward its original level. This effect is illustrated in Figure 1. Note in this figure that the arterial pressure was decreased to approximately 50% of normal and was kept at that level for the remainder of the experiment. The cardiac output at first also fell to 50% of normal but then returned to 87% of normal despite the fact that the arterial pressure remained at the low level. Note also in the second curve of Figure 1 that total oxygen consumption by the animal decreased markedly when the arterial pressure was first lowered. However, O_2 consumption returned even more rapidly and more completely than did cardiac output, returning so near to normal that one could not tell that oxygen consumption had been reduced after approximately 30 min.

Thus, even in an animal with no reflexes, the local tissue mechanisms for control of blood flow still function. Furthermore, when they function to control blood flow throughout the entire body, they also automatically control the cardiac output. That is, the blood flowing from the tissues enters the heart and is pumped back around the circulation again.

When the same experiment as that illustrated in Figure 1 is repeated in an animal with functioning nervous reflexes, the results are entirely different because the reflexes are geared toward trying to return the arterial pressure back to normal. The reflexes constrict the peripheral blood vessels more and more, and the cardiac output, instead of returning back toward normal, actually

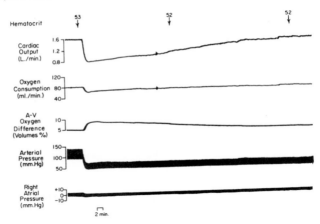

Figure 1. Autoregulation of cardiac output in a headless dog. Arterial pressure was decreased suddenly to one-half normal and held at this level by removing blood from the circulation. Note the essentially complete recovery of oxygen consumption and about 84% recovery of cardiac output. (Reprinted from Granger et al. (1969). Circ. Res. 25:379. By permission of the American Heart Association, Inc.)

becomes reduced still more (21). Also, oxygen consumption can suffer severely, as has been demonstrated in many studies of circulatory shock induced by bleeding an animal down to a fixed pressure level of 30–40 mm Hg and maintaining the pressure at that level. However, after a period of hours, and especially after several days, the local control factors begin to break through the reflexes so that in the long run the local control factors become dominant (22–25). Yet, for many hours the circulatory reflexes and the local control factors can remain in severe conflict, which often is severely damaging to an animal or human being in shock—causing such effects as ischemic necrosis of the kidneys, ischemic damage of the liver, and even ischemic damage of the heart itself (26–29).

Control of Arterial Pressure and Blood Volume in the Areflexic Circulation—Role of the Kidneys

The areflexic circulation can also control arterial pressure and blood volume. This is demonstrated in Figure 2 which compares the responses of the circulation in the normal and areflexic dogs to rapid infusion of whole blood equal to approximately 30% of the animal's control blood volume (9, 10, 30). Note that the increased blood volume caused increases in arterial pressure, cardiac output, and urinary output in both the normal and the areflex animals. However, two differences are readily apparent: First, the effects on arterial pressure were seven to eight times as great in the areflex animal as in the normal, indicating very poor acute control of arterial pressure in the areflex animal. Second, though the urinary output increased markedly in both types of animals, the output increased immediately in the areflex animal but rose to its maximum rate only after an hour's delay in the normal animal.

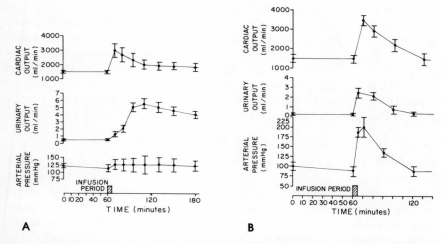

Figure 2. Changes in arterial pressure, cardiac output, and urinary output following infusion of 300 ml of autologous blood into (A) normal dogs and (B) headless dogs. Note readjustment of arterial pressure and cardiac output as fluid was lost from the circulation into the urine. Also note the tremendous increase in arterial pressure in the headless dogs. (Modified from W.A. Dobbs, Ph.D. dissertation, Jackson, Mississippi, University of Mississippi, 1970.)

Now, let us concentrate on the results in the areflex animal alone for a moment. The increase in blood volume obviously created a hyperdynamic circulatory state, with excess venous return to the heart, increased cardiac output, and increased arterial pressure. The increase in arterial pressure in turn had a direct and potent effect on the kidneys to increase urinary output, which has been demonstrated many times in isolated kidneys in many laboratories (31–35), and which is shown very potently in Figure 2B. The rapid excretion of urine decreased the body fluid volumes and along with these the cardiac output and arterial pressure as well. Note that as soon as the arterial pressure returned to normal in Figure 2B, so also had the urinary output returned to normal. Thus, even the areflexic circulation has the capability of controlling its own blood volume and thereby controlling its arterial pressure simultaneously.

The converse experiment can also be performed: A decrease in blood volume causes immediate decrease in urinary output. If there is a source of fluid for the animal, the fluid volume of the animal will continue to increase until the arterial pressure rises back to normal, at which time urinary output also returns to normal.

Now, returning to the normal animal, one sees in Figure 2A that the arterial pressure responses were greatly blunted, and the initial urinary response was blunted as well. These effects can be ascribed to the circulatory reflexes which are especially geared to prevent tremendous rises in arterial pressure. Since the arterial pressure rose only 17 mm Hg, one can suspect that the marked increase in urinary output was not due solely to the rise in pressure. The volume receptor reflexes described in detail by Gauer and Henry in Chapter 4 of this volume were presumably responsible for much of the urinary output increase in the normal animal. A special reason for believing this is that the initial rise in urinary output was considerably less than the rise after about 60 min. This could have resulted from diminished ADH production, one of the effects elicited by volume receptor reflexes. Since the ADH that had already been secreted must have continued to circulate for many minutes until it could be destroyed (36), one would have expected a delayed increase in urinary output after volume loading, much as that illustrated for the normal animal in Figure 2A.

Lack of Acute Stability of the Arterial Pressure in Areflex Animals

Thus far we have spoken of the capability of the areflex circulation to control local blood flow, cardiac output, arterial pressure, and blood volume. Now let us speak of one feature of the circulation that is decidedly abnormal in areflex animals: The arterial pressure in these animals is very labile (9, 10, 30). This has already been demonstrated in Figure 2 which illustrated that the arterial pressure increased seven to eight times as much in response to a blood infusion of 300 ml as in the normal animals. And the effect is demonstrated again in Figure 3 which shows a 100-min recording of arterial pressure in a normal awake dog compared with a recording under similar conditions in a dog several weeks after its baroreceptor reflexes had been completely abrogated by baroreceptor denervation. Note the extreme lability of the arterial pressure in the absence of the baroreceptor reflexes in comparison with the relatively stable arterial pressure in the normal animal. The changes in pressure occur when the animal stands up, sits down, hears a noise, sees a person entering a room, or so forth. Thus, one can see the importance of the nervous reflexes to "buffer" the arterial pressure changes and to keep these at a steady level throughout the day.

QUANTITATION OF HEMODYNAMIC
FUNCTIONS AND THEIR INTRINSIC CONTROLS

Quantitation of Autoregulation (Local Blood Flow Regulation)

Blood flow in most tissues of the body automatically increases 2- to 5-fold during states of high metabolic activity, a phenomenon called *autoregulation* (37–40). Indeed, in some tissues the local blood flow can increase as much as 20-fold or perhaps even more. For instance, during very heavy exercise in a well trained athlete (6) as well as in racing animals (41), the cardiac output can increase as much as 7-fold, essentially all of this increase occurring in the exercising muscles. Since the normal resting muscles have a blood flow equal to only about 25% of the total cardiac output, one can calculate that the total blood flow in the muscles increases from 25% to 625% of the total normal cardiac output, or an increase of 25-fold. Part of this increase is caused by an increase in arterial pressure, but the major portion is caused by local vasodilatation in the muscles.

Relation of Autoregulation to Oxygen Consumption Almost all research workers have demonstrated that either decreased availability of oxygen in the

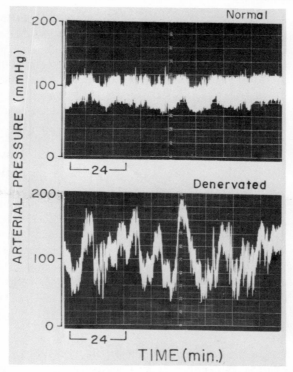

Figure 3. Arterial pressure record for 100 min in a normal awake dog and in a dog several weeks after denervation of the arterial baroreceptors. (Reprinted from Cowley et al. (1973). Circ. Res. 32:564.)

arterial blood or increased consumption of oxygen by a local tissue will increase local blood flow (42–48). This effect is illustrated in Figure 4 which shows progressive increase in blood flow in an isolated limb as the arterial oxygen saturation was decreased. Similar effects have also been demonstrated for blood flow in individual small arteries (49), cardiac output in the entire body (15), and contraction of isolated smooth muscle strips (50) when oxygen availability is decreased.

The mechanism of the oxygen effect is still in considerable dispute. The most prevalent belief is that diminished oxygen in each local tissue area causes some vasodilator substance to be released, and this vasodilator in turn diffuses to the arterioles to cause local arteriolar dilatation. One of the most widely discussed vasodilators in recent years has been adenosine (37, 38, 44). However, there are real problems with the adenosine vasodilator concept that have yet to be settled (45, 48), including the fact that isolated small arteries without associated parenchymatous tissue from which the adenosine could be derived still dilate in response to oxygen deficiency (49). Therefore, there is much reason to believe that simple lack of enough oxygen to supply energy to the

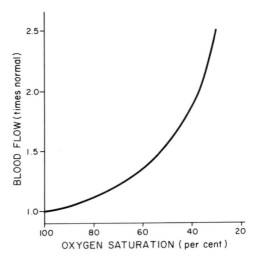

Figure 4. The effect of arterial blood oxygen saturation on blood flow through the isolated leg of the dog. (Drawn from data in Crawford et al. (1959). Amer. J. Physiol. 197:613.)

vascular wall to maintain contraction might be the most important stimulus for vasodilation (45).

However, for the present discussion, the precise mechanism of vasodilatation during oxygen deficiency is not of major importance. What is important is that whenever the tissues are deficient in oxygen the blood flow normally increases and thereby supplies the needed oxygen.

Relation of Autoregulation to Other Nutritional Factors Besides Oxygen Deficiency of oxygen is not the only nutritional deficiency that can cause local vascular dilatation. For instance, deficiency of several of the vitamins, especially of thiamine (51, 52), will cause progressive, long-term local vascular vasodilatation. Thus, local blood flow in a patient with beri beri is often as much as two times normal, and the cardiac output is also elevated by this amount despite the fact that the heart itself is in a weakened condition.

More recently, experiments on perfusion of isolated tissues have shown that deficiency of glucose alone will also cause vasodilatation (F. Haddy, personal communication). However, the vasodilatation that occurs following perfusion of a tissue with glucose deficient perfusate is slow to develop instead of the 20–40 s required in oxygen deficiency.

Therefore, a rather general statement of local blood flow control could be the following: Deficiency of almost any nutrient, whether it be oxygen, glucose, or many others, can perhaps cause local vasodilatation which, hopefully, causes subsequent delivery of adequate quantities of the deficient nutrient. Also, excesses of end products of metabolism—adenosine (44), potassium (53), carbon dioxide (54), acids (55), and many others—perhaps can all increase blood flow and thereby wash out the unwanted substances.

Special Attributes of Long-Term Autoregulation—Changes in Tissue Vascularity Thus far we have spoken only of the acute aspects of autoregulation. However, when a tissue remains ischemic for long periods of time, not only do the blood vessels dilate, but their basic physical sizes also enlarge, and in some instances they increase in number as well. In a series of rats, Folkow and his colleagues (22, 23) have recently demonstrated that partial occlusion of the femoral artery for several weeks caused marked enlargement of the vessels beyond the occlusion. Then, removal of the femoral restriction allowed return of the vessels to normal in another few weeks. These effects were not caused by simple acute vasodilatation, because measurements of resistance to flow during the different stages after blockade of vasomotor tone showed that the basic sizes of the vessels changed. Similar long-term vascular changes have been demonstrated by Patz (24) and Dollery et al. (25) in the retinal vessels.

This long-term effect for adjustment of peripheral vascularity is also seen in many clinical conditions. The most striking is coarctation of the aorta (56–58), in which the arterial pressure may be almost one and one-half times as great in the upper part of the body as in the lower body. Yet, measurements of blood flow per unit mass of tissue have demonstrated that the blood flow in both these areas of the body is nearly identical (although, unfortunately, the measuring techniques—mainly plethysmography—have not been the most satisfactory). If subsequent studies and more critical measurements of local tissue blood flow in coarctation of the aorta bear out these initial findings, then this is one of the most compelling reasons for believing that long-term autoregulation is a highly efficient control system, much more efficient than the acute autoregulation mechanism.

Figure 5 illustrates in quantitative terms the meaning of the difference between acute autoregulation and long-term local autoregulation. The solid curve of this figure shows acute changes in blood flow at different arterial pressure levels when the pressure is changed over its range in approximately 1 hour, a curve that has been measured in the hindleg of the dog (17). On the other hand, the dashed curve is based on the few available long-term autoregulation experiments (22–25) and on measurements that have been made in patients with coarctation of the aorta (56–58). This curve shows that over a wide range of arterial pressures the local tissue blood flow remains almost exactly constant after long-term autoregulation has functioned fully.

Quantitation of Cardiac Pumping—Cardiac Output Regulation

We do not need to spend much time characterizing the pumping capability of the heart because, of the many functions of the circulation, this is the most easily and most widely understood of all. Even so, it is strange that many well known physiologists have argued violently about the best means for quantitating the pumping ability of the heart. A vague quantity called "contractility" of the heart has often been equated with the heart's pumping ability, but, unfortunately, how to measure contractility itself has been of considerable dispute. Some have expressed a belief that contractility is closely related to dp/dt (59,

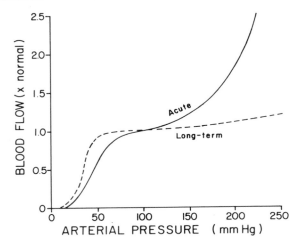

Figure 5. Relationship of arterial pressure to blood flow per unit mass of tissue, showing this relationship under acute conditions when the arterial pressure changes rapidly through its pressure range (———) and under chronic conditions when the arterial pressure is changed through its pressure range over a period of weeks or months (– – –).

60), which means the maximum rate of rise of ventricular pressure at the onset of systole. Others have used the term V_{max} which is a quantity related to the maximum rate of shortening of the cardiac musculature during contraction (61). Others have used various cardiac function curves that relate some expression of cardiac contraction to an initial condition of the ventricle, such as the curve that relates minute work output to mean left atrial pressure (62), or better still by three-dimensional plots of volume output, input pressure, and output pressure (63, 64).

But let us ask ourselves what function of the heart is most important to the overall operation of the circulation? The answer to this is the capability of the heart to pump blood—that is, to provide cardiac output. The next question that we need to ask is, what hemodynamic input factor to the heart is most important to function of the remainder of the circulation? And the almost certain answer to this is the amount of back pressure that the heart displays to the blood returning to it from the peripheral circulation. The average value of this back pressure is the mean right atrial pressure. Therefore, if we wish to express quantitatively the pumping capability of the heart as it relates to overall circulatory function, then we need to express cardiac output as a function of right atrial pressure. This is illustrated for the normal human heart under two separate conditions by the two curves of Figure 6. The solid curve shows the approximate cardiac output of the heart at increasing right atrial pressures for the normal, resting, unstimulated heart. Actually, no such curve has ever been recorded in the human being, but the curve was constructed from data derived from many sources, including isolated measurements of right atrial pressure and cardiac output during cardiac catheterization of human beings, and information

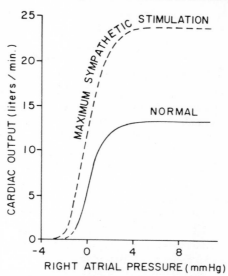

Figure 6. Relationship between right atrial pressure and cardiac output for the normal resting heart (——) and that for a heart strongly stimulated by the sympathetic nervous system (– – –). (Redrawn from a compilation of data in Guyton, Jones, and Coleman (1973). Circulatory Physiology: Cardiac Output and Its Regulation, 2nd Ed., Philadelphia, W.B. Saunders Co.)

calculated from other measurements of heart function (20, 65). In general, one can state that in the normal adult the cardiac output averages almost exactly 5 liters per minute when the right atrial pressure is zero mm Hg—somewhat greater than this in the young adult male, somewhat less in the older female, with other adults falling in between. When the heart is primed with additional amounts of venous return so that the right atrial pressure rises to + 4–8 mm Hg, the cardiac output of the *unstimulated* heart reaches a maximum plateau level which is about two and one-half to three times as great as the normal cardiac output. The most important lesson to be learned from this curve is that the only necessity required to increase the cardiac output is a slight increase in right atrial pressure. Or any factor that increases the venous return of blood to the heart will automatically increase the cardiac output by an amount equal to the increase in venous return.

However, the venous return mechanism for increasing the cardiac output is limited by the level of the plateau of the cardiac output curve. To attain cardiac outputs greater than three times normal, the pumping capability of the heart itself must be increased. In the normal circulation, this cannot be achieved except by nervous or hormonal stimulation of the heart, which will be discussed in more detail later in this chapter. Suffice it to say at the present time that nervous stimulation or norepinephrine and epinephrine released from the adrenal medulla can increase the plateau level of the heart about 100%, as illustrated by the dashed cardiac output curve of Figure 6 (66). It increases the level of the cardiac output curve by a combination of two separate mechanisms, the two of which potentiate each other: 1) an increase in heart rate and 2) an increase in the strength of cardiac muscle contraction with consequent increase in stroke volume.

Venous Return Mechanism for Cardiac Output Regulation To understand the normal mechanism of cardiac output regulation we need only to consider the combination of blood flow regulation in all areas of the body and the effect of right atrial pressure on cardiac output. Since each tissue area generally controls its own blood flow, it also follows that all of the tissues put together control the total blood flow through the peripheral circulation, and this is equal to the total venous return to the heart. Therefore, the normal hemodynamic mechanism for control of cardiac output is simply the sum of all the local blood flow controls in the peripheral circulation (14, 15, 18–20). If the blood flow through any single tissue area increases, this increases the venous return by an equal amount, which increases the right atrial pressure and therefore increases the cardiac output also by an equal amount. However, the effectiveness of this mechanism is limited by several special factors as follows:

First, the absolute limitation is the upper plateau level of the cardiac output curve as illustrated in Figure 6.

Second, as the venous return increases, the slight increase in right atrial pressure causes back pressure on the peripheral circulation, which impedes venous return a small amount (67–69). (Fortunately, it does not impede venous return very much because the increase in back pressure dilates the veins, an effect that enhances venous return; this enhancement nullifies most of the impeding effect at least for the first few mm Hg rise in right atrial pressure.)

Third, as the resistance to blood flow through the tissues decreases and blood flows more rapidly from the arteries to the veins, the arterial pressure falls (8, 9). This tends to keep the tissue blood flow from increasing as much as would occur should the arterial pressure not fall. In the areflex animal there is no acute hemodynamic mechanism to nullify this acute fall in arterial pressure. On the other hand, under long-term conditions, several hours to several days, the arterial pressure is returned to its normal value by automatic enhancement of blood volume, a mechanism that is discussed in a following section of this chapter. Therefore, the fall in arterial pressure is a limitation only under acute conditions.

Cardiac Hypertrophy as a Long-Term Hemodynamic Mechanism for Cardiac Output Regulation When the pumping load of the heart remains increased for several days to several weeks, caused either by excess arterial pressure against which the heart must pump or by long-term excess cardiac output, the heart hypertrophies. This changes the cardiac function curve in Figure 6 from the solid curve approximately to the dashed curve, sometimes enhancing the overall pumping capability of the heart by as much as 100% or more (as estimated from clinical studies). Unfortunately, here again, precise measurements of the enhancement of cardiac pumping by the hypertrophied heart have never been made in a human being and only crudely so in animals. Therefore, the quantitative enhancement of the cardiac output curve as shown in Figure 6 has been pieced together from many bits of information (20).

Thus, cardiac hypertrophy can be considered to be a long-term hemodynamic mechanism of cardiac output regulation.

Hemodynamic Regulation of
Arterial Pressure and Blood Volume—Role of the Kidneys

There is no intrinsic hemodynamic mechanism for *acute* regulation of arterial pressure nor for *acute* regulation of blood volume. But, over a period of hours or days the hemodynamic system, operating entirely independently of both nervous control and hormonal control, has a very effective mechanism for simultaneously regulating both arterial pressure and blood volume (1–5). This mechanism is based on a very simple phenomenon of kidney function called *pressure diuresis*. When the arterial pressure rises above normal, the output of fluid from the kidneys increases markedly. Conversely, when the arterial pressure falls below normal, the output becomes greatly reduced, normally falling to zero when the pressure has decreased to about 60 mm Hg. Though this pressure-diuresis mechanism is known and recognized by all renal physiologists, most circulatory physiologists have failed to understand its significance for regulation of arterial pressure and blood volume. Therefore, let us discuss the overall mechanism more completely.

Figure 7 gives a block diagram of the complete hemodynamic mechanism for long-term arterial pressure regulation (1). In this diagram a solid arrow indicates an increasing effect while a dashed arrow indicates a decreasing effect. Therefore, let us begin at the block labeled "arterial pressure" and explain each step.

1. Arterial pressure increases urinary output (the pressure diuresis phenomenon).
2. The increase in urinary output decreases extracellular fluid volume; but, balanced against this, the fluid intake increases extracellular fluid volume.
3. An increase in extracellular fluid volume increases blood volume (70).
4. Increased blood volume increases the circulatory filling pressure (71, 72).
5. Increased circulatory filling pressure increases venous return (72, 73).
6. Increased venous return increases cardiac output.
7. Increased cardiac output increases the total peripheral resistance because increased blood flow through tissues causes local tissue vascular constriction (the phenomenon of *autoregulation* discussed earlier).
8. Arterial pressure is increased in two ways, first, by the increase in cardiac output and, second, by the autoregulatory increase in total peripheral resistance.

Thus, when arterial pressure increases, pressure diuresis causes loss of fluid from the body, and this decreases the overall dynamic activity of the circulation until the arterial pressure falls back to normal. Conversely, a decrease in arterial pressure reduces urinary output, and the extracellular fluid volume increases because of continued intake of fluid; this increases the dynamic state of the circulation, thus raising the arterial pressure again back to normal.

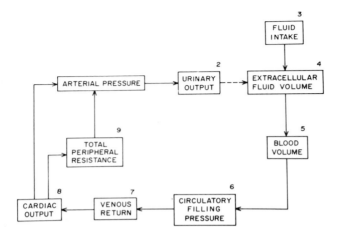

Figure 7. The hemodynamic system for long-term control of arterial pressure and blood volume. (A *solid arrow* means an increase, a *dashed arrow* means a decrease.)

The operation of this hemodynamic mechanism for regulating arterial pressure is illustrated very dramatically in the second panel of Figure 2 presented earlier in the chapter (9, 10). This figure shows the hemodynamic alterations of the circulation following massive infusion of blood into an animal, causing an acute increase in blood volume of 30%. Note in Figure 2B that the cardiac output increased immediately to almost double normal and the arterial pressure to more than double normal. The pressure in turn caused the urinary output to increase approximately 12-fold. Concurrently with the rapid loss of fluid in the urine, both the cardiac output and the arterial pressure returned toward normal. Note, also, that the arterial pressure did not stop falling back toward normal until the urinary output also returned to its original normal level, thus demonstrating almost perfectly the automatic hemodynamic feedback mechanism for regulation of arterial pressure.

Though this hemodynamic mechanism for arterial pressure regulation has been known for at least a hundred years, its importance as a pressure regulator has been seriously questioned by many if not most circulatory physiologists during the last half century. However, recent experiments have shown that the mechanism not only is a powerful one but indeed is the dominating regulator of the long-term level of arterial pressure. The reason that this mechanism is so dominating is that it is an "infinite gain" feedback pressure regulating mechanism (5, 74, 75). This is such an important concept that it requires special discussion as follows.

"Infinite Gain" Feature of Hemodynamic Mechanism for Regulation of Arterial Pressure Let us begin our discussion of the infinite gain feature of the hemodynamic mechanism for arterial pressure regulation by discussing the solid curve of Figure 8 labeled "normal renal function." This curve shows the relationship between arterial pressure and urinary output (5, 75, 76). It has been

recorded very precisely in dogs in the following way: the fluid intake of the dog is increased step by step over many days, and both the arterial pressure and the urinary output are recorded. The curve is almost identical in the normal animal whether the fluid infused be salt solution or pure water. But, in the case of abnormal renal function, the curve slopes much more to the right (to higher pressure values) when salt solution is infused (J. DeClue, unpublished results).

Now, let us see how the normal renal function curve operates when there is a steady intake of fluid. The lowermost line across the graph is labeled "normal intake," and this line intersects the normal renal function curve at point A. At this point the arterial pressure is 100 mm Hg, also depicting that both the intake and the output of fluid are approximately 1.5 liters per day for the human being. Next, let us see what happens when the arterial pressure rises to a value slightly over 100 mm Hg—for instance, to about 105 mm Hg. The renal function curve demonstrates that this rise in arterial pressure, if it remains at this level for a day or so, is associated with an increase in urinary output by about 3-fold. Yet, it does not change the intake of fluid at all. Therefore, more fluid is lost from the body than is gained; the dynamic state of the circulation diminishes because of progressively diminishing blood volume, and the arterial pressure falls back toward normal. The arterial pressure will not stop falling until the renal output falls to equal the intake of fluid, and this results only when the output and the intake become precisely equal. (Of course, fluid loss from the body in other ways besides through the kidneys must also be taken into consideration, but this will not affect the concept in any way.)

Now, to examine the mechanism in the reverse direction when the arterial pressure falls below normal: This causes the urinary output to decrease to a very low value while at the same time intake continues unabated. Therefore, the body fluid volume increases and the dynamic state of the circulation also increases until renal output once again precisely equals fluid intake. This will occur only when the arterial pressure has risen all the way back to point A.

Thus, when the arterial pressure rises too high or when it falls too low, it is eventually returned automatically back to the precise equilibrium point A. Furthermore, the adjustments never stop until the pressure reaches this precise point. That is, the capability of the system to return the pressure to its original control value is infinite, which gives rise to the "infinite gain" capability of this long-term hemodynamic mechanism for arterial pressure regulation (5, 74, 75).

Two Determinants of Long-Term Arterial Pressure Level There are only two ways in which the regulated pressure level of the hemodynamic control system can be changed. These are: 1) a change in the renal function curve, and 2) a change in the fluid intake level. Therefore, the renal function curve and the fluid intake level are called the *determinants* of the long-term arterial pressure level (5).

The role of each of these determinants in affecting the arterial pressure is illustrated in Figure 8. First, assume that the renal function curve has shifted far to the right as illustrated by the curve labeled "abnormal renal function." Such a shift occurs in many types of kidney disease (5, 75), and this is often the cause

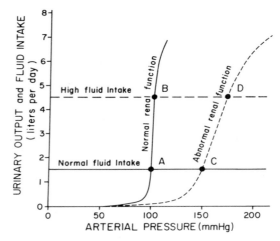

Figure 8. Relationship between arterial pressure, on the one hand, and fluid intake or urinary output, on the other hand. This figure shows the levels at which arterial pressure stabilizes (at the intersection points) with different combinations of 1) normal or abnormal renal function and 2) normal or high fluid intakes.

of hypertension, as we shall see. Note that the abnormal renal function curve crosses the normal intake line at point C. Therefore, with normal intake and with this abnormal function curve, the arterial pressure will always tend to reapproach the pressure level of point C, a level of 145 mm Hg—if the pressure rises above this, the mechanism will decrease the pressure with infinite feedback gain back toward 145 mm Hg; if the pressure falls below this value, the arterial pressure will increase back toward 145 mm Hg, again with infinite feedback gain.

The second determinant that can change the regulated pressure level in the hemodynamic control system is a change in fluid intake. The elevated dashed line of Figure 8 represents a 3-fold increase in fluid intake; this line intersects the normal renal function curve at point B, which represents an arterial pressure of approximately 106 mm Hg. Therefore, when there are high intake and normal renal function, the arterial pressure is controlled with infinite gain to a pressure level of 106 mm Hg rather than the normal value of 100 mm Hg. Note how little change occurs in the arterial pressure despite the three-fold increase in fluid intake.

Next, let us consider the combination of an abnormally shifted renal function curve and a high intake of fluid. In Figure 8 this is represented by the intersection of the dashed renal function curve with the dashed intake line at point D, depicting an arterial pressure level of 165 mm Hg. With these two inputs to the circulatory system, a high fluid intake and the shifted renal function curve, the mean arterial pressure is now regulated with infinite gain at a pressure of 165 mm Hg. Thus, the person has severe hypertension.

Factors That Can Shift Renal Function Curve to Abnormally High Pressure Levels One can see from Figure 8 that the pressure level to which the hemodynamic system will regulate the arterial pressure usually is determined far more by the renal function curve than by the fluid intake level. Therefore, it is important to understand the different factors that can cause the renal function curve to shift to high pressure levels as illustrated by the dashed curve in Figure 8. It will not be possible to go into all the different abnormalities of renal function that can cause such a shift in the renal function curve. However, we can list most of the important causes as follows (5, 75, 77, 78): 1) constriction of the renal arteries, as occurs with Goldblatt kidneys; 2) decreased glomerular filtration coefficient, as occurs in toxemia of pregnancy; 3) sympathetic constriction of the renal vasculature, as occurs in neurogenic and psychogenic hypertension; 4) hormonal constriction of the renal vasculature, as occurs when a pheochromocytoma secretes large quantities of norepinephrine and epinephrine, or as occurs when a renal tumor secretes large quantities of renin; 5) increased reabsorption of salt and water by the tubules, as occurs when excess aldosterone is secreted by a tumor of the adrenal cortex; 6) structural damage to the kidneys caused by renal disease.

It is important to note, however, that not all renal diseases shift the renal function curve to high pressure levels. Indeed, many instances of renal disease will cause lethal uremia long before they will cause hypertension (79, 80). This is particularly true of some patients with interstitial nephritis in which the kidneys actually lose increased quantities of salt and water at a normal arterial pressure. It is also particularly important to note that simple reduction of renal mass causes mainly uremia rather than hypertension (81) (also D. Manning and P. Baer, unpublished results). The reason for this is that reducing the number of nephrons causes far more retention of waste products of bodily metabolism than retention of salt and water. Even a slight rise in arterial pressure, only 5−10 mm Hg, is usually sufficient to cause kidneys with as much as 70% reduction in renal mass to excrete normal amounts of water and salt (81−84). Therefore, when there is reduced renal mass but the remaining nephrons are otherwise normal, the animal or human being usually dies of uremia before developing hypertension. On the other hand, if this animal or human being eats excessive quantities of salt and drinks commensurate amounts of water to go with the salt, the reduced kidney mass now cannot excrete the heavy load of salt and water, so that hypertension does ensue.

Time-Course for Function of Hemodynamic Mechanism for Arterial Pressure Regulation Though the hemodynamic mechanism for arterial pressure regulation is an extremely potent one and fully dominates the picture of arterial pressure regulation over the long term, this mechanism is of almost no importance in acute, minute by minute arterial pressure regulation. However, under appropriate conditions, as when there are a high rate of fluid intake and an abnormally high rate of urinary excretion, the mechanism can readjust the arterial pressure in as little as 30 min (D. Ferguson and R. Norman, unpublished results). But under other conditions, when the fluid intake is relatively low and the kidneys are already operating at a low rate of urinary output, the hemodynamic mechanism may take as long as several days to reach a new steady state pressure level (81–84). Therefore, it is important to have other pressure regulating mechanisms for acute pressure regulation. We shall see later in the chapter that this function is achieved especially by means of nervous reflexes and to some extent also by the renin-angiotensin-vasoconstrictor feedback mechanism for control of arterial pressure.

The Minute Quantity of Extracellular Fluid Volume Change That Is Required to Regulate Arterial Pressure   In long-term regulation of arterial pressure by the hemodynamic mechanism, the amount of fluid volume change that must occur to increase or decrease the arterial pressure as much as 50 mm Hg is very minute. This has been shown especially in the treatment of hypertensive patients with diuretic drugs (85–88). Following approximately a month of treatment, loss of less than one-half liter of fluid from the body of a patient will often reduce his arterial pressure back to normal. This illustrates that even minute changes in extracellular fluid volume, and, therefore, correspondingly minute changes in blood volume (in the order of 2–3% change) probably can, over a long period of time, cause a very large change in arterial pressure.

However, under acute conditions, the fluid volume must be changed tremendously to change the arterial pressure even a small amount (9, 10, 30, 89, 90). This effect can be seen by referring back to the second panel of Figure 2 which shows the effect of blood volume on arterial pressure, showing that in the headless animals without nervous reflexes the blood volume had to be increased approximately 30% to increase the arterial pressure from 100 mm Hg to over 200 mm Hg (9, 10). This compares with an estimated 4–5% change in blood volume required to effect this same change in arterial pressure under long-term chronic conditions. The question that we need to answer is, why this difference between the extreme long-term pressure effect of increased blood volume versus the relatively minor effect of short-term increase in blood volume? The answer to this probably lies in the other long-term adjustments of the circulation, especially long-term autoregulation of the circulation and hypertrophy of the heart.

Role of Long-Term Autoregulation in Hemodynamic Mechanism for Arterial Pressure Control   One of the most important reasons for the very potent long-term effect of small quantities of body fluid on arterial pressure is the long-term role of autoregulation in the hemodynamic pressure control system (5, 14, 15, 18, 19, 83, 84, 91–95). This may be explained as follows: when a small increase in blood volume causes a slight increase in cardiac output, the blood flow through all the tissues of the body also increases by the same amount. Remember that an acute increase in flow through a tissue delivers increased quantities of nutrients to the tissue and causes a natural tendency for the blood vessels to constrict, which is the phenomenon of acute autoregulation. And, if the flow through the tissue remains excessive over a long period of time, it causes long-term autoregulation in addition to acute autoregulation. The long-term autoregulation mechanism, on the basis of quantitative data from studies in patients with coarctation (56–58) as explained earlier in this chapter, is a much more potent mechanism than acute autoregulation. Therefore, over several weeks, even a slight increase in blood flow through the tissues causes marked vascular constriction. Since long-term autoregulation results from actual structural changes in the vasculature, one finds progressive diminishment of vascular dimensions (22–25). And this causes progressive increase in total peripheral resistance. Indeed, in experiments on volume loading hypertension, critical measurements have shown that the total peripheral resistance continues to increase for at least several weeks (83, 84). Also, these measurements show that most of the increase in total peripheral resistance probably results from the long-term mechanism of autoregulation rather than the short-term mechanism, illustrating again the far greater potency of the long-term autoregulatory mechanism than of the short-term.

Therefore, we can begin to understand why an acute increase in blood volume does not raise the arterial pressure greatly, though a chronic sustained increase in volume of only a slight amount causes a very great increase in pressure. Under acute conditions the increase in pressure results almost entirely from the effect of increased cardiac output to cause a direct increase in arterial pressure (83, 84). However, under long-term conditions the arterial pressure rises because of both increased cardiac output and increased total peripheral resistance. Furthermore, the results in volume loading hypertension experiments show that the pressure elevation derives much more from the long-term increase in total peripheral resistance rather than the increase in cardiac output (83, 84). Thus, the same amount of volume loading can probably increase the arterial pressure five to ten times as much under long-term chronic conditions as it can under acute conditions. Therefore, we can see why it is necessary to have a 30% increase in blood volume to double the arterial pressure under acute conditions while only a 4–5% increase can probably double or almost double the arterial pressure under long-term conditions.

Role of Cardiac Hypertrophy in Hemodynamic Mechanism for Arterial Pressure Control   Another factor that allows a minute increase in blood volume to increase arterial pressure markedly under long-term conditions is hypertrophy of the heart. Acutely, any attempt to raise the arterial pressure greatly causes a back-loading effect on the heart that to at least some extent diminishes the ability of the heart to achieve the high arterial pressure. However, as the heart hypertrophies, less and less atrial pressure is required to maintain the arterial pressure, which means that less and less fluid volume also is required to maintain the hyperdynamic state required to cause the hypertension.

"Cascade" Effect of Blood Volume on Arterial Pressure   In the above sections we have seen that a minute amount of increase in blood volume at first has relatively little effect to increase arterial pressure. However, over a period of days, weeks, and months, the effect on pressure becomes multiplied. This multiplication effect has been called the "cascade" effect of blood volume on the long-term level of arterial pressure. It begins with an initial increase in cardiac output that elevates the arterial pressure a slight amount. The next stage is long-term autoregulation that increases the total peripheral resistance and thereby multiplies the increase in arterial pressure as much as an additional 5- to

10-fold. The third stage is hypertrophy of the heart, occurring over a period of weeks and months, and multiplying the effect on arterial pressure still a further unknown amount.

Importance of Arterial Pressure Regulation for Function of Cardiac Output Regulating Mechanism We can now recall that the hemodynamic mechanism for regulating cardiac output is based on local autoregulation of blood flow in the peripheral tissues. When the metabolism of the peripheral tissues increases, the blood vessels dilate and venous return to the heart increases. But, acutely, the arterial pressure falls so that the increase in venous return is only about 30% as great as it would be if the arterial pressure did not fall (7, 8). Therefore, in the absence of an acute mechanism for arterial pressure regulation, the normal hemodynamic mechanism for cardiac output regulation is a relatively inefficient mechanism. Yet, in long-term states of decreased peripheral resistance lasting for more than a few hours, the hemodynamic arterial pressure regulating mechanism comes into play and returns the arterial pressure back to its normal level; and the cardiac output rises accordingly (96, 97). Therefore, one of the most important factors in cardiac output regulation is background regulation of arterial pressure. In the intact circulation, the arterial pressure is maintained at or near the normal level under acute conditions by the extrinsic controls of the circulation, mainly the nervous reflexes; in long-term states this role is played by the renal-fluid volume hemodynamic mechanism for pressure control.

Distribution of Fluid Volume between the Blood and the Interstitial Fluid

It is generally understood that an increase in extracellular fluid volume normally increases both blood volume and interstitial fluid volume. However, under some conditions almost all of the increase in extracellular fluid volume goes into the interstitial spaces rather than into the circulation (98). To understand why these different effects occur, it is especially important to look at the *quantitative* aspects of the mechanism for distributing fluid between the blood and interstitial fluid.

All physiologists know that an increase in capillary pressure increases the tendency for fluid to filter from the circulation into the interstitial spaces, while a decreasing capillary pressure allows osmosis of fluid in the opposite direction. Also, all understand that lymphatic flow is an important factor in draining fluid from the interstitial spaces back into the circulation. On the other hand, many physiologists are not familiar with the quantitative aspects of the pressure-volume curves of both the circulatory system and the interstitial spaces or the importance of these two curves on the distribution of fluid between the two compartments. Therefore, let us first discuss the two pressure-volume curves.

Pressure-Volume Curve of the Circulation Figure 9 illustrates two different relationships between the circulatory filling pressure and the blood volume. The circulatory filling pressure is the pressure in the circulation when the heart has been stopped and the blood in the circulation has been redistributed so that all pressures everywhere in the circulation are exactly the same value (71). The normal circulatory filling pressure (which has also been called "static pressure,"

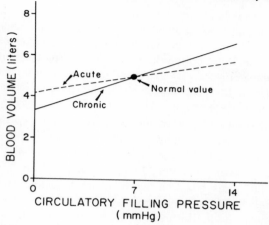

Figure 9. Relationship between the circulatory filling pressure and blood volume. The *dashed curve* shows the acute relationship when circulatory filling pressure is changed through a range of pressures in a period of about 15 s. (Drawn from data in Richardson et al. (1961). Amer. J. Physiol. 201:471.)

"mean circulatory filling pressure," or "mean circulatory pressure") is approximately 7 mm Hg.

The dashed curve of Figure 9 illustrates the acute relationship between circulatory filling pressure and blood volume. This curve was measured by Richardson by increasing or decreasing the blood volume acutely and then rapidly measuring the circulatory filling pressure after each acute change in blood volume (99). Note that at a circulatory filling pressure of 0 mm Hg, the blood volume is about 15% below normal. At a circulatory filling pressure of twice normal, 14 mm Hg, the blood volume is about 14% above normal. Thus, there is almost a linear relationship between circulatory filling pressure and blood volume.

Unfortunately, the curve relating circulatory filling pressure to blood volume under acute conditions does not apply to chronic long-term states because of the phenomenon of stress-relaxation of the circulation. Immediately upon increasing the blood volume, the vascular walls at first stretch elastically. Then they begin a prolonged period of elastic stretch called stress-relaxation (90, 100). This occurs rapidly for the first 3–5 min, then more slowly during the next hour, then extremely slowly over a period of days, weeks, and months thereafter. Therefore, the chronic curve relating blood volume to circulatory filling pressure is much steeper than the acute curve, the acute curve being caused only by elastic stretch of the circulation while the chronic curve is caused by the sum of the elastic stretch and the stretch resulting from stress-relaxation. Thus far, no one has critically measured the chronic curve relating blood volume to circulatory filling pressure. However, a study of the effect of stress-relaxation on the relationship of circulatory filling pressure to blood volume over a period of several hours was performed by Prather (90), and an approximate study of the long-term relationship between circulatory filling pressure and blood volume was made by Akutsu and his colleagues in a calf that was maintained for 23 days with an artificial heart and in which the blood volume was increased or decreased at various times during the procedure (T. Akutsu, personal communication). The curve in Figure 9 labeled "chronic" is an approximate extrapolation of the results from these studies to show the chronic relationship between circulatory filling pressure and blood volume.

Relationship between Circulatory Filling Pressure and Capillary Pressure To understand the effect of the circulatory filling pressure-blood volume curve on exchange of fluid through the capillary membrane, it is necessary to understand the relationship between circulatory filling pressure and capillary pressure. This is a complex relationship based on many factors, including arteriolar resistance, cardiac pumping capability, arterial pressure, and others—as well as the circulatory filling pressure itself. However, in most states, one can probably consider the capillary pressure to increase approximately linearly with an increase in circulatory filling pressure. Therefore, an increase in circulatory filling pressure tends to increase the rate at which fluid will leave the circulation and enter the interstitial fluid spaces.

Pressure-Volume Curve of the Interstitial Spaces In contrast to the linear relationship between circulatory filling pressure and blood volume, the relationship between interstitial fluid pressure and interstitial fluid volume is very nonlinear (101, 102). This is illustrated by the curve labeled "total fluid volume" in Figure 10, which shows the effect of increasing interstitial fluid pressure on interstitial fluid volume. This curve has been measured by the following procedure. Small perforated capsules were implanted in the limbs of dogs for use in measuring the interstitial fluid pressure approximately 4 weeks prior to measuring the pressure-volume curve. At the time of measurement, a limb was removed from the animal and placed on a scale to record changes in fluid volume in the leg. Then the leg was perfused with fluids having different oncotic pressures and different hydrostatic pressures so that the fluid was either absorbed by osmosis into the capillaries from the interstitial spaces when the oncotic pressure was high, or filtered out of the capillaries into the spaces when the hydrostatic pressure was high. At the same time, the interstitial fluid pressure was measured by measuring the pressure of the fluid in the implanted perforated capsule. The normal interstitial fluid pressure in these experiments averaged −7 mm Hg. As small amounts of fluid filtered into the tissues, the interstitial fluid pressure rose rapidly to zero while the interstitial fluid volume increased approximately 50%. Then with further filtration the interstitial fluid volume suddenly increased another 600% while the interstitial fluid pressure increased only to +3−4 mm Hg. Thus, it was clear that an increase in interstitial fluid pressure to only a few mm Hg above atmospheric pressure (zero pressure) is associated with marked increase in interstitial fluid volume. On further analysis of this pressure-volume curve, the following interesting relationship has been found (101−105).

1. Under normal conditions, when the interstitial fluid pressure is less than atmospheric pressure (negative interstitial fluid pressure), there is essentially no free fluid in the tissues. Instead, the fluid is in a semi-gel state because of a hyaluronic acid and collagen reticulum matrix throughout the fluid.

2. When the interstitial fluid pressure rises above atmospheric pressure (that is, into the positive interstitial fluid pressure range), the gel fluid of the tissue increases about 50% (105), and any additional fluid beyond this amount is free fluid rather than gel fluid. This allows free mobility of the fluid in the tissues, which gives rise to the phenomenon of "pitting" edema. That is, in the normal tissues the gel fluid is not highly mobile; but, in the edematous state that results from positive interstitial fluid pressure, the tissue fluid flows readily through the tissue spaces.

3. So long as the interstitial fluid pressure remains negative, interstitial fluid volume does not change greatly (101). To express this quantitatively, one can state that the compliance of the interstitial spaces is relatively low so long as the interstitial fluid pressure is negative (subatmospheric). On the other hand, when the interstitial fluid pressure rises into the positive pressure range, the compliance of the interstitial spaces suddenly increases at least 25-fold in the loose areolar tissues of the body such as in the subcutaneous tissues beneath the skin. Therefore, very slight further increases in interstitial fluid pressure then cause tremendous increases in the volume of the tissue spaces. This explains the tremendous edema that occurs when the interstitial fluid pressure rises above a certain critical pressure value, which in most instances is atmospheric pressure.

It is important to understand why the pressure-volume curve of the interstitial spaces changes so abruptly as the interstitial fluid pressure rises above atmospheric pressure (101, 102). The reason for this is simply the following: normally, the interstitial fluid is continually pumped back into the circulation by way of the lymphatics, and additional fluid is absorbed by osmosis into the capillaries due to the oncotic pressure of the capillaries. These two effects cause a continuous dehydrating state of the tissue spaces, which compacts the tissue elements. The cells are compacted against each other and the reticular filaments of the hyaluronic acid and collagen gel are also compacted. Because of this compacted state, the walls of the normal tissue spaces exhibit very little change in elastic distortion as the interstitial fluid pressure changes. On the other hand, when the interstitial fluid pressure rises into the positive pressure range, there is now excess fluid in the interstitial spaces. This enlarges the spaces and eliminates the compaction of the tissue elements. Now the only limiting restraints to change in interstitial fluid volume are the tensile elements of the tissues. In loose areolar tissue, such as the subcutaneous spaces, the tensile elements are very weak in comparison with the elastic strength of the compaction elements (cells, collagen fibers, hyaluronic acid gel). Therefore, in the positive pressure range, the compliance becomes at least 25 times as great as it is in the negative pressure range. To state this another way, the elastic volume deformation of the tensile elements is at least 25 times as great for a given pressure change as is the elastic volume compression of the compaction elements. In other tissues of the body this great difference between the negative and the positive pressure ranges is not so apparent. For instance, in the kidneys (106), which have a very strong elastic tensile capsule, there is not this sudden change in compliance between the negative and the positive pressure ranges. Likewise, in muscles, which lie in strong sheaths, the difference also is not so great.

Effect of Vascular and Interstitial Pressure-Volume Curves on Distribution of Fluid between the Blood and the Interstitial Spaces Figure 9 illustrated that there is almost a linear relationship between circulatory filling pressure and blood volume. On the other hand, Figure 10 illustrated that there is marked nonlinearity between interstitial fluid pressure and interstitial fluid volume. Therefore, as the total volume of the combined vascular interstitial space system (the extracellular fluid volume) increases, one would also expect a non-linear distribution of fluid between the vascular space and the interstitial spaces. This indeed does occur, as illustrated in Figure 11 (107). Note that as the blood volume increases from its normal value of about 5 liters in the human being up to approximately 7 liters, the interstitial fluid volume rises from approximately 12 liters up to 17 liters. Then, with almost no further increase in blood volume, the interstitial fluid volume continues to increase very greatly. This is a well

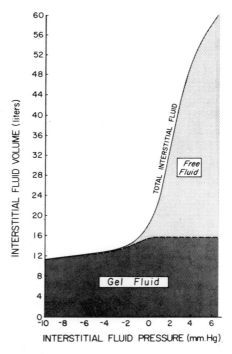

Figure 10. Relationship between interstitial fluid pressure and interstitial fluid volume, showing also the volumes of gel fluid and free fluid at different interstitial fluid pressures. (Modified from Guyton et al. (1971). Physiol. Rev. 5:527.)

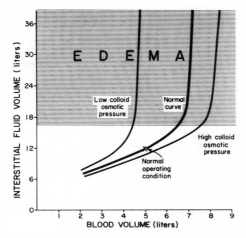

Figure 11. Relationship between blood volume and interstitial fluid volume (*heavy curve*) as the extracellular fluid volume is increased. This curve shows a relatively linear relationship between blood volume and interstitial fluid volume until the blood volume reaches about 7 liters. Above that point the interstitial fluid volume increases massively while the blood volume hardly changes at all. Note also the effects of high and low plasma colloid osmotic pressures (*light curves*). (Modified from Guyton, Taylor, and Granger (1975). Circulatory Physiology II: Dynamics and Control of the Body Fluids. Philadelphia. W.B. Saunders Co.)

known phenomenon that has especially been observed clinically when fluids are administered to patients. A small amount of fluid infusion increases the blood volume a moderate amount at the same time that interstitial fluid volume is increased. However, infusion of large amounts of fluid increases the blood volume only to a critical value, and then essentially all of the remaining fluid passes into the interstitial spaces.

The nonlinearity of the relationship between blood volume and interstitial fluid volume is explained, for obvious reasons, by the nonlinearity of the pressure-volume curve of the interstitial spaces in Figure 10.

Overflow Reservoir Function of the Interstitial Spaces   The nonlinear relationship between blood volume and interstitial fluid volume is one of the most remarkable features of the circulatory hemodynamic system. It actually describes an overflow reservoir function of the interstitial spaces for the circulation (102, 107). That is, when the circulation becomes overly filled with fluid, there comes a critical point at which it cannot be filled still more. At this point, all excess fluid that enters the circulation, provided it does not enter too rapidly, will be dumped out of the circulation into the interstitial spaces. This causes peripheral edema, but fortunately peripheral edema is not a lethal condition. On the other hand, excessive filling of the circulation is a lethal condition because it can cause acute failure of the heart or, even more important, can cause fulminating pulmonary congestion followed by death within minutes. (Fortunately, the dumping of fluid into the *peripheral tissues* almost always occurs before fluid begins to dump into the *pulmonary tissues,* except in the case of left heart failure (108).)

Thus, the interstitial fluid spaces function as an overflow reservoir for the circulation. Furthermore, the nonlinear relationship between blood volume and interstitial fluid volume, as illustrated in Figure 11, represents an indispensable safety valve relief system for the circulation.

Relationship between Blood Volume and Interstitial Fluid Volume under Normal Nonedematous Conditions From the point of view of normal circulatory hemodynamics, the portion of the curve in Figure 11 that occurs prior to the steep increase in interstitial fluid volume is exceedingly important. This portion of the curve shows that for each increase in blood volume there is approximately a 2-fold increase in interstitial fluid volume. In other words, in this range of operation of the fluid distributing mechanism, the extracellular fluid volume is distributed approximately two-thirds in the interstitial fluid spaces and one-third in the vascular space (70). Thus, in the normal range of function, intake of fluid increases the blood volume as well as the interstitial fluid volume.

Role of Lymphatic System in Determining Distribution of Fluid between the Vascular Space and the Interstitial Spaces The volume of fluid returned to the circulation by way of the lymphatics each day is only 2—3 liters (109). Yet, probably 10 to 30 times this much fluid actually filters into the interstitial spaces through the capillaries and is reabsorbed by the capillaries each day. Therefore, from a quantitative point of view, the return of fluid volume to the circulation by way of the lymphatics is of small dimensions in relation to the total fluid movement out of and then back into the circulation. Therefore, what is the real importance of a lymphatic system to the overall hemodynamics of the circulation? The answer to this lies in a unique function of the lymphatic system: It is only by way of the lymphatics that the protein which leaks into the tissues can return to the circulation. The slow flow of lymph drains the leaked protein out of the tissue spaces and back into the circulation. This maintains a high colloid osmotic pressure differential between the plasma and the interstitial fluid. And this colloid osmotic pressure causes continual osmosis of fluid into the capillaries.

EXTRINSIC CONTROLS OF THE CIRCULATION

The extrinsic controls of the circulation—the nervous reflex circulatory controls and the hormonal controls—need little emphasis in this chapter because they have been discussed and analyzed in great detail time and again. However, it is important that we discuss the *quantitative* importance of each of these controls and their overall role in the function of the circulation.

Control of the Circulation by the Nervous System

When one recognizes the high degree of circulatory control that persists in an animal whose nervous system has been completely destroyed as discussed earlier in this chapter, and when one understands the potency of the intrinsic controls of the circulation, he can begin to see that many of the controls that in the past have been ascribed to the nervous system simply are not true. For instance, studies in animals with the nervous system either totally destroyed or partially destroyed show that local blood flow, cardiac output, blood volume, and long-term level of arterial pressure can all be controlled to a high degree of precision in the total absence of the nervous system. Yet, there are four very important functions of the nervous controls that cannot be performed by the intrinsic controls of the circulation. These are: 1) control of fluid and electrolyte intake; 2) acute control of arterial pressure; 3) acute control of blood volume; and 4) acute control of pumping capacity of the heart.

Nervous Control of Fluid and Electrolyte Intake The only absolutely essential role of the nervous system in the control of the circulation is to provide the nervous drive for fluid and electrolyte intake. It is not highly important that this intake be at a precise level but merely that there be an intake of both water and electrolytes within a reasonable range. Indeed, referring back to Figure 8, one can see that changing the fluid and electrolyte intake from as little as one-third normal up to as high as three to four times normal has very little effect on the control of arterial pressure because most of this control is achieved by the control of water and electrolyte excretion through the kidneys (81–84). And the same can be said for control of the body fluid volumes. Therefore, so far as the normal circulation is concerned, it is not *control* of water and electrolyte intake by the nervous system that is important; instead, what is important is the normal basic drive that makes an animal each day drink some water and eat some electrolytes.

Yet, we shall see shortly that the thirst mechanism—the mechanism for controlling water intake—operates as part of a complex system that includes the various hormonal controls of the circulation, especially aldosterone and ADH, to regulate the composition of the body fluids.

Acute Control of Arterial Pressure by the Nervous System

Baroreceptor Reflexes Of all of the controls of the circulation, the barore-ceptor reflexes have been the most widely studied and discussed. They have been accorded the capability of controlling arterial pressure both acutely and chroni-cally, controlling cardiac output, controlling heart pumping both acutely and chronically, controlling blood volume, and, indeed, controlling almost any and all aspects of circulatory function. Such omnipotence is almost certainly not true for several reasons. First, any attempt to control all of these different factors at the same time by the same reflex would lead to serious confusion as to which factor should be controlled. For instance, in vasomotor collapse of the circulation, the cardiac output is frequently high while the arterial pressure is low. On the other hand, in hemorrhagic shock, both arterial pressure and cardiac output are low. Therefore, which of the two, arterial pressure or cardiac output, is the baroreceptor reflex going to correct? The response must be exactly opposite for one of these conditions versus the other. Also, the baroreceptors have been shown to adapt (reset) over a period of several hours to several days

(110–114), and thus far no one has proved an exception to this. Therefore, it is doubtful that the baroreceptor reflexes can modify circulatory function for longer than several hours to several days.

Since it is arterial pressure that the arterial baroreceptors detect, and since the efferent limb of the baroreceptor reflex is especially organized to correct abnormalities of arterial pressure, the baroreceptor system plays a very important role in the acute control of arterial pressure. On the other hand, it seems to play a role in cardiac output control only when arterial pressure is a factor in providing tissue blood flow. During acute hemorrhage the baroreceptor system is exceedingly important to prevent serious decrease in arterial pressure (115–117). Also, when the baroreceptor system is functioning, the arterial pressure changes far less during postural changes than when the baroreceptors are absent (118, 119). If the reader will refer back to Figure 3, he will see recordings of arterial pressure in normal awake dogs both before and after the arterial baroreceptor system had been completely denervated (119). Note the tremendous variation in pressure in the denervated state in comparison with a much more stable pressure in the normal state. Thus, the baroreceptor system "buffers" the arterial pressure, which gives rise to the name "buffer nerves" often used to describe the carotid sinus nerves.

Figure 12 illustrates frequency-distribution curves for the arterial pressure over a period of 24 hours for dogs (a) before baroreceptor denervation and (b) after denervation (119). In each instance the dog was awake, free to move in its cage, and fully healed for several weeks after any surgery. Note that the variation in arterial pressure was two to three times as great in the denervated state versus the normal state.

Another feature that can be discerned from Figure 12 is that the mean arterial pressure over the 24-hour period was almost identical in the denervated state as in the normal state. In other words, the process of denervation did not change the long-term level of arterial pressure. Chronic pressure recordings and frequency distribution curves similar to those shown in Figure 12 have now been made on 34 denervated dogs by Cowley et al. (119), and an additional 16 by Liard et al. (120). The average arterial pressure in all of these animals in the normal state was 106.5 ± 4.5 mm Hg and in the denervated state 107.2 ± 4.7 mm Hg. Thus, the data at present do not show a statistical difference in the basal arterial pressure level whether the baroreceptors be present or not.

Therefore, we can probably come to the conclusion that the truly important function of the baroreceptor system is to "buffer" the acute changes in arterial pressure, to keep the arterial pressure from varying drastically up and down from the mean value.

We need also to consider the feedback gain of the arterial baroreceptor system. This has been variously stated to be as low as 1–2 (121–124) or as high as 30 (125). (A gain of 1 corrects abnormal pressures by 1/2; 2 = 2/3 correction; 30 = 30/31 correction.) However, the high values have generally been estimated from circumstantial evidence rather than direct measurements, while direct measurements of gain have been about 1–2 in the dog. The gain can also be

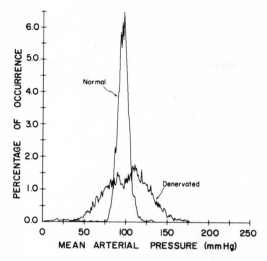

Figure 12. Frequency distribution curves for minute by minute arterial pressures over a period of 24 hours in a normal awake dog and in a dog several weeks after baroreceptor denervation. (Reprinted from Cowley et al. (1973). Circ. Res. 32:564. By permission of the American Heart Association, Inc.)

estimated from the frequency-distribution curves shown in Figure 12 (119). The deviations in pressure in Figure 12 were approximately two times as great in the denervated state as in the normal state. Mathematically, this would give a feedback gain for the baroreceptor system of 1.0, which fits well with the direct measurements of feedback gain in the awake dog.

Therefore, the arterial baroreceptor system is not by itself an excessively powerful mechanism for control of arterial pressure, though even a gain of 1.0 is important to prevent wild excursions of arterial pressure during the course of the day.

Low Pressure Baroreceptor System   The low pressure baroreceptors include the stretch receptors of the pulmonary arteries as well as those of the heart, especially of the two atria. The atrial receptors are probably the same receptors that play a role in acute volume regulation as discussed by Gauer and Henry in Chapter 4 of this volume. However, these low pressure baroreceptors probably also are important in buffering the arterial pressure as well. One of the major reasons for believing this is that arterial pressure is far better buffered against one particular type of circulatory stress than against all others: This is stress caused by increased blood volume. An increase in blood volume has a special propensity for increasing the pressure in both the low pressure areas of the circulation and in the systemic arteries at the same time, thus strongly exciting both the low and the high pressure baroreceptors simultaneously. On the other hand, postural changes of the body, changes in peripheral resistance, and other types of circulatory stresses that affect mainly the systemic circulation are likely to have large effects on the systemic arterial pressure while not affecting the pressure in the low pressure areas. Therefore, it is significant that the reflex feedback gain for control of arterial pressure in response to volume changes averages 6–7 (9, 10, 126) rather than the 1.5 that occurs in response to systemic arterial pressure changes alone; this difference perhaps can be explained by the stimulation of the low-pressure receptors by the volume expansion.

Central Nervous System Ischemic Response   When the arterial pressure to the brain falls to a very low level, ischemia of the brain elicits a powerful sympathetic response to both the peripheral arteries and the heart, thereby elevating the arterial pressure (127–130). This is another reflex feedback mechanism for control of arterial pressure. However, this mechanism is not excited to a significant extent until the arterial pressure falls below 50 mm Hg, and it becomes maximally developed only when the arterial pressure falls into the range of 20–30 mm Hg (128–130). This is in contrast to the pressure range for maximum effectiveness of the arterial baroreceptor system between approximately 60 and 160 mm Hg. However, when the arterial pressure does fall into the pressure range of 20–30 mm Hg, the feedback gain of the CNS ischemic response can be extremely high, having been measured as high as 11 in the dog (114) and as high as 20 in the rabbit (130). Thus, these gains are several times as great as the maximum feedback gains recorded for the arterial baroreceptor feedback system.

Therefore, the CNS ischemic response can be classified as an emergency system to prevent further fall in arterial pressure once it has fallen into the low pressure, lethal range below 50 mm Hg. This can be an especially important life saving event in the case of severe hemorrhage.

Importance of Chemoreceptors for Pressure Control   As much as the chemoreceptors have been studied, unfortunately we still know very little about their capability to help control arterial pressure. Heymans and Neil emphasize that these receptors do have the propensity to help prevent further fall in arterial pressure when the arterial pressure has fallen into the pressure range below 80–100 mm Hg (115). That is, low pressure causes diminished delivery of blood to the chemoreceptors and therefore diminished oxygen and enhanced carbon dioxide in the chemoreceptor cells. The resulting excitation of the chemoreceptors then elicits reflex effects equivalent to those elicited by the baroreceptors. At present we do not know the feedback gain of this mechanism, but one can suspect it to be no greater than that of the baroreceptor system, because when both the baroreceptors and the chemoreceptors are functioning together one cannot demonstrate better pressure stability in the 50–80 mm Hg pressure range than in the normal pressure range where the baroreceptors seem to operate alone.

Acute Control of Blood Volume by the Nervous System   From discussions earlier in the chapter it has been emphasized that very precise long-term blood volume control can be achieved by the pressure diuresis mechanism even in the absence of the nervous system. This is especially well demonstrated in patients whose spinal cords have been destroyed and, therefore, who have lost most of their nervous link between the brain and the body. Also, the blood volume can be controlled over long periods of time within a few per cent of normal without ADH (as occurs in patients with diabetes insipidus (131)) or in the presence of great excesses of ADH (as occurs in the clinical condition called inappropriate ADH syndrome (132)). However, under acute conditions, nervous control of blood volume is probably important, a subject recently reviewed by Goetz et al. (133) and Gauer et al. (134). Also, in Chapter 4 of this volume Gauer and Henry have summarized these acute controls and have emphasized that when they are fully operative they can either correct or help to correct abnormal blood volumes within a few hours, perhaps before the long-term volume control mechanism has time to function fully. On the other hand, the importance of the long-term renal-hemodynamic volume control system is that any residual abnormality of blood volume that fails to be corrected by the acute nervous controls can still be corrected by these long-term mechanisms. Furthermore, after the nervous controls "adapt" during the chronic state (to be discussed below), the long-term renal-hemodynamic blood volume control system still maintains the normal blood volume day in and day out and month in and month out.

Acute Nervous Control of Pumping Capacity of the Heart The intrinsic controls of the heart are capable of increasing cardiac output acutely up to a maximum of about 3-fold (20), this resulting simply from increased venous return to the heart that invokes the Frank-Starling mechanism to drive the heart to a greater pumping level. If the increased input to the heart continues for weeks at a time, the heart will hypertrophy and will eventually increase its pumping level perhaps as much as 6- to 7-fold. However, there are times when even the normal, unenlarged heart needs to be driven to pumping levels as great as five to six times normal, especially during heavy exercise. It is well known that the circulatory reflexes, as well as direct drive of the sympathetic nervous system by signals originating in the brain during exercise, can excite the heart to a pumping capacity far above normal. Though there have been almost no direct measurements to determine precisely how much the nervous controls can increase pumping capacity in the human being, indirect information from lower animals indicates this to be about 100% (20).

Adaptation of the Nervous Receptors of the Circulation and Its Importance to Nervous Control of the Circulation Several different studies of the circulation have now shown that the baroreceptors adapt completely within a few hours to several days (110–114)—as little as 6 hours under some conditions in the rat and perhaps as long as 2 days in the dog. The degree of adaptation of the other nervous receptors is still unknown. It is presumed that the low pressure receptors (including the volume receptors) adapt in approximately the same way as the baroreceptors. One of the most compelling evidences for this is the fact patients with congestive heart failure having very high atrial pressures do not exhibit reflex loss of fluid through the kidneys but, instead, actually exhibit the opposite effect, a tendency to retain fluid. Therefore, if the baroreceptor and volume receptor mechanisms are to be beneficial as a volume control mechanism, it is important that they make their corrections of pressure and blood volume during the first few hours after stimulation and before adaptation takes place. Following this time, the arterial pressure and blood volume must be adjusted and maintained by other mechanisms, presumably by the long-term, intrinsic, combined pressure and volume control mechanism discussed earlier.

We do not know whether or how much the chemoreceptors or the CNS ischemic response adapt over a long period of time. However, since these detect chemical changes rather than stretch, it is possible that they do not exhibit the complete adaptation that seems to occur for the stretch receptors. If this be true, then long-term nervous controls of the circulation could be effected through the chemoreceptor reflex or through the CNS ischemic response. Unfortunately, neither of these has been shown to have a high feedback gain for control of the circulation in the normal pressure range (115, 128–130).

Therefore, in view of adaptation of the stretch receptors, and in view of the very low gain of the chemoreceptors and the CNS ischemic receptors in the normal pressure range, there is at present no compelling evidence that the nervous system normally exerts a significant degree of long-term feedback control over the circulation (though it does exert marked acute control). This, obviously, would explain why the patient with a severed spinal cord controls his arterial pressure, his cardiac output, and other aspects of his circulation at almost exactly the same level under long-term conditions as does the normal person.

Summary of Nervous Control of the Circulation Putting all of the above quantitative aspects of nervous control of the circulation together, we can state that the nervous system is highly important for acute modification of circulatory function. The one indispensable need for the nervous system in circulatory control is to provide a nervous drive for water and electrolyte intake, which is a long-term and continuing necessity. If this is achieved, the intrinsic controls of the circulation can effect relatively normal long-term circulatory control. In addition, the nervous system provides the following very important acute circulatory regulations that allow the circulation to adapt to important stresses of the body.

1. Acute nervous drive to the heart can double the pumping effectiveness of the heart.

2. The nervous system acutely buffers the arterial pressure, decreasing the normal excursions of pressure to one-half or less of those that would occur without these controls. This effect is especially potent in response to blood volume changes, in which case the nervous controls reduce the arterial pressure excursions as little as one-seventh what would occur without the controls (9, 10, 126).

3. The acute nervous controls help to return the blood volume to normal whenever it becomes abnormal (Gauer and Henry, Chapter 4).

Hormonal Control of the Circulation

The basic concepts of hormonal control of the circulation, like those of nervous control, have been widely discussed and in general are understood by physiologists everywhere. For instance, it is well known that norepinephrine and epinephrine act as mediators for a major share of the control of the circulation by the sympathetic nervous system and that loss of direct sympathetic control can usually be compensated by these circulating catecholamines. It is also widely understood that renin is an important vasoconstrictor agent and that it plays a major role in at least certain types of hypertension. Finally, all physiologists know that both ADH and aldosterone play important roles in the extracellular fluid economy of the body.

On the other hand, the *quantitative* aspects of hormonal control of the circulation are generally less well understood and, therefore, deserve special consideration here. The two hormonal systems that especially require discussion are 1) the renin-angiotensin system and 2) the ADH and aldosterone system for overall control of the body fluids.

The Renin-Angiotensin System

Feedback Control of Arterial Pressure by the Renin-Angiotensin-Vasoconstrictor System   When the arterial pressure falls below normal, renin is secreted; this in turn causes the formation of angiotensin. The angiotensin then causes peripheral vasoconstriction and return of the arterial pressure back toward the normal level. This is an important mechanism for maintaining the arterial pressure. But, let us quantitate this mechanism: in normal dogs, decreasing the arterial pressure from 100 mm Hg down to 90 mm Hg has very little effect to activate the renin-angiotensin system. However, further decreases down to about 60 mm Hg promote full activation (11, 135). The peripheral vasoconstriction becomes progressively more intense for almost exactly 20 min, at which time it reaches its peak. This delay in buildup of activity presumably results from progressive accumulation of circulating renin rather than accumulation of angiotensin, because angiotensin has a half-life of only about a minute and therefore cannot accumulate over long periods of time.

The feedback gain of the renin-angiotensin-vasoconstrictor system in the dog has been measured to be approximately 1.6 (11, 136), which means that after acute reduction of the arterial pressure to 60 mm Hg this system acting by itself can return the arterial pressure to approximately 85 mm Hg. Therefore, the gain of this system is almost the same as that of the arterial baroreceptor system which has a feedback gain of 1 to 2, as was discussed earlier in this chapter.

Unfortunately, the renin-angiotensin-vasoconstrictor system does not function to a significant extent to control arterial pressure when the pressure rises above normal (11). That is, this control system is primarily a unidirectional one, helping to elevate arterial pressure when it falls too low but not playing a significant role in readjusting arterial pressure when it rises too high.

Control of Renal Output by the Renin-Angiotensin System   Recent studies have shown that angiotensin has a very potent direct effect on the kidneys to decrease the rates of excretion of both salt and water (137–139). Therefore, over several days, the renin-angiotensin system can increase the arterial pressure through a fluid retention mechanism. It is possible, indeed likely, that this means for controlling arterial pressure is much more potent than is the direct vasoconstrictor effect of angiotensin. For instance, Dickinson (140) a number of years ago and Cowley (141) more recently have both demonstrated that infusion of very minute quantities of angiotensin into animals for several weeks, though causing essentially no rise in arterial pressure because of acute vasoconstriction at the onset of infusion, will, nevertheless, cause marked chronic rise in pressure. There is reason to believe that a large share of this chronic rise results from the renal fluid retention effects of the angiotensin.

Possible Role of the Renin-Angiotensin System to Enhance Aldosterone Secretion Beginning with the early studies of Davis (142) it has become widely believed and taught that angiotensin causes marked increase in aldosterone secretion and that this is one of the major mechanisms for long-term control of aldosterone secretion. However, recent studies in nine separate series of dogs by Cowley (142a) and McCaa (143) demonstrated that this angiotensin effect to elicit aldosterone secretion was mainly a transient one lasting only about 24 hours. Only under a few conditions did angiotensin cause a mild chronic increase in aldosterone secretion. Therefore, from a quantitative point of view, it is dubious that angiotensin has an outstandingly potent long-term effect to stimulate aldosterone secretion. On the other hand, as noted above, angiotensin has a direct effect on the kidneys to cause much the same effect as increased aldosterone secretion, namely, salt and water retention.

The Renal-ADH-Thirst-Aldosterone System for Controlling Blood Volume, Sodium Concentration, and Potassium Concentration Very potent roles have been ascribed to both ADH and aldosterone for controlling blood volume and extracellular fluid volume. However, in the case of each of these hormones, this is very doubtful for the following simple reasons: in patients with either no ADH secretion (diabetes insipidus) (131) or massive secretion (inappropriate ADH syndrome) (132), the blood volume and extracellular fluid volume generally range no more than approximately ± 10% from the normal. Certainly a weak effect such as this could not play a potent role in the control of either blood volume or extracellular fluid volume. Similarly, in patients with primary aldosteronism the blood and extracellular fluid volumes increase only a few per cent (144), and such patients only rarely exhibit edema or other signs of serious hypervolemia. In Addison's disease, the blood volume and extracellular fluid volume do decrease (145), but it is not clear whether this is caused by lack of the effect of aldosterone on the kidneys or caused by other effects such as gastrointestinal disturbances. At any rate, aldosterone secretion can decrease considerably without altering the blood volume more than a few per cent.

Yet, it is true that direct studies of kidney function demonstrate very potent effects of both ADH and aldosterone on the excretion of fluid volume. How do we reconcile these potent direct effects on renal fluid volume excretion with the relatively small changes in body fluid volumes when the concentrations of the hormones themselves become greatly abnormal? The answer to this seems to reside in a complex control system that involves several different mechanisms operating simultaneously, including: 1) the renal pressure diuresis mechanism; 2) ADH stimulation of water reabsorption by the kidneys; 3) control of water intake by the thirst mechanism; and 4) stimulation of sodium reabsorption and potassium secretion in the renal tubules by aldosterone. This system seems to work in the following way:

First, the most potent of all of these mechanisms for control of blood volume, according to quantitative measurements made by DeClue (unpublished observation), Coleman (83), and Manning (84), is the renal *pressure diuresis mechanism*. Even a slight rise in arterial pressure causes marked increase in urine output, as was illustrated in Figure 8 earlier in this chapter. Therefore, any time that either ADH or aldosterone attempts to increase the blood volume, within the next few days the arterial pressure rises a few mm Hg and causes enough pressure diuresis to oppose fully the fluid retention propensities of the hormone (146, 147). This seems to be the mechanism of the phenomena that have been called "aldosterone escape" and "ADH escape." Thus, the renal pressure diuresis mechanism seems to play a prepotent role in control of the blood volume and at the same time controls extracellular fluid volume as well.

Second, another potent factor in this overall control system is the combined *thirst/ADH mechanism*. Thirst controls the intake of water, and ADH controls the output of water. The two of these acting together can increase or decrease the water content of the body and, therefore, can also increase or decrease the osmolality of the body fluids. As already well understood by all physiologists, strong feedback mechanisms act through both the thirst (148) and ADH mechanisms (149) to achieve very precise control of body fluid osmolality.

On the other hand, many physiologists have overlooked the fact that body fluid osmolality is determined almost entirely by the sodium concentration of the extracellular fluid (150). The reason for this is that sodium constitutes about 94% of the cation osmols in the extracellular fluid, and this sodium also controls the quantity of anion osmols in the extracellular fluid at the same time—that is, an increase in sodium ions has an automatic effect, acting through renal tubular mechanisms, to adjust the anion concentration to equal the cation concentration. The importance of this relationship between sodium ion concentration and osmolality of the extracellular fluids is the following: whenever the thirst/ADH mechanism affects extracellular fluid osmolality, it simultaneously affects the sodium ion concentration approximately an equal amount.

Experiments have also shown that the thirst mechanism, and presumably the ADH mechanism as well, responds much more to sodium ion changes in the extracellular fluid than to changes of other osmotic agents (149). Therefore, conditions are ideal for a feedback regulatory mechanism for control of extracellular fluid sodium ion concentration: that is, an increase in sodium ion concentration increases both thirst and ADH secretion, which together increase the water content of the body. This decreases the sodium ion concentration back toward normal. It also increases the blood volume and arterial pressure which then elicit pressure diuresis until the pressure returns to normal, washing out additional quantities of sodium from the extracellular fluid. Therefore, this is an extremely potent sodium ion control mechanism; indeed, measurements by Young indicate that it can control sodium ion concentration about ten times as potently as can the aldosterone feedback mechanism (147, 151).

Third, we must now answer the question, if the thirst/ADH system is the really potent hormonal mechanism for control of sodium ion concentration, then what is the importance of and function of aldosterone? The answer to this seems to be that aldosterone is mainly important for control of potassium ion concentration. When all other factors are controlled and aldosterone is infused into a dog at rates only three to four times the normal rate of aldosterone secretion, the potassium ion concentration often falls so low that it is lethal (D. Young, unpublished observations). Conversely, in low aldosterone states, an elevated potassium can also be lethal. Furthermore, recent potassium infusion experiments have shown that extremely minute increases in potassium ion concentration in the extracellular fluid can cause marked increase in aldosterone secretion—for instance, 0.8 mEq/liter increase in potassium gives 200% increase in aldosterone (143). Furthermore, this increased secretion continues indefinitely. Thus, the aldosterone mechanism constitutes a very powerful feedback system for control of potassium ion concentration: that is, an increase in potassium ion concentration increases aldosterone secretion, and this in turn reduces the potassium ion concentration back toward normal by increasing potassium excretion.

To compare the ability of the aldosterone mechanism to control sodium ion concentration with its ability to control potassium ion concentration, the following experiment was performed: when the aldosterone level in the body fluids was fixed so that it could neither rise nor fall (by removing the adrenals and infusing aldosterone), sodium ion concentration changed no more than 2% despite sodium intake changes as great as 20-fold (151); this same amount of change also occurred in normal animals. On the other hand, potassium ion concentration increased five to eight times as much following potassium loading as in the normal animal.

SUMMARY

In this chapter we have emphasized especially the intrinsic controls of the circulation, such as the autoregulation mechanism for control of local blood flow, automatic control of cardiac output, long-term control of arterial pressure, long-term control of blood volume, and automatic distribution of fluids between the circulation and the interstitial spaces. The reasons for emphasizing these mechanisms are several: first, many experiments have now shown that the intrinsic mechanisms can provide highly stable long-term control of the circulation. Second, the value of the nervous and hormonal controls have probably been greatly overemphasized in the past. And, third, there are special complexities of the intrinsic controls—such as nonlinearities, delay in responses, and other effects—that have made these difficult to understand; it is probably these difficulties that have led to their underemphasis.

However, we have not meant to take from the nervous and hormonal systems their true importance in circulatory control. For instance, intrinsic mechanisms have almost no capability for acute arterial pressure control (only for long-term control), and they have no mechanism for providing the drive necessary to make the animal ingest water and electrolytes. These require the nervous controls. Also, nervous reflexes are important in enhancing the effectiveness of blood volume control and control of cardiac pumping.

Among the hormonal mechanisms, the renin-angiotensin system can provide a modest degree of arterial pressure control when the pressure falls below normal by eliciting a vasoconstrictor response in the peripheral blood vessels. However, this system seems to have an even more important renal function, a direct effect on kidneys to cause fluid retention; this in turn increases the body fluid volume and in this way increases the arterial pressure.

Finally, the roles of ADH and aldosterone in the control of blood volume have probably been greatly overemphasized. On the other hand, both clinical experience and experimental studies are beginning to demonstrate that the thirst/ADH system is probably by far the most potent mechanism that we have for control of extracellular fluid sodium ion concentration. On the other hand, the aldosterone mechanism seems to be our primary control system for maintaining a normal extracellular fluid concentration of potassium.

REFERENCES

1. Guyton, A.C., and Coleman, T.G. (1967). Long-term regulation of the circulation; Interrelationships with body fluid volumes. *In* E.B. Reeve and A.C. Guyton (eds.), Physical Bases of Circulatory Transport: Regulation and Exchange, pp. 179–201. W.B. Saunders Co., Philadelphia.
2. Guyton, A.C., and Coleman T.G. (1969). A quantitative analysis of the pathophysiology of hypertension. Circ. Res. 24:1.
3. Guyton, A.C., Coleman, T.G., and Granger, H.J. (1972). Circulation: Overall regulation. Annu. Rev. Physiol. 34:13.
4. Guyton, A.C., Coleman, T.G., Cowley, A.W., Jr., Liard, J.F., Norman, R.A., Jr., and Manning, R.D., Jr. (1972). Systems analysis of arterial pressure regulation and hypertension. Ann. Biomed. Engin. I:254.
5. Guyton, A.C., Coleman, T.G., Cowley, A.W., Jr., Manning, R.D., Jr., Norman, R.A., and Ferguson, J.D. (1974). A systems analysis approach to understanding long-range arterial blood pressure control and hypertension. Circ. Res. 35:159.
6. Rowell, L.B. (1974). Human cardiovascular adjustments to exercise and thermal stress. Physiol. Rev. 54:75.
7. Askar, E., and Hamilton, W.F. (1963). Cardiovascular response to graded exercise in the sympathectomized-vagotomized dog. Amer. J. Physiol. 204:291.
8. Banet, M., and Guyton, A.C. (1971). Effect of body metabolism on cardiac output: role of the central nervous system. Amer. J. Physiol. 220:662.
9. Dobbs, W.A.J. (1970). Relative importance of nervous and intrinsic mechanical factors in cardiovascular control systems. Ph.D. thesis, University of Mississippi School of Medicine.
10. Dobbs, W.A., Prather, J.W., and Guyton, A.C. (1971). Relative importance of nervous control of cardiac output arterial pressure. Amer. J. Cardiol. 27:507.
11. Cowley, A.E., Jr., Miller, J.P., and Guyton, A.C. (1971). Open-loop analysis of the renin-angiotensin system in the dog. Circ. Res. 28:568.
12. Guyton, A.C., Batson, H.M., Jr., and Smith, C.M., Jr. (1951). Adjustments of the circulatory system following very rapid transfusion or hemorrhage. Amer. J. Physiol. 164:351.
13. Guyton, A.C., Batson, H.M., Jr., Smith, C.M., Jr., and Armstrong, G.G. (1951). Method for studying competence of body's blood pressure regulatory mechanism and effect of pressoreceptor denervation. Amer. J. Physiol. 164:360.

14. Granger, H.J., and Guyton, A.C. (1969). Autoregulation of the total systemic circulation following destruction of the central nervous system in the dog. Circ. Res. 25:379.
15. Shepherd, A.P., Granger, H.J., Smith, E.E., and Guyton, A.C. (1973). Local control of tissue oxygen delivery and its contribution to the regulation of cardiac output. Amer. J. Physiol. 225:747.
16. Guyton, A.C., Douglas, B.H., Langston, J.B., and Richardson, T.Q. (1962). Instantaneous increase in mean circulatory pressure and cardiac output at onset of muscular activity. Circ. Res. 11:431.
17. Walker, J.R., and Guyton, A.C. (1967). Influence of blood oxygen saturation on pressure-flow curve of dog hindleg. Amer. J. Physiol. 212:506.
18. Liedtke, A.J., Urschel, C.W., and Kirk, E.S. (1973). Total systemic autoregulation in the dog and its inhibition by baroreceptor reflexes. Circ. Res. 32:673.
19. Conway, J. 1966. Hemodynamic consequences of induced changes in blood volume. Circ. Res. 18:190.
20. Guyton, A.C., Jones, C.E., and Coleman, T.G. (eds.) (1973). Circulatory Physiology: Cardiac Output and Its Regulation, 2nd Ed. W.B. Saunders Co., Philadelphia.
21. Rothe, C.F., Love, J.R., and Selkurt, E.E. (1963). Control of total vascular resistance in hemorrhagic shock in the dog. Circ. Res. 12:667.
22. Folkow, B., Hallback, M., Lundgren, Y., and Weiss, L. (1970). Background of increased flow resistance and vascular reactivity in spontaneously hypertensive rats. Acta Physiol. Scand. 80:93.
23. Folkow, B., Gurevich, M., Hallback, M., Lundgren, Y., and Weiss, L. (1971). The hemodynamic consequences of regional hypotension in spontaneously hypertensive and normotensive rats. Acta Physiol. Scand. 83:532.
24. Patz, A. (1965). Effect of oxygen on immature retinal vessels. Invest. Ophthalmol. 4:988.
25. Dollery, C.T., Henkind, P., Patterson, J.W., et al. (1966). Ophthalmoscopic and circulatory changes in focal retinal ischemia. Brit. J. Opthalmol. 50:285.
26. Weaver, D.Q., Henson, E.C., Crowell, J.W., Arhelger, R.B., and Brunson, J.G. (1972). Structural alterations produced in dogs in sublethal hemorrhagic shock. Arch. Pathol. 93:155.
27. Crowell, J.W., and Guyton, A.C. (1962). Further evidence favoring a cardiac mechanism in irreversible hemorrhagic shock. Amer. J. Physiol. 203:248.
28. Harvey, I.W., Weaver, D.Q., and Crowell, J.W. (1971). Hypotensive necrosis of the heart. J. Miss. Acad. Sci. 16.
29. Henson, E.C., Lockard, V.G., Crowell, J.W., Arhelger, R.B., and Brunson, J.C. (1973). Recovery from a usually lethal period of hypotension. Arch. Pathol. 95:73.
30. Cevese, A., and Guyton, A.C. "Isohemic" blood volume expansion in normal and headless dogs. Submitted for publication.
31. Selkurt, E.E., Hall, P.W., and Spencer, M.P. (1949). Influence of graded arterial pressure decrement on renal clearance on creatinine, p-amino hippurate and sodium. Amer. J. Physiol. 159:369.
32. Selkurt, E.E. (1951). Effects of pulse pressure and mean arterial pressure modification on renal hemodynamics and electrolyte and water excretion. Circulation 4:541.
33. Thompson, D.D., and Pitts, R.F. (1952). Effects of alterations of renal arterial pressure on sodium and water excretion. Amer. J. Physiol. 168:490.
34. Thurau, K., and Deetzen, P. (1962). Diuresis in arterial pressure increases. Pflügers Arch. 294:567.
35. Navar, L.G., Uther, J.B., and Baer, P.G. (1971). Pressure diuresis in dogs with diabetes insipidus. Nephron 8:97.
36. Reeve, E.B., Kulhanek, L. (1967). Regulation of body water content. In E.B. Reeve and A.C. Guyton (eds.), Physical Bases of Circulatory Transport, p. 151. W.B. Saunders Co., Philadelphia.
37. Johnson, P.C. (1964). Review of previous studies and current theories of autoregulation. Circ. Res. 14 and 15 (Suppl. 1):I-2.
38. Johnson, P.C. (1974). The microcirculation and local and humoral control of the circulation. In A.C. Guyton (ed.), International Review of Physiology, Vol. 1, p. 163. University Park Press, Baltimore.
39. Stainsby, W.N. (1962). Autoregulation of blood flow in skeletal muscle during increased metabolic activity. Amer. J. Physiol. 202:273.

40. Stainsby, W.N. (1973). Local control of regional blood flow. Annu. Rev. Physiol. 35:151.
41. Donald, D.E., and Shepherd, J.T. (1963). Response to exercise in dogs with cardiac denervation. Amer. J. Physiol. 205:393.
42. Crawford, D.G., Fairchild, H.M., and Guyton, A.C. (1959). Oxygen lack as a possible cause of reactive hyperemia. Amer. J. Physiol. 197:613.
43. Ross, J.M., Fairchild, H.M., Weldy, J.F., and Guyton, A.C. (1962). Autoregulation of blood flow by oxygen lack. Amer. J. Physiol. 202:21.
44. Berne, R.M. (1964). Regulation of coronary blood flow. Physiol. Rev. 44:1.
45. Guyton, A.C., Ross, J.M., Carrier, O., Jr., and Walker, J.R. (1964). Evidence for tissue oxygen demand as the major factor causing autoregulation. Circ. Res. 14:60.
46. Jones, R.D., and Berne, R.M. (1964). Intrinsic regulation of skeletal muscle blood flow. Circ. Res. 14:126.
47. Jones, R.D., and Berne, R.M. (1964). Local regulation of blood flow in skeletal muscle. Circ. Res. 15 (Suppl. I):I-30.
48. Fairchild, H.M., Ross, J.M., and Guyton, A.C. (1966). Failure of recovery from reactive hyperemia in the absence of oxygen. Amer. J. Physiol. 210:490.
49. Carrier, O., Walker, J.R., and Guyton, A.C. (1964). Role of oxygen in autoregulation of blood flow in isolated vessels. Amer. J. Physiol. 206:951.
50. Detar, R., and Bohr, D.F. (1968). Oxygen and vascular smooth muscle contraction. Amer. J. Physiol. 214:241.
51. Burwell, C.S., and Dexter, L. (1947). Beri-beri heart disease. Trans. Assoc. Amer. Physicians 60:59.
52. Lahey, W.J. (1953). Physiologic observations on a case of beriberi heart disease, with a note on the acute effects of thiamine. Amer. J. Med. 14:248.
53. Kjellmer, I. (1965). The potassium ion on a vasodilator during muscular exercise. Acta Physiol. Scand. 63:460.
54. Kontos, H. (1971). Role of hypercopnic acidosis in the local regulation of blood flow in skeletal muscle. Circ. Res. 28 (Suppl. I):I-98.
55. Carrier, O., Jr., Cowsert, M., Hancock, J., and Guyton, A.C. (1964). Effect of hydrogen ion changes on vascular resistance in isolated artery segments. Amer. J. Physiol. 207:169.
56. Patterson, G.C., Shepherd, J.T., and Whelan, R.F. (1957). The resistance to blood flow in the upper and lower limb vessels in patients with coarctation of the aorta. Clin. Sci. 16:627.
57. Wakim, K.G., Slaughter, O., and Clagett, O.T. (1948). Studies on the blood flow in the extremities in cases of coarctation of the aorta: determination before and after excision of the coarctate region. Proc. Mayo Clin. 23:347.
58. Fries, E.D. (1960). Hemodynamics of hypertension. Physiol. Rev. 40:27.
59. Baker, D., Ellis, R.M., Franklin, D.L., and Rushmer, R.F. (1959). Some engineering aspects of modern cardiovascular research. Proc. Inst. Radio Engrs. 47:1917.
60. Mason, D.T. (1969). Usefulness and limitations of the rate of rise of intraventricular pressure (dp/dt) in the evaluation of myocardiac contractility in man. Amer. J. Cardiol. 23:516.
61. Sonnenblick, E.H. (1970). Contractility of cardiac muscle. Circ. Res. 27:479.
62. Sarnoff, S.J. (1955). Myocardial contractility as described by ventricular function curves; observations on Starling's law of the heart. Physiol. Rev. 35:107.
63. Sagawa, K. (1967). Analysis of the ventricular pumping capacity as a function of input and output pressure loads. In E.B. Reeve and A.C. Guyton (eds.), Physical Bases of Circulatory Transport, p. 141. W.B. Saunders Co., Philadelphia.
64. Herndon, C.W., and Sagawa, K. (1969). Combined effects of aortic and right atrial pressure on aortic flow. Amer. J. Physiol. 217:65.
65. Bishop, V.S., Stone, H.L., and Guyton, A.C. (1964). Cardiac function curves in conscious dogs. Amer. J. Physiol. 207:677.
66. Cowley, A.W., Jr., and Guyton, A.C. (1971). Heart rate as a determinant of cardiac output in dogs with arteriovenous fistula. Amer. J. Cardiol. 28:321.
67. Guyton, A.C. (1965). Determination of cardiac output by equating venous return curves with cardiac response curves. Physiol. Rev. 35:123.
68. Guyton, A.C., Satterfield, J.H., and Harris, J.W. (1952). Dynamics of central venous resistance with observations on static blood pressure. Amer. J. Physiol. 169:691.
69. Guyton, A.C., Lindsey, A.W., Abernathy, J.B., and Richardson, T.Q. (1957). Venous

382   Guyton et al.

return at various right atrial pressures and the normal venous return curve. Amer. J. Physiol. 189:609.

70. Shires, T., Williams, J., and Brown, F. (1960). Simultaneous measurement of plasma volume, extracellular fluid volume and red cell mass in man, utilizing I^{131} $S^{35}O_4$ and Cr^{51}. J. Lab. Clin. Med. 55:776.
71. Guyton, A.C., Polizo, D., and Armstrong, G.G. (1954). Mean circulatory filling pressure measured immediately after cessation of heart pumping. Amer. J. Physiol. 179:261.
72. Guyton, A.C., Lindsey, A.W., Kaufmann, B.N., and Abernathy, J.B. (1957). Effect of blood transfusion and hemorrhage on cardiac output and on the venous return curve. Amer. J. Physiol. 194:263.
73. Guyton, A.C., Lindsey, A.W., and Kaufman, B. (1955). Effect of mean circulatory filling pressure and other peripheral circulatory factors on cardiac output. Amer. J. Physiol. 180:463.
74. Guyton, A.C., Cowley, A.W., Jr., and Coleman, T.G. (1972). Interaction between the separate control systems in normal arterial pressure regulation and in hypertension. In J. Genest and E. Kiow (eds.), Hypertension '72, pp. 384–398. Springer-Verlag, New York.
75. Guyton, A.C., Coleman, T.G., Cowley, A.W., Jr., Scheel, K.W., Manning, R.D., Jr., and Norman, R.A., Jr. (1972). Arterial pressure regulation: overriding dominance of the kidneys in long-term regulation and in hypertension. Amer. J. Med. 52:584.
76. Guyton, A.C., Coleman, T.G., Fourcade, M.J., and Navar, L.G. (1969). Physiological control of arterial pressure. Bull. N.Y. Acad. Med. 45:811.
77. Guyton, A.C., Cowley, A.W., Jr., Coleman, T.G., DeClue, J.W., Norman, R.A., Jr., and Manning, R.D., Jr. 1974. Hypertension, a disease of abnormal circulatory control. Chest 65:328.
78. Guyton, A.C., Cowley, A.W., Coleman, T.G., Liard, J.F., McCaa, R.E., Manning, R.D., Jr., Norman, R.A., Jr., and Young, D.B. (1974). Pretubular versus tubular mechanisms of renal hypertension. In M.P. Sambhi (ed.), Mechanisms of Hypertension, pp. 15–30. American Elsevier, New York.
79. Scribner, B.H., Fergus, E.G., Boen, S.T., and Thomas, E.D. (1965). Some therapeutic approaches to chronic renal insufficiency. Annu. Rev. Med. 16:285.
80. Ulvila, J.M., Kennedy, J.A., Lamberg, J.D., and Scribner, B.H. (1972). Blood pressure in chronic renal failure. JAMA 220:233.
81. Langston, J.B., Guyton, A.C., Douglas, B.H., and Dorsett, P.E. (1963). Effect of changes in salt intake on arterial pressure and renal function in nephrectomized dogs. Circ. Res. 12:508.
82. Douglas, B.H., Guyton, A.C., Langston, J.B., and Bishop, V.S. (1964). Hypertension caused by salt loading. II: Fluid volume and tissue pressure changes. Amer. J. Physiol. 207:669.
83. Coleman, T.G., and Guyton, A.C. (1969). Hypertension caused by salt loading in the dog. III. Onset transients of cardiac output and other circulatory variables. Circ. Res. 25:152.
84. Manning, R.D., Jr. (1973). Hemodynamic and humoral changes during the initial phase of salt-induced, renoprival hypertension. Ph.D. thesis, University of Mississippi.
85. Wilson, I.M., and Fries, E.D. (1959). Relationship between plasma and extracellular fluid volume depletion and the antihypertensive effect of chlorothiazide. Circulation 20:1028.
86. Conway, J., and Lauwers, P. (1960). Hemodynamic and hypotensive effects of long-term therapy with chlorothiazide. Circulation 21:21.
87. Ulrych, M., Frohlich, E.D., Dustan, H.P., et al. (1968). Immediate hemodynamic effects of beta-adrenergic blockade with propranolol in normotensive and hypertensive man. Circulation 37:411.
88. Tarazi, R.C., Dustan, H.P., and Frohlich, E.D. (1970). Long-term thiazide therapy in essential hypertension. Circulation 47:709.
89. Guyton, A.C., Lindley, J.E., Touchstone, R.N., Smith, C.M., Jr., and H.M. Batson, Jr. (1950). Effects of massive transfusion and hemorrhage on blood pressure and fluid shifts. Amer. J. Physiol. 163:525.
90. Prather, J.W., Guyton, A.C., and Taylor, A.E. (1969). Effect of blood volume, mean circulatory pressure, and stress relaxation on cardiac output. Amer. J. Physiol. 216:467.

91. Ledingham, J.M., and Cohen, R.D. (1962). Circulatory changes during the reversal of experimental hypertension. Clin. Sci. 22:69.
92. Ledingham, J.M., and Cohen, R.D. (1963). Role of the heart in the pathogenesis of renal hypertension. Lancet ii:979.
93. Borst, J.G.G., and Borst-De Geus, A. (1963). Hypertension explained by Starling's theory of circulatory homeostasis. Lancet i:677.
94. Ledingham, J.M., and Cohen, R.D. (1964). Changes in the extracellular fluid volume and cardiac output during the development of experimental renal hypertension. Can. Med. Assoc. J. 90:292.
95. Ledingham, J.M. (1971). Blood pressure regulation in renal failure. J. R. Coll. Physicians Lond. 5:103.
96. Holman, E. (1937). Arteriovenous Aneurysm: Abnormal Communication Between Arterial and Venous Circulations. Macmillan Co., New York.
97. Warren, J.V., Nickerson, J.L., and Elkin, D.C. (1951). The cardiac output in patients with arteriovenous fistulas. J. Clin. Invest. 20:210.
98. Gilligan, D.R., Altschule, M.D., and Volk, M.C. (1938). The effects on the cardiovascular system of fluids administered intravenously in man. I. Studies of the amount and duration of changes in blood volumes. J. Clin. Invest. 17:7.
99. Richardson, T.W., Stallings, J.O., and Guyton, A.C. (1961). Pressure-volume curves in live, intact dogs. Amer. J. Physiol. 201:471.
100. Porciuncula, C.I., Armstrong, G.G., Jr., Guyton, A.C., and Stone, H.L. (1964). Delayed compliance in external jugular vein of the dog. Amer. J. Physiol. 207:728.
101. Guyton, A.C. (1975). Interstitial fluid pressure: II. Pressure-volume curves of interstitial space. Circ. Res. 16:452.
102. Guyton, A.C., Taylor, A.E., Granger, H.J., and Coleman, T.G. (1971). Interstitial Fluid Pressure. Physiol. Rev. 51:527.
103. Guyton, A.C. (1963). A concept of negative interstitial pressure based on pressures in implanted perforated capsules. Circ. Res. 12:399.
104. Guyton, A.C., Scheel, K.W., and Murphree, D. (1966). Interstitial fluid pressure: III. Its effect on resistance to tissue fluid mobility. Circ. Res. 19:412.
105. Guyton, A.C., Granger, H.J., and Taylor, A.E. (1972). Compliance of the interstitial space. Pflügers Arch. 336 (Suppl.):51.
106. Ott, C.E., Cuche, J.L., and Knox, F.G. Measurement of renal interstitial fluid pressure and polyethylene matrix capsules. J. Appl. Physiol. In press.
107. Guyton, A.C., Taylor, A.E., and Granger, H.J. (1975). Circulatory Physiology II: Dynamics and Control of the Body Fluids. W.B. Saunders Company, Philadelphia.
108. Staub, N.C. (1974). Pulmonary edema. Physiol. Rev. 54:678.
109. Landis, E.M., and Pappenheimer, J.R. (1963). Exchange of substances through the capillary walls. In Handbook of Physiology, Sec. 2, Vol. 2, p. 961. The Williams & Wilkins Co., Baltimore.
110. McCubbin, J.W., Green, J.H., and Page, I.H. (1956). Baroreceptor function in chronic renal hypertension. Circ. Res. 4:205.
111. McCubbin, J.W. (1958). Carotid sinus participation in experimental renal hypertension. Circulation 17:791.
112. Kezdi, P., and Wennemark, J. (1958). Baroreceptor and sympathetic activity in experimental renal hypertension. Circulation 17:785.
113. Kezdi, P., and Spickler, W. (1967). The evidence for resetting of the baroreceptors in hypertension. In P. Kezdi (ed.), Baroreceptors and Hypertension. Pergamon Press, Oxford.
114. Kreiger, E.M. (1970). Time course of baroreceptor resetting in acute hypertension. Amer. J. Physiol. 218:486.
115. Heymans, C., and Neil, E. (1958). Reflexogenic Areas of the Cardiovascular System. (J. and A. Churchill, Ltd. London.)
116. Kumada, M., and Sagawa, K. (1970). Aortic nerve activity during blood volume changes. Amer. J. Physiol. 218:961.
117. Kumada, M., Schmidt, R.M., Sagawa, K., and Tan, K.S. (1970). Carotid sinus reflex in response to hemorrhage. Amer. J. Physiol. 219:1373.
118. Smith, E.E., and Guyton, A.C. (1963). Center of arterial pressure regulation during rotation of normal and abnormal dogs. Amer. J. Physiol. 204:979.
119. Cowley, A.W., Jr., Liard, J.F., and Guyton, A.C. (1973). Role of the baroreceptor

reflex in daily control of arterial blood pressure and other variables in dogs. Circ. Res. 32:564.

120. Liard, F.F., Cowley, A.W., Jr., McCaa, R.E., McCaa, C.S., and Guyton, A.C. (1974). Renin-aldosterone, body fluid volumes and baroreceptor reflex in the development and reversal of Goldblatt hypertension in conscious dogs. Circ. Res. 34:549.

121. Scher, A.M., and Young, A.C. (1963). Servoanalysis of carotid sinus reflex effects on peripheral resistance. Circ. Res. 12:152.

122. Sagawa, K., and Watanabe, K. (1965). Summation of bilateral carotid sinus signals in the barostatic reflex. Amer. J. Physiol. 209:1278.

123. Allison, J.L., Sagawa, K., and Kumada, M. (1969). An open-loop analysis of the aortic arch barostatic reflex. Amer. J. Physiol. 217:1576.

124. Sagawa, K., Kumada, M., and Schrann, L.P. (1974). Nervous control of the circulation. In A.C. Guyton (ed.), International Review of Physiology, Vol. 1, p. 197. University Park Press, Baltimore.

125. Scher, A.M., Franz, G.N., Ito, C.S., and Young, A.C. (1967). Studies on the carotid sinus reflex. In E.B. Reeves and A.C. Guyton (eds.), Physical Bases of Circulatory Transport, pp. 113–120. W.B. Saunders Co., Philadelphia.

126. Brown, D.R., and Taylor, A.E. (1972). Arterial pressure response to sinusoidal blood volume changes in unanesthetized control and sino-aortic denervated dogs. Fed. Proc. 31:367.

127. Guyton, A.C. (1948). Acute hypertension in dogs with cerebral ischemia. Amer. J. Physiol. 154:45.

128. Sagawa, K., Taylor, A.E., and Guyton, A.C. (1961). Dynamic performance and stability of cerebral ischemic pressor response. Amer. J. Physiol. 201:1164.

129. Sagawa, K. (1967). Analysis of the CNS ischemic feedback regulation of the circulation. In E.B. Reeves and A.C. Guyton (eds.), Physical Bases of Circulatory Transport, p. 129. W.B. Saunders Company, Philadelphia.

130. Uther, J.B., and Guyton, A.C. (1973). Cardiovascular regulation following changes in central nervous perfusion pressure in the unanesthetized rabbit. Aust. J. Exp. Biol. Med. Sci. 51:295.

131. Leaf, A. (1975). Diabetes insipidus. In P.B. Beeson and W. McDermott (eds.), Textbook of Medicine, 14th Ed., p. 1700. W.B. Saunders Company, Philadelphia.

132. Bartter, F.C., and Schwartz, W.B. (1967). Syndrome of inappropriate secretion of antidiuretic hormone. Amer. J. Med. 42:79.

133. Goetz, K.L., Bond, G.C., and Bloxham, D.D. (1975). Artrial receptors and renal function. Physiol. Rev. 55:157.

134. Gauer, O.H., Henry, J.P., and Behn, C. (1970). The regulation of extracellular fluid volume. Annu. Rev. Physiol. 32:547.

135. Cowley, A.W., Jr., and Guyton, A.C. (1972). Quantification of intermediate steps in the renin-angiotensin-vasoconstrictor feedback loop in the dog. Circ. Res. 30:557.

136. Brough, R.D., Jr., Cowley, A.W., Jr., and Guyton, A.C. 1975. Quantitative analysis of the acute response to hemorrhage of the renin-angiotensin-vasoconstrictor feedback loop in areflexic dogs. Cardiovasc. Res. 9:722.

137. Yeyati, N.L., Taquini, C.M., and Etcheverry, J.C. (1972). A possible role of angiotensin in the regulation of glomerulotubular balance for sodium. Medicina 32:68.

138. Waugh, W.H. (1972). Angiotensin II: Local renal effects of physiological increments in concentration. Can. J. Physiol. Pharmacol. 50:711.

139. Fagard, R. (1975). Renal responses to slight elevations of renal arterial plasma angiotensin II concentration. Submitted for publication.

140. Dickinson, C.J., and Yu, R. (1967). Mechanism involved in the progressive pressor response to very small quantities of angiotensin. Circ. Res. 21:157.

141. Cowley, A.W., Jr., and Guyton, A.C. (1974). Baroreceptor reflex contribution in angiotensin II induced hypertension. Circulation 50:61.

142. Davis, J.O., Carpenter, C.C.J., Ayers, C.L., Holman, J.E., and Bahn, R.C. 1961. Evidence for secretion of an aldosterone-stimulating hormone by the kidney. J. Clin. Invest. 40:684.

142a. Cowley, A.W., Jr., and McCaa, R.E. (1974). Effects of chronic infusions of small doses of angiotensin II on aldosterone secretion. Physiologist 17:201.

143. McCaa, R.E., McCaa, C.S., and Guyton, A.C. (1975). Role of angiotensin II and

potassium in the long-term regulation of aldosterone secretion in intact conscious dogs. Circ. Res. (Suppl. 1) 36 and 37:57.

144. August, J.T., Nelson, D.H., and Thorn, G.W. (1958). Response of normal subjects to large amounts of aldosterone. J. Clin. Invest. 37:1549.

145. Liddle, G.W. 1975. Addison's disease. In P.B. Beeson and W. McDermott (eds.), Textbook of Medicine, 14th Ed., p. 1736. W.B. Saunders Company, Philadelphia.

146. Norman, R.A., Jr., Coleman, T.G., Wiley, T.L., Jr., Manning, R.D., Jr., and Guyton, A.C. The separate roles of sodium ion concentration and fluid volumes in salt-loading hypertension in sheep. Amer. J. Physiol. In press.

147. Young, D.B., McCaa, R.E., Pan, Y.J., and Guyton, A.C. (1975). Influence of ADH and plasma Na on renin and aldosterone. Fed. Proc. 34:311.

148. Fitzsimmons, J.T. (1972). Thirst. Physiol. Rev. 52:468.

149. Cowley, A.W., Jr. (1975). Role of thirst and vasopressin in control of body fluid osmolality and volume. In A.C. Guyton, A.E. Taylor, and H.J. Granger (eds.), Dynamics and Control of the Body Fluids, p. 274. W.B. Saunders Company, Philadelphia.

150. Guyton, A.C., and Young, D.B. (1975). Relative unimportance of hormonal mechanisms for volume control: their all-importance for extracellular electrolyte concentration regulation. In A.C. Guyton, A.E. Taylor, and H.J. Granger (eds.), Dynamics and Control of the Body Fluids, p. 349. W.B. Saunders Company, Philadelphia.

151. Young, D.B., and Guyton, A.C. (1974). The role of the aldosterone system in control of fluid volume and plasma electrolytes. Fed. Proc. 33:340.

Index